Lecture Notes in Economics and Mathematical Systems   577

Kai Rudolph

Bargaining Power Effects in Financial Contracting

A Joint Analysis of Contract Type and
Placement Mode Choices

With 25 Figures
and 99 Tables

 Springer

Dr. Kai Rudolph

Im Tiergarten 58
8055 Zurich
Switzerland
E-mail: kai.rudolph@gmail.com

ISBN-10 3-540-34495-0   Springer Berlin Heidelberg New York
ISBN-13 978-3-540-34495-7   Springer Berlin Heidelberg New York

Springer is a part of Springer Science+Business Media
springeronline.com

© Springer-Verlag Berlin Heidelberg 2006
Printed in Germany

Typesetting: Camera ready by author
Cover: Erich Kirchner, Heidelberg
Production: LE-TEX, Jelonek, Schmidt & Vöckler GbR, Leipzig

SPIN 11760870   Printed on acid-free paper – 88/3100 – 5 4 3 2 1 0

Preface

This work was accepted as a dissertation by the University of Muenster, Germany, in 2004. It was written while I was a teaching and research assistant at the Department of Banking.

I own many debts – personal and intellectual – to Professor Dr. Andreas Pfingsten, my doctoral adviser, for his overall support while I was writing my dissertation. This thesis benefited much from his constructive criticism. I am also grateful that Professor Dr. Klaus Röder, Department of Finance, University of Regensburg, Germany, acted as my second advisor.

Furthermore, I want to thank Dr. Alistair Milne, Sir John Cass Business School, City University, Great Britain, since he assisted me during a crucial phase of my dissertation project while I was staying as a visiting scholar at the Marie Curie Training Site in Corporate Finance, Capital Markets and Banking at Cass. This five month visit in London was financially supported by the European Commission (Fellowship Ref. No. HPMT-GH-01-00330-04).

I am also indebted to my previous colleagues, mainly Dr. Hendrik Hakenes and Dr. Markus Ricke, for their encouragement and helpful discussions throughout my whole dissertation project. The dissertation also benefited from comments on a first working paper about the dissertation's topic by participants of research seminars at the Universities of Constance, Tuebingen and Osnabrueck, Germany, and at the 2003 Annual Meeting of the European Economic Association, Stockholm, Sweden.

Thanks to the editors of Springer's "Lecture Notes in Economics and Mathematical Systems" series for publishing my work. Financial support of the Gesellschaft zur Foerderung der Bankwirtschaftlichen Forschungsstelle at the University of Muenster is greatefully acknowleged.

And finally, I owe more than I can say to the support and assistance of my girlfriend Melanie Probst and my family. They helped me to cope with my mood swings while writing the dissertation and understood that I had less time for them during the final phase of the dissertation.

Thank you all very much.

Kai Rudolph, Muenster, May 2006

Contents Overview

Contents

Symbols / Notation

AC	$\in [0, \infty)$	Agency costs
B		Borrower
D°		Public debt
$D^\bullet$		Private debt
DP	$\in [0, 1]$	Default probability
E°		Public equity
$E^\bullet$		Private equity
f_y	$\in [0, 1]$	Project return distribution
h	$\in [0, y_{max}]$	Fixed repayment obligation (debt financing)
h^*	$\in [0, y_{max}]$	Optimal fixed repayment obligation (debt financing)
h_j^*	$\in [0, y_{max}]$	Optimal fixed repayment obligation given that j optimizes contracts' conditions (debt financing)
$h^*_{j,\ i\text{-type con-tractual part-ner}}$	$\in [0, y_{max}]$	Optimal fixed repayment obligation given that j optimizes contracts' conditions conditioned on the i-type contractual partner (debt financing)
IR	$\in [0, \infty)$	Informational rent
L		Lender
LPM_h^1	$\in [0, \infty)$	Lower partial moment 1
mc	$\in [0, \infty)$	Monitoring costs (equity financing)
nc	$\in [0, \infty)$	Negotiation costs (private placement)
$\overline{P}_j$	$\forall j \in \{L, B\}$	Investment/financing alternative of j
P_j^+	$\forall j \in \{L, B\}$	Positive investment/financing alternative of j
PB	$\in (-\infty, \infty)$	Borrower's expected profit
$\widetilde{PB}$	$\in (-\infty, \infty)$	Borrower's expected profit when his project is financed by the lender

PL	$\in (-\infty, \infty)$	Lender's expected profit
$\widetilde{PL}$	$\in (-\infty, \infty)$	Lender's expected profit when the borrower's project is financed
q	$\in [0,1]$	Proportional return participation (equity financing)
q^*	$\in [0,1]$	Optimal proportional return participation (equity financing)
q_j^*	$\in [0,1]$	Optimal proportional return participation given that j optimizes contracts' conditions (equity financing)
$q^*_{j,\ i\text{-type con-tractual part-ner}}$	$\in [0,1]$	Optimal proportional return participation given that j optimizes contracts' conditions conditioned on the i-type contractual partner (equity financing)
$S(\widetilde{y})$	$\in [0,\infty)$	Spread of return distribution
vc	$\in [0,\infty)$	Verification costs (debt financing)
$\mathrm{Var}(\widetilde{y})$	$\in [0,\infty)$	Variance of project return
$\widetilde{y}$		Risky project return
y	$\in [y_{min}, y_{max}]$	Project return realization
y_{max}, y_{min}	$\in [0,\infty)$	Maximum, minimum project return
$\widehat{y}$	$\in [y_{min}, y_{max}]$	Project return proclaimed by the borrower
ΔPB		Change in the borrower's expected profit due to a bargaining power redistribution
ΔPL		Change in the lender's expected profit due to a bargaining power redistribution
λ_j	$\forall j \in \{L, B\}$	Probability of the A-type contractual partner (j)
$\widehat{\lambda}_j$	$\forall j \in \{L, B\}$	Expected probability of the A-type contractual partner (j)
$\lambda_j\vert x$	$\forall j \in \{L, B\}$	Conditional probability of the A-type contractual partner (j)
$\widehat{\lambda}_j\vert x$	$\forall j \in \{L, B\}$	Conditional expected probability of the A-type contractual partner (j)
μ	$\in [0,\infty)$	Expected project return
ζ	$\in [0, \zeta_{max}]$	Relative return disparity/risk
ζ_{max}	$\in [0,1]$	Maximum relative return disparity/risk
ζ_{crit}	$\in [0,1]$	Critical relative return disparity/risk
$\wedge$		And
$\vee$		Or
$\max_j$		Maximization with respect to j
s.t.		Subject to
$\vert_{X=x}$		For $X = x$

$Prob(X = x)$		Probability of $X = x$
x	$\in [x_1, x_2)$	$x_1 \leq x < x_2$
x	$\in [x_1, x_2]$	$x_1 \leq x \leq x_2$
x	$\in \{x_1, x_2\}$	$x = x_1$ or $x = x_2$
$\forall$		For all
$\in$		Element of
$x \succ y$		Strict preference for x over y
$x \prec y$		Strict preference for y over x
$\sim$		Indifference between x and y

Abbreviations

AOB	Alternating offer bargaining
ARD	American Research and Development
A-type	Party with a positive profit/financing alternative
B	Borrower
BCOC	Borrower's contract offer condition
BCOC(NL)s	Borrower's contract offer condition satisfied for N-type lender
BCOC(AL)s/ns	Borrower's contract offer condition for A-type lender satisfied / not satisfied
BM	Basic model
BPC	Borrower's participation constraint
BPCb/nb	Borrower's participation constraint binding / non-binding
BS^a	Borrower scenario with absolute power
BS^r	Borrower scenario with restricted power
C	Candidate
Ct	Contract type
Cts	Contract types
Ct/pm	Contract type / placement mode
CDr/nr	Case distinction between risky / riskless debt
EM	Extended model
FCC	Feasible contracting constraint
IMF	International Monetary Fund
L	Lender
LCOC	Lender's contract offer condition
LCOC(NB)s	Lender's contract offer condition satisfied for N-type borrower
LCOC(AB)s/ns	Lender's contract offer condition for A-type borrower satisfied / not satisfied
LPC	Lender's participation constraint
LS^a	Lender scenario with absolute power
LS^r	Lender scenario with restricted power

NoN	Non-negativity constraint
N-type	Party without a profit/financing alternative
PC	Parameter constellation
Pm	Placement mode
Pms	Placement modes
S1	Strategy 1
S2	Strategy 2
SCCE	Set of consistent conditional expectations

1

Introduction

Bargaining power plays an important role in financial contracting when determining which firms / projects are financed and how the value created is distributed between firms and their financiers (cf., e. g., in case of venture capital financing Fairchild (2004)). A bargainer possesses bargaining power if he can affect the bargaining outcome in a way desirable for him.

The aim of this dissertation is to examine bargaining power effects in financial contracting. In particular power effects on firms' choices of contract type (debt vs. equity) and placement mode (public offering vs. private placement) are considered.

The motivation for this research is presented in Section 1.1, while Section 1.2 discusses the methodology applied and describes the structure of the dissertation.

1.1 Motivation

In their pioneering work on capital structure Modigliani and Miller (1958) showed that in a simple world, with no taxes, no incentive or information problems, and a given profitability, the way a firm is financed is irrelevant for the value of the firm. However, subsequent empirical research revealed that in reality capital structure does matter.[1]

---

[1] For example, Rajan and Zingales (1995), Hovakimian et al. (2001), Antoniou et al. (2002), and Frank and Goyal (2003a) discover systematic factors which influence firms' debt-equity choice.

According to Harris and Raviv (1991) and Hart (2001) the forces driving firms' financing decisions basically belong to one of the following categories.[2] These are the desire to

- minimize the burden due to corporate and personal taxes;
- influence corporate control contests;
- convey private information to capital market participants to mitigate adverse selection difficulties (the asymmetric information approach);
- ameliorate conflicts of interests among various claim holders to the firm's resources (the agency approach) and
- affect competition in the product/input market.

However, the empirical validity of these forces, respectively of their underlying factors, has been intensively discussed. For example, while Titman and Wessels (1988, p. 17) find that,

> "results do not provide support for an effect on debt ratios arising from non-debt tax shields, volatility, collateral value, or future growth,"

Harris and Raviv (1991, p. 334) observe that

> "leverage increases with fixed assets, non-debt tax shields, growth opportunities, and firm size and decreases with volatility, advertising expenditure, research and development expenditure, bankruptcy probability, profitability and uniqueness of the product."

Myers (2003, p. 217) concludes so far[3]

> "There is no universal theory of capital structure, and no reason to expect one. There are useful conditional theories, however [...] Each factor could be dominant for some firms or in some circumstances, yet unimportant elsewhere."

Now recently, Röell (1996) and Pagano et al. (1998) revealed that a primary reason for a firm's initial stock market listing, i. e. for a firm's decision to go public, is to increase competition among potential fund suppliers. Hence,

---

[2] Of course, the following broad categorization is far from being free of overlaps among the categorized forces since many interdependencies among the forces driving firms' financing decisions exist. However, the categorization helps to illustrate the wide range of forces on which firms' capital structure choices depend.

[3] As the notion in Myers (1984) shows, his opinion nearly did not change over the last nineteen years, see Myers (1984, p. 575):

> "How do firms choose their capital structures? [...] the answer is 'We do not know'."

going public improves a firm's bargaining position with fund suppliers.[4] They empirically observe that a firm's strategic decision to go public actually increases competition among banks providing loans and other potential sources of finance. Pagano et al. (1998) find that firms after their initial stock market listing borrow from more banks, obtain better terms of loans even after controlling for leverage effects due to their increased equity base, and reduce the concentration of their borrowings. Such bargaining power effects have to be considered in a firm's decision to go public or not, as otherwise the immense costs of such a decision can not be justified (cf. Ellingsen and Rydqvist (1997)).[5]

Independently, if bargaining power considerations possess such a big impact on a firm's decision to go public or not, they also affect a firm's financing decision in general.

The significance of bargaining power considerations in a firm's financing decision is also confirmed by Gompers and Lerner (1996). They find for the United States that venture capital funds can reduce the number of restrictive covenants imposed by investors when the supply of venture capital is high. Moreover, Inderst and Müller (2004) show that the distribution of bargaining power between start-up firms and venture capitalists affects the financial contracts negotiated, the final contract agreement probability as well as the value created in the start-up firm.

Inspired by Röell (1996) and Pagano et al. (1998), this dissertation examines how bargaining power affects a firm's choices of contract type (debt vs. equity) and placement mode (public offering vs. private placement). Inderst and Müller's venture capital focus is relaxed by incorporating the firm's market segment choice in the decision process. The dissertation deals with firms which already have access to private as well as public debt and equity markets.

The primary objective of this dissertation is to demonstrate that a firm's financing decision depends among other things on bargaining power considerations and to illustrate potential reasons for this dependency. Thereby, the importance of bargaining power considerations for corporate finance is emphasized. Firms planning to raise funds have to be aware about their current bargaining position when negotiating with potential fund suppliers. Firms should know whether and how they can improve their bargaining position.

---

[4] Further benefits of going public are, for example, to overcome borrowing constraints (e. g. credit rationing by banks), to decrease agency problems between shareholders and management by market discipline (e. g. managers have to fear takeovers by superior rivals) and to obtain the option to change corporate control (e. g. sell the company), see Röell (1996) and Pagano et al. (1998).

[5] Ritter (1987) and Barry et al. (1991) estimate that twenty cents per dollar raised at a stock market introduction is a reasonable measure for an average company's total costs.

Additionally, bargaining power considerations might help to reduce the gap between theoretical capital structure predictions and empirical evidence.

The dissertation highlights, for example, that

- the advantages of debt financing increase with a firm's bargaining power since e. g. the (agency) costs of debt financing decrease due to reductions in the firm's repayment obligation while the (agency) costs of equity financing are unaffected by variations in the lenders' proportional return participation,
- the favorability of private placements in comparison to public offerings increases with a firm's bargaining power since the cost difference between a public and a private placement is lower from the firm's than from the lenders' perspective,[6]
- a firm's contract type and placement mode choice are interrelated and must be treated jointly and not separately; from our analysis it becomes obvious that the cost differences of the available placement modes depend on the underlying contract type choice which implies that the optimal placement mode choice depends on the contract type choice and vice versa,[7]
- in the presence of an ex-ante informational asymmetry about the firm's financing and the lenders' profit alternatives the contract agreement probability depends (in a non-monotonous way) on the firm's bargaining power.

1.2 Methodology and Structure

Bargaining problems between agents about splitting gains from trade have a long history in economic analysis, see, for example, the contract curve in the Edgeworth box (cf. Edgeworth (1881)). But due to the difficulty of these problems the assumption of perfect competition is widely used in economic theory (cf. Roth (1985b)). Perfect competition represents an idealized case in which strategic aspects of economic interactions are reduced to negligible proportions by market discipline. However, in typical bargaining situations the interaction between the bargainers is essential for the bargaining outcome and the market simply determines the range of possible outcomes (cf., e. g., Canning (1989)).

---

[6] In our model private placements are assumed to be more expensive than public offerings despite all occurring transaction costs like road shows and investment banking fees. This assumption is empirically confirmed by Hertzel and Smith (1993) and Wu (2004) for equity and by Best and Zhang (1993) as well as Krishnaswami et al. (1999) for debt financing, see p. 40 for details.

[7] Gomes and Phillips (2005) empirically support that a firm's choices of contract type and placement mode are interrelated. In their sample they observe that conditional on issuing in the public market the pecking order of contract types as developed by Myers and Majluf (1984) and Myers (1984), see p. 11 for details, holds. However, in the private market they find a reversal of this pecking order.

Basically, two different approaches have emerged to cope with bargaining problems, i. e. to predict their outcome. Both are based on the game theoretic foundations laid by von Neumann and Morgenstern (1944), showing that each game consists of the following elements: (i) the number of players; (ii) the structure of the game, i. e. who moves when and which actions are available;[8] (iii) the payoffs related to the players' actions / strategies[9] and (iv) the distribution of information among the players (complete vs. incomplete; symmetric vs. asymmetric). The *first approach*, i. e. the *non-cooperative (or strategic)* game theoretic method, is due to Nash (1950, 1951). This approach assumes the absence of coalitions among the bargainers; each bargainer acts independently without communicating with the others. To determine bargaining outcomes each player's strategy to maximize his own profit given the others' strategies are held fixed must be examined. Potential bargaining outcomes are found when *each* player's strategy is optimal against those of the others. The *second approach*, i. e. the *co-operative (or coalition)* game theoretic method, is due to Nash (1953). This approach allows the players / bargainers to discuss their situation and agree on a rational enforceable joint plan of action. In this case bargaining outcomes can be determined by a particular set of postulates or axioms about the relationship of the bargainers' predicted utility outcomes to the set of feasible utilities.[10]

In this dissertation we apply the first approach to explore how bargaining outcomes depend on the bargaining power of the involved parties. We are interested in the strategic choices of the involved parties under a given set of bargaining rules (contract negotiation game). The explicit structure of the contract negotiation game assigns bargaining power to the bargainers. Our analysis is set in a principal-agent setting with a lender (principal) and a borrower (agent) bargaining how to finance the borrower's risky project. The lender and the borrower are risk neutral and are maximizing their expected profits. Informational asymmetries cause difficulties in financial contracting since the lender anticipates adverse selection and moral hazard problems. We distinguish between ex-ante, interim and ex-post uncertainty.

- There is an *ex-ante informational asymmetry* if information relevant for financial contracting is private knowledge of either the lender or the borrower prior to contract agreement (hidden characteristics).

---

[8] An action refers to a particular choice available to a player at a specific (decision) point in the game.

[9] A strategy is a complete listing of the player's actions to be taken at every (decision) point in the game, i. e. the player's contingency plan.

[10] An alternative solution mechanism for co-operative games is proposed by Nash (1951). He proclaims that nearly all co-operative games can be transformed in non-cooperative games (the "Nash Program") and hence latter's' solution techniques become available. However, up to now the "Nash Program" works only in particular, restrictive circumstances (cf. Serrano (2004)).

- There is an *interim informational asymmetry* if the borrower's behavior during the duration of the contract is unobservable for the lender (hidden action).
- Finally, in the case of an *ex-post informational asymmetry* the borrower's project return cannot be observed by the lender at zero cost (costly state verification).

The contract negotiation game is analyzed for *four altenative bargaining power scenarios* while power effects are determined by a pair-wise (binary) comparison of the optimal contract choices in the alternative bargaining power scenarios. It seems reasonable to employ our bargaining power analysis in an asymmetric information framework as informational asymmetries are crucial for bargaining outcomes (cf. Crawford and Sobel (1982)).[11]

The remainder of this dissertation is divided into five chapters.

Related literature is reviewed in *Chapter 2*. Section 2.1 focuses on a firm's contract type / placement mode-choice in the presence of an ex-ante, interim or ex-post informational asymmetry. Section 2.2 demonstrates in which areas of financial contracting bargaining power aspects have previously been considered. Section 2.3 concludes this chapter by summarizing the contributions of the current study for research in the areas of financial contracting under asymmetric information or under alternative bargaining power scenarios.

The basic model to examine bargaining power effects is developed and analyzed in *Chapter 3*. Section 3.1 describes the basic principal-agent setting where a lender (principal) and a borrower (agent) bargain about how to finance the borrower's risk project when interim and ex-post uncertainty cause moral hazard difficulties. No ex-ante informational asymmetry is considered. The contract negotiation game is assumed to be a three stage (sequential) game where firstly the contract type (debt vs. equity) and the placement mode (public offering vs. private placement) choices have to be made before the respective contract conditions can be determined. Finally, at the last stage the contract conditions can either be accepted or rejected. If they are accepted the borrower's project is financed. Otherwise, the game ends and the lender as well as the borrower are left with their outside options, i. e. their profit / financing alternative (if they have any). For this contract negotiation game four alternative bargaining power scenarios are defined where the power to determine contract type / placement mode and the right to set contracts' conditions are either assigned jointly or separately to the lender or to the borrower. The resulting expected profits are defined in Section 3.2. However, before bargaining power effects can be determined by a pair-wise (binary) comparison of the optimal contract choices in the alternative bargaining power scenarios, see Section 3.5, the contract negotiation game has to be solved for the alternative power scenarios (Sections 3.3 and 3.4). Accor-

---

[11] For bargaining under asymmetric information see, e. g., Roth (1985a), Chatterjee (1985), Fudenberg et al. (1985) and Canning (1989).

ding to the backward-induction principle, firstly all feasible contract type /
placement mode-choices, i. e. satisfying the lender's and the borrower's participation constraints, are defined in Section 3.3 before a profit comparison of
all feasible choices reveals the optimal choice for the alternative bargaining
power scenarios (Section 3.4).

The basic model is extended in *Chapter 4* by an ex-ante informational
asymmetry about the lender's profit and the borrower's financing alternative.
The lender, respectively the borrower, possesses either a positive outside option (A-type) or not (N-type). Hence, the lender and the borrower are unaware
of the contractual partner's outside option when negotiating about financing.
Such an ex-ante uncertainty limits the lender's, respectively the borrower's,
rent extraction opportunity resulting from the assigned bargaining power.
The chapter is structured in accordance with the previous chapter. In Section
4.1 the ex-ante informational asymmetry is defined. In Sections 4.2 and 4.3
the contract negotiation game is solved for the alternative bargaining power
scenarios before bargaining power effects are determined in Section 4.4. Due
to the ex-ante uncertainty about the contractual partner's outside option the
feasible contract type / placement mode-choices now depend on the lender's,
respectively the borrower's, (ex-ante) private information since they know
their own type but are unaware of the partner's type. The robustness of the
derived bargaining power effects is examined in Section 4.5.

Chapter 5 copes with possible methodological concerns. Section 5.1 states
the main concerns while Section 5.2 presents our methodological justification.

Finally, *Chapter 6* concludes this dissertation by summarizing the obtained
bargaining power effects in Section 6.1 and discussing their economic relevance
in Section 6.2.

2

A Review of Related Research

In this chapter we review studies on financial contracting under asymmetric information (Section 2.1) and show in which areas of financial contracting bargaining power aspects have previously been analyzed (Section 2.2). To demonstrate the main ideas, we concentrate on the most prominent studies in these areas.

More comprehensive reviews can be found, for example, in Harris and Raviv (1991) and Hart (2001), while Schmid-Klein et al. (2002) focus on informational aspects in a firm's financing decision. Graham and Harvey (2001) even present evidence about the significance of informational and bargaining power aspects in a firm's financing decision from the chief financial officer's perspective.

At the end of this chapter we point out how this dissertation is related to the studies conducted so far in the areas of financial contracting under asymmetric information and under alternative bargaining power constellations (Section 2.3).

2.1 Financial Contracting Under Asymmetric Information

Assuming a basic financial contracting problem, i. e. a firm (the borrower) has a project but insufficient funds to finance the project while an investor (the lender) has funds to invest but no project, our aim is to examine whether a contract will be agreed upon and, if so, how the project will be financed. The outcome of the project is uncertain. Informational asymmetries between the lender and the borrower can exist ex-ante, interim and/or ex-post, causing difficulties for financial contracting. Since the objective of this section is to become familiar with difficulties caused by different types of informational asymmetry, each type is considered separately.

Depending on the type of informational asymmetry between the lender and the borrower when bargaining if and how to finance the borrower's risky project, either "private information" or agency problems are caused for financial contracting. "Private information" problems arise when the borrower possessing private information e. g. about his investment opportunities tries to raise funds from potential lenders, i. e. uninformed capital market participants.[1] In such a situation "private information" problems create adverse selection difficulties for financial contracting. Apart from "private information" problems, agency problems, i. e. problems due to a conflict of interest among the lender (principal) and the borrower (agent), arise when the consequences of the borrower's (agent's) actions are mainly borne by the lender (principal).[2] The latter problems are essential elements of the contractual view of the firm, developed by Coase (1937), Jensen and Meckling (1976), and Fama and Jensen (1983). The problems for financial contracting due to informational asymmetries and their implications are discussed in more detail in Sections 2.1.1 and 2.1.2.

As the aim of this dissertation is to examine how the (ex-ante) distribution of bargaining power affects a firm's financing decision, we do not determine a (new) contract which copes optimally with a particular informational asymmetry, but base our analysis on a set of predefined contracts. Therefore, we restrict the following review to studies about these contracts, i. e. to debt and equity contracts which are either publicly or privately placed. The implications of informational asymmetries for a firm's contract type choice are discussed in Section 2.1.1, while Section 2.1.2 deals with a firm's placement mode choice.

The empirical evidence on theoretical predictions in financial contracting under asymmetric information is far from conclusive, i. e. not consistently supporting one theory (cf. Schmid-Klein et al. (2002)). Hence, we only state exemplary findings which we take at face value. We do not question e. g. the methodologies or the data sets used.

2.1.1 A Firm's Contract Type Choice Under Asymmetric Information

This section focuses on a firm's contract type choice, i. e. debt vs. equity financing. The distinction is motivated by the observation that debt financing typically involves a fixed repayment obligation to outside investors which is supported by a conditional right to liquidate the firm's assets if the obligation is not fully satisfied. On the other hand, equityholders are promised a fractional return participation after debtholders' obligations are satisfied. This

---

[1] Harris and Raviv (1991) refer to this approach to financial contracting as the asymmetric information approach.

[2] Harris and Raviv (1991) refer to this approach to financial contracting as the agency approach.

fractional right is supported by an unconditional right to engage in the firm's decision making (cf. Myers (2000)).[3]

2.1.1.1 Contract Type Choice in the Presence of an Ex-ante Informational Asymmetry

Ex-ante asymmetric information between lender and borrower has distinct effects on a firm's (the borrower's) debt-equity choice depending on whether the firm's investment decision is considered to be endogenous (see, e. g., Myers and Majluf (1984)) or exogenous (see, e. g., Ross (1977)).

For example, Myers and Majluf (1984) assume a manager needs capital to finance a project. The project can be valuable, i. e. possessing a positive net present value. The manager has private knowledge about the firm's value and the project's prospects and acts in the interest of the current shareholders. Therefore, he only issues new shares to finance the valuable project

- if the firm is overvalued since a new issue transfers wealth from the new to the current (old) shareholders, since the new shareholders pay more for their rights than they are actually worth, or
- if the firm is undervalued, which implies that a new issue transfers wealth from the old to the new shareholders, only equity financing is possible, and the current (old) shareholders' wealth loss due to the transfer is overcompensated by their project return participation.

Otherwise the valuable project is not financed by equity. On the other side, a project with a negative net present value is only financed if the firm is overvalued and the wealth transfer due to the equity issue overcompensates current (old) shareholders' loss of wealth due to the bad project. Anticipating this behavior potential investors interpret an equity issue announcement as a bad signal and adjust their expectations since they assume being exploited. They can not separate information about the new project from information about whether the firm is under- or overvalued. This expectation adjustment affects the price investors are willing to pay and therefore the firm's issue-investment decision.

To avoid such difficulties, Myers (1984) suggests the pecking order for external finance. Projects should be financed by retained earnings and riskless debt before issuing (risky) equity to avoid the dilemma of either passing up valuable projects or issuing undervalued shares. Therefore, capital structure emerges dynamically over time to mitigate inefficiencies in a firm's investment decision, i. e. as a solution to the problem of underinvestment due to the dilution costs of selling underpriced securities.

A different approach is illustrated by Ross (1977). He assumes that the firm's investment decision is fixed and capital structure is designed to signal (private) insider information to investors to optimize market valuation.

---

[3] See Titman and Wessels (1988) for a study considering further contract type distinctions such as short-term, long-term and convertible debt.

In Ross's model, managers know the true distribution of the firm's returns while investors do not (ex-ante uncertainty) and managers benefit from high market valuation of the firm's securities. However, managers are penalised if the firm goes bankrupt. Hence, the managers try to signal the firm's prospects by their capital structure choice. A large debt level is taken as a signal for good firm prospects (high quality). Firms with poorer prospects have higher marginal expected bankruptcy costs for any debt level and, therefore, managers do not imitate higher quality firms by issuing more debt, i. e. a separating equilibrium exists.

Both models provide rich sets of empirically testable hypotheses, see, e. g., Harris and Raviv (1991). The results of the hypotheses' tests are mixed. In total neither the pecking order theory nor signaling models seem to be a good description of reality (cf. Schmid-Klein et al. (2002)).

For example, according to Myers and Majluf (1984) and Myers (1984)

- leverage should increase with reductions in the firm's free cash flow; this hypothesis is empirically confirmed by Chaplinsky and Niehaus (1993),
- dividend-paying firms are expected to hold more debt since dividends are part of the firm's financing deficit (cf. Shyam-Sunder and Myers (1999)); this hypothesis is rejected by Frank and Goyal (2003b),
- leverage should be negatively related to the firm's profitability; this hypothesis is widely confirmed, see, e. g., Titman and Wessels (1988), Rajan and Zingales (1995), and Frank and Goyal (2003a,b).

According to Ross (1977)

- a firm's stock price should increase on the announcement of a debt issues; this hypothesis is empirically confirmed by Masulis (1983),
- leverage is supposed to be positively related with firm value; Israel et al. (1989) confirm this prediction,
- contrary to Myers and Majluf (1984) leverage should increase with a firm's profitability; this hypothesis is rejected by Titman and Wessels (1988), Rajan and Zingales (1995), and Frank and Goyal (2003a,b).

Finally, both models share the prediction that

- a firm's stock price decreases on the announcement of an equity issue; Masulis and Korwar (1986) confirm this expectation.

2.1.1.2 Contract Type Choice in the Presence of an Interim Informational Asymmetry

Interim informational asymmetry can exist between bondholders and shareholders and between shareholders and managers. Therefore, two moral hazard problems can arise since the agent, e. g. the manager, maximizes primarily his own profit which implies in the presence of market imperfections like asymmetric information that the principals, e. g. the bond- and/or shareholders, have to incur agency costs to keep the agent in line, e. g. by monitoring his behavior or by designing particular incentive schemes (cf. Jensen and Meckling (1976)). Agency costs arise due to a conflict of interest among agent and principal. Firstly, in this section we focus on agency problems between shareholders (agents) and bondholders (principals), neglecting potential difficulties between shareholders and managers. I. e. we assume that the managers' interests are fully aligned with the shareholders' interests. This assumption is then relaxed and we focus, in particular, on agency problems between shareholders (principals) and managers (agents). Overall, we see that the value of the firm is not fixed and depends on the ownership, i. e. the capital structure, of the firm.

In the context of an interim informational asymmetry between bondholders and shareholders, firstly, Myers (1977) demonstrates that the resulting agency problems affect a firm's investment decisions and thereby the value of the firm. He shows that (long-term) debt financing provides shareholders with a potential incentive to surpass positive net present value projects since shareholders only receive the profit remaining after debtholders' obligations are satisfied. This behavior results in an underinvestment problem by shareholders, which bondholders anticipate and hence demand an appropriate compensation.

Another incentive problem arises from interim uncertainty among bond- and shareholders due to the firm's limited liability in debt financing. This property of debt financing can encourage the firm's shareholders to unobservably increase the firm's risk, i. e. engage in asset substitution (cf. Jensen and Meckling (1976)).[4] Asset substitution can be attractive to shareholders since they fully participate from the return's upside potential while the downside potential is limited. Anticipating the shareholders' adverse behavior the bondholders demand compensation (ex-ante) for the costs incurred to mitigate this problem, e. g. writing restrictive covenants and monitoring the firm's investment policy.

Therefore, debt financing can cause different kinds of agency problems, i. e. costs in the presence of interim uncertainty among bond- and shareholders.

So far any potential interim informational asymmetry between shareholders and managers has been neglected, but such an informational asymmetry can cause agency costs of equity financing when managers (agents) are able to maximize their utility at the the expense of the shareholders (principals). Therefore, shareholders have to monitor the managers to ensure that they act in their interest. Jensen and Meckling (1976), for example, assume that managers have the opportunity of consuming "non-pecuniary" benefits (perks) like fancy offices, etc. These benefits are attractive to managers while the costs

---

[4] Alternatively, one can refer to the asset substitution problem as debt financing's risk-shifting incentive. While the former considers the case in which the borrower can unobservably choose among different risky projects in the latter case the borrower sticks to the same project but can unobservably vary the risk of the project.

of these benefits are born at least partly by others, i. e. shareholders. These benefits are not in the interest of shareholders since they reduce the value of the firm.

Jensen and Meckling show that the optimal debt-equity mix, i. e. the firm's capital structure is determined at the point where the marginal benefits of keeping the manager from taking perks is offset by the marginal costs of causing risky behavior.

Furthermore, Jensen (1986) finds that if a firm has sufficient "free cash flow", i. e. its managers may "build empire" or make inefficient investment decisions, debt financing with its fixed obligation helps to discipline firms' management.

The benefit of debt financing is therefore to mitigate any potential conflict between equityholders and managers.

The empirical evidence concerning all hypotheses from these theoretical models is again non conclusive (cf. Harris and Raviv (1991), Schmid-Klein et al. (2002)).

However, empirically confirmed are the following exemplary selected predictions by Jensen and Meckling (1976),

- leverage should increases with lack of growth opportunities (see Titman and Wessels (1988), Chaplinsky and Niehaus (1993)), and
- covenants prohibit asset substitution in bond financing (see Smith and Warner (1979)),

while the hypothesis by Jensen (1986) that

- leverage increases with increases in "free cash flow" has not been confirmed by Chaplinsky and Niehaus (1993).

2.1.1.3 Contract Type Choice in the Presence of an Ex-post Informational Asymmetry

Finally, the most obvious moral hazard problem occurs in the presence of ex-post informational asymmetry between lender and borrower since the borrower always has the opportunity to understate the firm's, i. e. the project's, return in order to minimize his repayment obligation.

Assuming that the decision to verify the return stated by the borrower follows a deterministic scheme[5], debt contracts dominate equity contracts (cf. Townsend (1979), Gale and Hellwig (1985) and Williamson (1986)). In debt financing verification is only necessary in the proclaimed default state while equity financing requires that the lender always verifies the proclaimed return since cheating is always profitable for the borrower.

---

[5] A verification scheme is called deterministic when before the borrower's return announcement the lender has to define under which circumstances, e. g. return announcements, he will verify the borrower's proclamation.

However, studies by Bernanke and Gertler (1989) and Mookherjee and Png (1989) question the "classical" debt contract's optimality. They show that stochastic verification mechanisms are Pareto-superior, i. e. reduce the verification costs and still prevent the borrower from cheating.[6] Stochastic mechanisms also cope with the deterministic verification mechanism's shortcoming of lacking subgame completeness, i. e. to verify the project return (ex-post) even if the potential gains exceed the costs.

Furthermore, by relaxing the assumption that debt and equity contracts cause identical verification costs recently Hvide and Leite (2002) integrate equity contracts in the costly state verification setting and show that an optimal debt-equity mix exists. Hvide and Leite argue that the verification costs of equity financing are lower than the costs for debt financing since debt- and equityholders possess different control rights which implies that their incentives to invest in "cheap" monitoring technology ex-ante differ. According to Habib and Johnsen (2000) the verification costs differ since equityholders care about the firm in its primary use, i. e. the firm's operating value, while debtholders are concerned about the firm's alternative use, i. e. the firm's liquidation value.

Hvide and Leite (2002) point out that the implications drawn from such extended costly state verification frameworks are consistent with observed empirical regularities like strategic defaults of debt obligations and low debt ratios for high risk projects.

2.1.2 A Firm's Placement Mode Choice Under Asymmetric Information

In this section a firm's placement mode decision under asymmetric information is analyzed. We focus on the choice of public offering vs. private placement. The distinction is motivated by the fact that typically a borrower can either negotiate privately with a small group of investors (private placement) or issue the securities publicly to a large number of dispersed investors (public offering).

Firms with the intention to raise debt can either issue a corporate bond publicly or place a bond privately by selling the debt contract directly to a small group of investors (cf. Kwan and Carleton (2004)). Private lenders are mainly institutional investors such as commercial banks and insurance companies which are specialized in the assessment of credit quality before a debt issue and in monitoring the firm's performance after an issue (cf. Krishnaswami et al. (1999)).[7] We abstract from further placement mode distinctions, i. e.

---

[6] A verification scheme is called stochastic when the lender can decide after the borrower's return announcement whether to verify the return or not.

[7] Best and Zhang (1993) empirically confirm the expectation that private lenders have in comparison to public lenders superior firm-specific information.

we treat a bank loan like a privately placed corporate bond since in both circumstances the borrower and the lender(s) negotiate personally.

Firms with the intention to raise equity can also choose between a public offering or a private placement. The choice is even more complex since different types of public equity offerings exist. A firm can e. g. choose between a regular offering or an (uninsured) rights offering.[8, 9] Further alternatives are an (insured) underwritten rights offering (an underwriter has a standby commitment to purchase any unsubscribed shares) or a firm-commitment underwritten rights offering (the underwriter agrees to purchase all new shares for resale to public).[10] However, all public offerings have in common that the securities are offered publicly to a large number of dispersed investors, while in a private equity placement new shares are sold to a small group of current or new investors.

The structure of this section is related to the structure of the previous section. Firstly, we examine the implications of a potential ex-ante informational asymmetry between lender and borrower for a firm's placement mode choice before considering the effects of an interim asymmetry.[11] The factors affecting debt's placement decision are similar to the factors influencing equity's placement mode choice while differences arise due to equityholders' right to control the firm in the non-default state, i. e. when debtholders' claims are satisfied.

2.1.2.1 Placement Mode Choice in the Presence of an Ex-ante Informational Asymmetry

2.1.2.1.1 Debt's Placement Mode

Each firm's debt placement mode choice is affected by ex-ante informational asymmetries between lender and borrower, since each borrowing decision causes adverse selection difficulties for financial contracting (cf. Leland and Pyle (1977)).

These difficulties can arise since the managers, acting in the interest of the current shareholders, only issue securities to new investors if investors

---

[8] See, e. g., Heron and Lie (2004) for a comparison of a regular and an (uninsured) rights offering's information content.

[9] In a rights offering current shareholders obtain a short term option, i. e. right, to purchase the new shares issued on a pro rata basis at the exercise price. Rights offerings are used to prevent any potential wealth transfers from current to new shareholders.

[10] See, e. g., Cronqvist and Nilsson (2003).

[11] The effects considered in this section are separated according to their origin. For interdependencies among the stated effects see, e. g., Diamond (1991). Diamond examines a firm's choice between bank loan and public debt in the presence of ex-ante and interim uncertainty.

are willing to pay as much or more than the managers believe the securities are worth. The managers try to prevent any wealth transfer from current shareholders to new investors. Therefore, the managers, i. e. respectively the firms, care about who provides their funding because different providers have different information and expectations about the firm's prospects. Hence, their valuations of the securities offered differ. Investors who believe they have poorer information than managers anticipate the managers' behavior and pay less for new securities than better informed investors do since they assume to be exploited (cf. Mackie-Mason (1990)). Thus the firm prefers to obtain funds from investors which are better informed and do not require a large premium as compensation for potential adverse selection.[12]

Additionally, Ramakrishnan and Thakor (1984) demonstrate that private lenders have a comparative advantage to public lenders in producing firm-specific information since they possess information which is not publicly available. Hence, firms with greater ex-ante uncertainty such as younger firms will issue more private than public debt, since private lenders mitigate the contracting costs due to adverse selection difficulties.

This hypothesis is empirically confirmed by Blackwell and Kidwell (1988). They find that firms with only private debt are significantly younger and have higher levels of ex-ante informational asymmetry than firms which issue public debt. Furthermore, Mackie-Mason (1990) reveals that problems due to ex-ante informational asymmetry are a significant determinant of a firm's placement mode choice even after controlling for the security type (debt or equity).

However, Rajan (1992) damps the placement mode prediction obtained by the consideration of potential adverse selection difficulties. He points out that private lenders can obtain information monopolies about their borrowers in the long-run due to their monitoring and relationship lending. Such information monopolies increase private lenders' bargaining power in financial contracting enabling the lenders to extract excess returns from project financing. In such circumstances, Rajan argues, public debt is beneficial to reduce the private lenders' ability to extract rents.

Another aspect is put forward by Dhaliwal et al. (2003). The authors argue that public offerings are not only costly due to potential adverse selection difficulties. In addition, the public disclosure of information may convey information to the firm's competitors harming the firm's prospects. Of course, an optimal level of public disclosure exists balancing the costs and the benefits of

---

[12] This consequence of an ex-ante informational asymmetry has firstly been illustrated by Akerlof's "market for lemons" (cf. Akerlof (1970)). Akerlof demonstrated that the market for used cars can collapse if the sellers are better informed about the quality of their cars than the potential buyers since the buyers anticipate their disadvantage and are only willing to pay a price for an expected average quality car. Hence, the sellers of high quality cars leave the market since their cars are worth more than the buyers are willing to pay. Again this is anticipated by the buyers. Finally, only cars with the poorest quality remain in the market.

public disclosure (cf. Verrecchia (1983)). However, if public disclosure has to be increased above this level further information relevant to the firm's competitors is revealed and causes the firm to lose its competitive advantage or bargaining power (cf. Admati and Pfleiderer (2000)). Any information disclosure above the optimal level is harmful for the firm. Yosha (1995) finds that private placements help to protect valuable firm-specific information from being publicly disclosed. Additionally, private lenders are less likely to reveal their private information since they are concerned about the borrower's success.

2.1.2.1.2 Equity's Placement Mode

Similar to debt financing, the presence of ex-ante informational asymmetry between a firm, i. e. respectively its managers, and potential investors causes adverse selection problems if equity is issued regularly to outside investors.

This is shown by Myers and Majluf (1984). As already mentioned, they demonstrate that managers only issue equity if the firm is overvalued and any potential wealth loss due to a negative net present value project is overcompensated by the wealth transfer from new to current shareholders; or if the firm is undervalued while the wealth transfer from current to new shareholders is overcompensated by shareholders' project return participation. This issuing behavior is anticipated by the investors. Therefore, investors fear to be exploited and demand an appropriate compensation which in turn affects the firm's issue-investment decision. An underinvestment problem occurs.

One solution to this underinvestment problem is suggested by Eckbo and Masulis (1992) and Eckbo and Norli (2005). They suggest to allow current shareholders to participate in public equity placements by rights offerings. However, the adverse selection and hence the underinvestment problem prevails when the firm expects that less that 100% of their shareholders participate in the equity issue. In such situations where the expected participation rate is below 100% only underwritten rights issues can ease the adverse selection difficulties. Underwriters perform a certification role to mitigate any informational asymmetry between managers and potential investors. Based on these ideas and bearing in mind that underwriting services are costly, Frank and Goyal (2005) derive a pecking order of equity floatation method choices: Firms expecting low shareholder participation in an (uninsured) rights issue prefer underwritten rights offerings while firms expecting high shareholder participation favor (uninsured) rights offerings.

An alternative solution to this underinvestment problem is put forward by Hertzel and Smith (1993). They incorporate the possibility of private placements in the Myers and Majluf's framework. Private placements enable the investor to reveal the true value of the firm at some costs. Therefore, undervalued firms prefer private placements as long as the net present value of their investment opportunity exceeds the costs of information production borne by the private investors.

Cronqvist and Nilsson (2003) empirically confirm that adverse selection costs affect a firm's equity placement structure. They observe that firms which are potentially undervalued depend significantly on private financing.

As for debt financing, the aspect of costly information disclosure, put forward by Dhaliwal et al. (2003), is relevant for the firm's equity placement mode choice. Increased public disclosure above the optimal level might decrease the firm's competitive advantage and/or bargaining power (cf. Admati and Pfleiderer (2000)). Therefore, private equity placements become preferable to conceal valuable information from the firm's competitors.

An argument against private equity financing in the presence of an ex-ante informational asymmetry à la Rajan's argument against private debt financing (cf. Rajan (1992) or Section 2.1.2.1.1) does not exist since shareholders are the residual claimholders anyway. However, in the presence of an ex-ante informational asymmetry between the firm's management and its shareholders similar considerations have to be taken into account.

2.1.2.2 Placement Mode Choice in the Presence of an Interim Informational Asymmetry

2.1.2.2.1 Debt's Placement Mode

Similar to the effects of an interim uncertainty on a firm's contract type choice, the lender's potential interim uncertainty about the borrower's, i. e. the firm's, behavior can cause two moral hazard problems which affect the cost of financial contracting and, thereby, the firm's preferred debt placement structure (see, e. g., Krishnaswami et al. (1999)).

Firstly, the asset substitution problem might arise since the borrowing firm's limited liability causes an adverse incentive for the firm's risk taking behavior. Due to limited liability the firm's shareholders and managers have an incentive to undertake riskier projects than previously agreed since they fully participate from the project return's upside potential but not from the return's downside potential (cf. Jensen and Meckling (1976)). Galai and Masulis (1976) illustrate the same difficulty by considering a leveraged firm's equity as a call option on the firm's underlying assets. An increase in the firm's cash flow risk therefore increases the value of the firm's equity while decreasing the value of the firm's debt. Debtholders who are unable to monitor the firm's behavior (interim uncertainty) anticipate this (adverse) debt financing response and demand a compensation for the firm's potential risk adjustment.

The second moral hazard problem, the problem of underinvestment, results from the fact that the firm's shareholders only receive the cash flows which remain after debt financing's obligations are satisfied. Therefore, a firm with debt outstanding only undertakes projects when returns exceed the face value of debt instead of considering all projects with a positive net present value (cf. Myers (1977)). The lenders anticipate this behavior and want an appropriate compensation.

Monitoring the firm's behavior and restricting it by covenants are options to mitigate both moral hazard problems. Therefore, privately placed debt is preferable to public debt since private lenders generally possess a stronger incentive to monitor the firm's behavior since the number of lenders is smaller which implies that the average default risk is higher (cf. Nakamura (1993)). Additionally, private lenders have a comparative advantage to public lenders in enforcing bond covenants since formal bankruptcy proceedings are not always optimal to cope with covenant violations (cf. Smith and Warner (1979)). Alternative private renegotiations are easier among a small number of private lenders than among a large number of dispersed public investors (cf. Chemmanur and Fulhieri (1994)).[13]

Krishnaswami et al. (1999) observe that firms with greater moral hazard problems hold higher proportions of private debt. They conclude that better monitoring incentives and stricter covenants of privately placed debt help to mitigate the agency costs of debt financing.[14] Denis and Mihov (2003) find as well that private lenders' concentrated holdings and superior access to information constrain managerial discretion.

2.1.2.2.2 Equity's Placement Mode

As it has been put forward by Cronqvist and Nilsson (2003) a new equity issue obviously affects a firm's control structure and thereby the firm's controlling parties' ability to use their discretion over the firm's decisions to extract private benefits. Furthermore, a private investor is probably a "large" shareholder with the incentive to monitor the firm's management and their use of the proceeds (cf. Shleifer and Vishny (1986)). Additionally, the usually lower liquidity of privately placed equity further increases the investor's incentive to monitor (cf. Maug (1998)). Therefore, private equity placements improve monitoring incentives even if the firm's ownership structure remains nearly unaffected. Hence, a private placement can reduce the value reducing managerial discretion.

The emerging hypothesis that corporate control considerations are a significant determinant of a firm's equity placement mode choice is empirically confirmed by Cronqvist and Nilsson (2003). They find that large controlling shareholders who enjoy significant private benefits of control are "control-dilution averse". A firm's ownership concentration is negatively related to the probability of a private placement. Current controlling shareholders are in particular control dilution averse if their control margin is small.

Another potential moral hazard problem affecting a firm's equity placement mode choice is related to strategic alliances in the product market.

---

[13] Private renegotiations and in particular the number of creditors become important in financial distress, see e. g. Bolton and Scharfstein (1996).

[14] This finding is supported by the observation that regulated firms tend to have lower private debt proportions than non-regulated since the former have alternative monitoring mechanisms (cf. Krishnaswami et al. (1999)).

Suppose, for example, the positive net present value of a joint venture is only realized if both partners invest in the project and no credible commitment opportunity exists. Contracts are incomplete. A private equity placement to the strategic partner can therefore increase the firm's incentive to invest in the joint venture, lowering contracting costs (cf., e. g., Aghion and Tirole (1994)).

In their sample of private and public equity placements Cronqvist and Nilsson (2003) find evidence that monitoring considerations in product market partnerships help to explain equity's placement mode choice. They show that strategic partners employ equity ownerships to reduce contracting costs.

2.2 Bargaining Power Considerations in Financial Contracting

In this section we review studies of financial contracting where bargaining power aspects have been considered. Since financial contracting is quite a broad field of research, see, e. g., Hart (2001) for a recent review, we focus on the in our opinion most interesting studies to demonstrate the different areas where bargaining aspects have been taken into account. This section is divided into three parts. Part one (Section 2.2.1) and two (Section 2.2.2) both review studies of financial contracting where firm's capital structure is used as a bargaining device, i. e. the firm's capital structure is chosen to optimize the firm's bargaining position in future negotiations. While in Section 2.2.1 only claim holders' cash flow rights matter, Section 2.2.2 focuses on their voting and control rights, i. e. on corporate control considerations. Finally, in Section 2.2.3 a quite new area of research is presented, examining the implications of capital market competition for firms' financing decisions. The studies in this new area are so far limited in the sense that they simply focus on one particular capital market segment, i. e. either bank or venture capital financing, and ignore all other possible market segments. The principal-agent model developed in this dissertation (see Chapters 3 and 4) to examine how the (ex-ante) distribution of bargaining power between a lender and a borrower (a firm) affects the firm's choices of contract type (debt vs. equity) and placement mode (public offering vs. private placement) belongs to this new research area while the bank / venture capital financing focus is relaxed.

Of course, further studies of bargaining power in financial contracting exist, in particular, focusing on sovereign-debt renegotiations.[15] However, in this

---

[15] See, for example, Fernández and Özler (1999) and Klimenko (2002). The former study analyzes how debt concentration - the proportion of the country's debt held by large banks relative to small banks - affects the bargaining power of the creditors in debt renegotiations as well as the maximum penalty they might impose on the debtor, while the latter examines the effects of the debtor country's trade pattern as well as the IMF's and the World Bank's sovereign debt policy on debt reschedulings.

section we only review corporate finance related studies since this dissertation examines how bargaining power affects a firm's financing decision.[16]

2.2.1 Leverage as a "Bargaining Tool"

In this section we present studies which focus on bargaining situations between different stakeholders of the firm where the firm's capital structure is used as a bargaining device. Studies about managerial compensation are reviewed in Section 2.2.1.1. Sections 2.2.1.2 and 2.2.1.3 concentrate on bargaining situations with firm "outsiders" belonging either to the firm's input or product market.

2.2.1.1 Managerial Power and Compensation

Managers hired by the firm's shareholders to maximize their wealth typically possess some managerial discretion in their behavior, e. g. due to informational asymmetries. Additionally, the interests of the shareholders (principals) and of the managers (agents) are not perfectly aligned (cf. Bebchuk et al. (2002)). Hence an agency problem exists and managers use their discretion to positively affect their compensation. Osano (2003), for example, suggests that managers can influence the bargaining process to determine managerial compensation by influencing the independency of the board members and their monitoring ability. Alternatively, the managers can affect the principal agent problem underlying their relationship with shareholders and thereby increase their compensation (cf. Stoughton and Talmor (1999)). Basically, managers' objective is to achieve a high level of compensation and a secure job. However, this opposes the shareholders' interests and they will improve their bargaining position by altering the firm's capital structure since this affects the firm's free cash flow and the managers' incentives (cf. Kose et al. (1993)).[17] Additionally, Berkovitch et al. (2000) demonstrate that a firm's leverage ratio influences the probability of managerial replacement and the managers' negotiated compensation schemes.

2.2.1.2 Bargaining with Suppliers and in Particular Employees

A firm is often confronted with situations in which it has to make up-front investments in firm-supplier relations while the suppliers' response is unclear.

---

[16] Sovereign debt differs in many aspects from corporate debt (cf., e. g., Bolton and Jeanne (2005)). One important difference is that if a sovereign defaults, there is *no* possibility for the lender to initiate formal bankruptcy proceedings in order to seize the borrower's assets, while for corporate defaults this opportunity exists. Hence, the corporate and sovereign debt markets are characterized by different incentive problems.

[17] The argument is based on the observation that "hard claims", i. e. debt contracts, help to restrict managerial discretion (cf. Hart and Moore (1995)).

Up-front investments in such situations weaken the firm's bargaining position in subsequent negotiations (cf. Subramaniam (1996)). The profits generated by these relations are exposed to expropriation by the supplier. The inability to write precommitment contracts causes therefore an underinvestment problem. However, this problem can be mitigated by debt financing since debt financing shields wealth from the suppliers (cf. Dasgupta and Sengupta (1993)).

Furthermore, an advantage of debt financing is that debt financing strengthens the bargaining position of the shareholders, respectively their managers, in dealing with input suppliers and employees since bondholders bear the main costs of any bargaining failure but do not fully participate from its gains (cf. e. g. Sarig (1998)). Bondholders partly insure equityholders against bargaining failures. Therefore, increases in the firm's leverage increase the equityholders' insurance and hence their bargaining position. Debt financing positively affects firm value.

Consequently, a firm should hold more debt the greater the bargaining power and/or the market alternatives of its suppliers are. Sarig predicts that highly unionized firms have more debt. Hanka (1998) empirically confirms this prediction. Bronars and Deere (1991) even find that firms use debt financing to protect wealth from the threat of unionization, i. e. the formation of a collective bargaining unit.

2.2.1.3 Bargaining with Rival Firms, Regulators and Prosecutors

Brander and Lewis (1986) propose another way in which a firm's production and financing decision are interwined. They demonstrate that a substantial amount of debt financing can credibly signal rival firms that the firm is not willing to reduce production. Instead the firm pursues an aggressive output policy. Thereby, the firm's bargaining position against its rivals is improved due to (strategic) debt financing.

Dasgupta and Nanda (1993) show that for regulated firms their capital structure choice, i. e. its leverage, is an effective bargaining device to induce the regulator to set higher prices for the firm's output. Equityholders' bargaining advantage arises since the main costs of any bargaining failure are borne by the bondholders. The equityholders are protected by limited liability. Dasgupta and Nanda's model predicts that firms prefer higher debt levels in harsh regulated environments than in not so harsh regulated environments. This prediction is consistent with cross sectional evidence the authors present for US electric utilities.

Finally, a firm's level of debt also affects the firm's bargaining position in settlements of civil claims when the civil judgement might force the firm into bankruptcy (cf., e. g., Spier and Sykes (1998)). A high level of debt makes shareholders tougher bargainers in pretrail negotiations by narrowing the settlement range. The usefulness of debt as a bargaining device depends on the bankruptcy priority rules. If the civil plaintiff does not receive top priority in bankruptcy debt financing directly dilutes the value of the civil claims. But even if the civil plaintiff receives top priority in bankruptcy the costs of large civil judgements are partly borne by the debtholders.

2.2.2 Corporate Control Considerations

The main reason why investors provide external financing to firms is that they receive control rights in exchange (cf. Shleifer and Vishny (1997)). Shareholders, for example, obtain the right to vote on corporate matters, such as the election of the board of directors, while creditors can liquidate the firm's assets if the firm defaults on its debt or participate in the company's reorganization.

Therefore, the firm's prospects depend on its capital structure since different claim holders, i. e. shareholders and creditors, have different control rights in different situations. Furthermore, the structure of debt- and equity-holders is important since it is not sufficient to assign rights to a particular party. The party, i. e. shareholders or creditors, must also be able to execute these rights and not be hampered e. g. by coordination problems among a large number of dispersed investors.

In this section we illustrate where bargaining power aspects have been analyzed in corporate control considerations. Section 2.2.2.1 focuses on equity's voting right structure, Section 2.2.2.2 on aspects of debt financing, while Section 2.2.2.3 deals with the determination of the optimal debt-equity mix. See Shleifer and Vishny (1997) and Becht et al. (2003) for more comprehensive corporate control reviews.

2.2.2.1 Equity's Voting Right Structure

To demonstrate the presence of bargaining power considerations in the determination of equity's optimal voting right structure, we focus on studies examining the active market for corporate control (takeover models).[18] In these models the optimal allocation of cash flow and voting rights assigned to equity securities is determined by its effects on rivals' behavior to obtain control of the firm from an incumbent management. The allocation of voting rights determines the securities a party must acquire to win control about the firm, while the assignment of income claims to the same securities determines the costs of acquiring these voting rights.

Grossman and Hart (1988), for example, assume that an entrepreneur draws a corporate charter for his firm in order to maximize the value of the firm. He can create different classes of shares, where the share of votes and the share of dividends assigned to a class of shares can vary. The entrepreneur only cares about the value of the firm (private optimization). He anticipates

---

[18] For alternative studies considering, for example, the impact of large blockholders, executive compensation or board models see Shleifer and Vishny (1997) and Becht et al. (2003).

that the securities will be widely held and that the firm is run by an incumbent management. A rival management, which may or may not be able to manage the firm better than the incumbent, can try to obtain control by bidding for the securities to which the voting rights are assigned. But before the security holders may decide whether to accept the offer or not, the incumbent management can make a counteroffer. The crucial assumption of Grossman and Hart is that the actual management obtains private benefits from controlling the firm.

They show that under these circumstances the optimal allocation of voting rights and dividends depends on the absolute and relative sizes of the private benefits accruing to the incumbent and the rival management. When private benefits are negligible, the allocation of control is unimportant. However, if the private benefits of control are one-sided, such that, e. g., the incumbent has no private benefits of control while the rival does, one share / one vote and the majority rule are optimal since maximizing the amount the rival management is willing to pay to obtain control.[19]

Harris and Raviv (1988b) apply a similar framework. Additionally, they consider social optimality, i. e. they take the private benefits accruing to the incumbent and to the rival management into account. The authors find that one share / one vote is socially optimal since it ensures that the management which generates the greatest total surplus (payout to shareholders and private benefits to managers) controls the firm. However, from the original owner's perspective (private optimization) it is preferable to issue two classes of extreme securities, i. e. one with all voting rights and one with all cash flow rights. This implies that the privately optimal voting right structure is not socially desirable. This deviation occurs since one share / one vote is socially optimal since the management with the highest willingness to pay gains control, while the current owner, i. e. the security holders, can extract as much as possible of the managers' private benefits due to competition between both managements (private optimization).

Obviously, the optimality of one share / one vote and of the majority rule depends significantly on the underlying assumptions. E. g. the private optimality hinges crucially on the assumption of asymmetric private benefits between the incumbent and the rival management. However, so far no clear justification for this asymmetry is known. Normally one would expect symmetric benefits, since the limitations of managers' private benefits such as corporate law etc. are identical for the incumbent and the rival. Differences, still, may occur when the parties have different outside options.

This concern has encouraged further research on private benefits of control, see for example, Bebchuk (1999) and Shleifer and Wolfenson (2002). While the existence of private benefits is empirically confirmed (cf. Shleifer and Vishny

---

[19] In such a constellation it is inefficient for the entrepreneur to issue preferred shares without voting rights since this would enable the rival management to aquire the voting majority without bidding for at least half of the issued capital.

(1997)), a recent study by Nenova (2003) reveals that private benefits vary widely across countries due to differences in the legal environment, investor protection and takeover regulations. An alternative way to proceed apart from further research on private benefits is offered by Mason et al. (2003). The authors model the actual takeover mechanism more precisely. However, the optimality of equity's voting right structure needs to be further investigated.

2.2.2.2 Aspects of Debt Financing

Of course, since debtholders obtain control about the firm when their obligations are not fully satisfied, corporate control considerations matter in the default state, see Section 2.2.2.2.1. Additionally, relationship lending has to be considered since such relationships significantly affect corporate control, see Section 2.2.2.2.2.

2.2.2.2.1 Renegotiation in the Default State

If a borrower defaults on his debt obligations held by a large number of creditors, renegotiations with his creditors are probably extremely difficult and the borrower might be forced into bankruptcy (cf. Gertner and Scharfstein (1991) and Bolton and Scharfstein (1996)). In contrast, to renegotiate with a bank or a single investor should be easier.

Bolton and Scharfstein (1996), for example, show that a firm's optimal debt structure, i. e. the number of creditors, the allocation of security interests and voting covenants, depends on post-default bargaining considerations. Assuming two possible types of default, liquidity and strategic defaults, they show that debt contracts reduce managers' incentives for strategic defaults since in default creditors obtain the right to liquidate the firm's assets.[20] However, liquidation results in inefficiencies following liquidity defaults. According to Bolton and Scharfstein an optimal debt structure should balance the benefits of deterring strategic defaults and the costs of inefficient liquidation in case of a liquidity default. A firm can maximize its liquidation value, i. e. minimize the costs of a potential inefficient liquidation, by just borrowing from one creditor, by giving only one creditor a security interest, or by adopting voting rules that make it easier to complement debt restructurings. On the other hand, strategic defaults are best prevented by borrowing from multiple creditors, by giving each creditor security interests and by adopting voting rules that allow some creditors to block restructurings.

A related paper by Berglöf and von Thadden (1994) also examines a firm's optimal debt structure taking post-default debt renegotiations into account. In a framework where the firm cannot commit to future payouts but assets

---

[20] In a liquidity default the firm's earnings are insufficient to meet its obligations while in a strategic default managers want to divert cash to themselves while the firm's earnings are sufficient to meet the firm's obligations.

can be contracted upon, Berglöf and von Thadden show that a debt structure with multiple investors, where some specialize in short-term and others in log-term claims, is superior to a structure with only one type of claims or with investors holding both claims. They assume that the firm's assets exhibit a certain degree of firm specificity, which makes the outside value of these assets less than the value within the firm. This strengthens the firm's incentive to default strategically. However, the firm's incentive is weakened by the separation of ownership, i. e. investors hold either short or long-term claims, which strengthens investors' bargaining power in renegotiations. An investor with short and long-term claims partly internalizes the impact of his short-run behavior and thereby weakens his short-run ex-post bargaining position. Similar to Bolton and Scharfstein the firm's debt structure is determined by a tradeoff between the desire to discourage the firm from strategic defaults and the wish to limit inefficiencies in liquidity defaults. This tradeoff determines endogenously the costs of financial distress. Berglöf and von Thadden demonstrate how investors can reduce the need for inefficient liquidations in liquidity defaults while maintaining the disincentives of strategic defaults by separating their claims.

Furthermore, Gorton and Kahn (2000) show that debt contract's (initial) conditions are not set to price default risk, instead they are determined to balance the firm's and the lender's bargaining power in later renegotiations. In their framework one lender, i. e. a bank, and one borrower renegotiate the terms of the debt contract after the arrival of new information. The borrower's bargaining power results from his risk-shifting (asset substitution) opportunity while the bank's power rests on its credible threat to eventually liquidate the borrower's project.

Finally, the strategic importance of collateral in debt renegotiations is empirically confirmed by Elsas and Krahnen (2000). They find that banks, i. e. relationship lenders, try to accumulate collateral to improve their bargaining position when the firm experiences financial distress.

2.2.2.2.2 Relationship Lending

Another aspect firms have to consider when determining their optimal debt structure is whether to engage in relationship lending, i. e. establish an implicit long-term contract with a single investor, e. g. a bank, or not.[21]

In the short-run relationship lending can be beneficial for the firm since being tied to a single lender might enable the lender to reduce the firm's initial interest payments and thereby the firm's potential incentive to engage in asset substitution (cf. Petersen and Rajan (1995)). The lender can reduce the firm's initial payments since he is able to backload interest payments over time, subsidizing the firm in the short-run and extracting rents later. This solution is beneficial for the lender and the borrower in total since the low

---

[21] See Boot (2000) for a recent review.

initial interest payments reduce the borrower's asset substitution incentive and any associated loss such as (extra) agency costs occurring otherwise.

However, Rajan (1992) argues that in such relationships the lender obtains bargaining power over the firm's profits in the long-run, e. g. due to informational monopolies, etc.[22]

Therefore, firms have to tradeoff the benefits and the costs of relationship lending when determining their optimal debt structure.

2.2.2.3 The Optimal Debt-Equity Mix

As seen in the previous sections a firm's optimal equity and debt structure is affected by corporate control considerations and therefore by bargaining power aspects. This section now illustrates corporate control implications for the firm's capital structure, i. e. its debt/equity choice. Similar to Section 2.2.2.1 on equity's voting right structure, we focus on takeover models.

For example, Harris and Raviv (1988a) show the relevance of a firm's capital structure for takeover contests taking the security structure of voting equity and debt as given. In their model the incumbent management possesses shares of the own company and has the ability to increase their voting/bargaining power by replacing outstanding equity through debt. However, any increase in the incumbent's voting power causes the firm's default probability to increase. In case of default the incumbent management loses the benefits it derives from corporate control. The firm's optimal capital structure balances the costs and the benefits of debt financing.

The arising hypothesis that

- a firm's leverage is positively correlated with the extent of managerial equity ownership (see Harris and Raviv (1988a)) is empirically confirmed by Amihud et al. (1990).

However, Harris and Raviv (1991) point out that the predictions based on such takeover models are only valid in the short-term in response to imminent takeover threats.

Harris and Raviv (1989) take a different approach. They examine the optimal allocation of voting and cash flow rights when the firm can issue debt and equity in order to maximize firm value (private optimization). The entrepreneur, who initially owns the firm, designs the securities in a way that any incumbent management team with private benefits is replaced by a superior rival. A rival management is superior when they can generate a higher dividend stream. Therefore, the costs of resisting a takeover by a superior rival have to be maximized. As in Harris and Raviv (1988b) one share / one vote among voting securities is found to be optimal. This implies that corporate

---

[22] Note: Lenders with higher bargaining power in the long-run, i. e. better ability to extract the firm's surpluses in the long-run, possess more flexibility to cut short-run interest rates initially (cf. Petersen and Rajan (1995)).

control cannot be acquired cheaply by the party with private benefits. When non-voting securities are sold they should be riskfree debt. The private optimality of one share / one vote hinges again on the asymmetric distribution of private benefits.

A further alternative is presented by Bergman and Callen (1991). They model debt renegotiations as a sequential bargaining process between creditors and shareholders and show that shareholders can enforce ex-post concessions from the firm's creditors by threatening to run down the firm since they are still in charge of the firm's investment decisions. The creditors anticipate the bargaining outcome and therefore limit the amount of debt financing.

Hege and Mella-Barral (2000) also find that bargaining power in post-default debt renegotiations is a major determinant of the firm's leverage. In their model, the firm's capital structure tries to balance debts' tax shield advantage and possible liquidation costs in the default case. Due to the bankruptcy costs the creditors might be willing to forgive parts of their outstanding claims. A debtor with some bargaining power in renegotiations will opportunistically try to extract parts of the renegotiation surplus, which is in turn anticipated by the creditors and they adjust their interest demand.

2.2.3 Competition in the Capital Market and Implications for Financial Contracting

In this section studies are presented examining the implications of capital market competition for financial contracting. Studies focusing on bank financing are described in Section 2.2.3.1, while Section 2.2.3.2 concentrates on studies focusing on venture capital financing. To our knowledge, so far no study exists examining bargaining power effects for financial contracting without restricting the analysis to one particular capital market segment.

2.2.3.1 Competition in Banking and Implications

Competition in banking is expected to affect the availability of corporate loans and the interest rates charged by banks; but so far theory has not yet provided an unambiguous prediction about the relationship between bank competition and loan supply (cf. Bonaccorsi di Patti and Dell'Ariccia (2004)).

Firstly, the theory of industrial organization predicts that any increase in banks' market power results in a lower supply of corporate loans at higher costs. This prediction is often called the structure-performance hypothesis (cf., e.g., Beck et al. (2004)) and implies that any deviation from perfect competition causes less loan availability at higher costs.[23] This hypothesis is based on models of oligopolistic behavior showing that collusive arrangements are

---

[23] Banks with market power face downward sloping demand curves and their loan supply is determined by the standard requirement of equality between the Lerner indices and the inverse elasticities (cf. Freixas and Rochet (1997)). Any reduction

less costly to maintain in an oligopolistic than in a fully competitive markets (cf. Goldberg and Rai (1996)). In line with this prediction, for example, Guzman (2000) shows that credit rationing is more likely to occur in a banking monopoly than in a fully competitive banking market.

Secondly, theories taking potential informational asymmetries between lender and borrower into account predict a positive relationship between the degree of banks' market power and the supply of corporate loans (cf. Zarutskie (2005)). This prediction is often called the information-based hypothesis. These theories show that due to the informational asymmetry adverse selection, moral hazard, and/or hold-up problems can arise which increase with the degree of bank competition.[24] In such circumstances any increase in banks' market power might be useful to improve corporate loan availability since such an increase reduces the difficulties for financial contracting caused by informational asymmetries (cf. Bonaccorsi di Patti and Dell'Ariccia (2004)). Stiglitz and Weiss (1981), for example, find that in an asymmetric information setting credit rationing can occur as an equilibrium outcome in a fully competitive banking market. Also Petersen and Rajan (1995) show that banks possessing market power might be beneficial for certain borrowers. These banks might have a strong incentive to establish long-term relationships with new unknown borrowers because of their ability to participate from the borrowers' future surpluses when they turn out to be successful. Banks without market power refuse to lend to new unknown borrowers when their potentially short-term relationship is unprofitable for them. In Petersen and Rajans' model banks can only sustain the cost of lending funds to risky unknown borrowers in the first place if their market power allows them to benefit at later stages from their successful borrowers' surpluses.

Finally, Beck et al. (2004) point out that also other characteristics of the banking sector, such as the ownership structure and the regulatory environment, can affect the relationship between the degree of banks' market power and the supply of corporate loans. The ownership structure of banks can affect this relationship since domestically owned banks probably posses better information about domestic borrowers than foreign owned banks do and, therefore, domestically owned banks are more willing to lend to domestic opaque borrowers. The regulatory structure of the banking system might also have an impact because high regulatory entry barriers can reduce the contestability and thus competitiveness of the banking system, thereby affecting the relationship between banks' market power and loan availability.

So far, empirical studies examining the implications of bank competition for loan supply have derived conflicting results, not uniquely supporting one

---

in banks' market power now increases the elasticity of the demand for loans and hence reduces banks' ability to extract rents from their customers by charging them higher interest rates (cf. Bonaccorsi di Patti and Dell'Ariccia (2004)).

[24] See Bonaccorsi di Patti and Dell'Ariccia (2004) for examples of such problems between lenders and borrowers.

of the presented theories / hypotheses. Studies find positive as well as negative relationships between competition in banking and corporate loan availability (cf., e. g., Beck et al. (2004)).

2.2.3.2 Competition in the Venture Capital Market and Implications

Young, high-potential companies typically receive finance from venture capitalists. The venture capitalists try due to due diligence, intensive monitoring and direct assistance to create companies that eventually go public (cf. Gompers and Lerner (1999)). For example, Compaq, Intel and Microsoft received venture capital. American Research and Development (ARD), founded in 1946, is the first modern venture capital firm. ARD is structured as a publicly traded closed-end fund to provide institutional as well as private investors with the opportunity to invest in such young and high potential companies.[25]

The venture capital market is part of the private equity market, being highly cyclical with persistent changes in capital supply and demand (cf. Gompers and Lerner (2000)). Therefore, the question arises in which way such changes in the venture capital market affect the value created for entrepreneurs and for venture capitalists themselves.

Inderst and Müller (2004), for example, demonstrate that the value created in the start-up firm and the contract agreement probability between entrepreneur and venture capitalist depend strongly on capital market characteristics like the level of capital supply and capital market competition. Inderst and Müller examine financial contracting in a market where entrepreneurs and venture capitalists bargain how to finance the entrepreneur's venture. In their model they assume that the bargaining power of the involved parties is determined by the relation of capital supply and demand. E. g. if capital supply grows an entrepreneur's alternative funding opportunities increase improving his outside option in bilateral bargaining with the venture capitalist. According to Inderst and Müller in a situation with many investors and entrepreneurs there is no reason why the latter should capture all rents.

Inderst and Mülller find, holding capital demand fixed, that for different levels of capital supply different contract agreements, i. e. allocations of financial claims, are optimal. Optimal contracts are efficient, i. e. they balance the incentive problems of the entrepreneur and the venture capitalist (effort of both parties is required to create a successful venture), while the actual division of the claims is determined by bargaining. During periods of either extremely high or extremely low capital supply bargaining power is strongly asymmetrically distributed among the entrepreneur and the venture capitalist. Consequently, the net present value created in a start-up firm is a hump-shaped function of capital supply (cf. Inderst and Müller (2004)).

---

[25] See, e. g., Gompers and Lerner (1999).

A further example illustrating the importance of bargaining power considerations in venture capital contracting is Fairchild (2004). He analyzes the bargaining process between a venture capitalist and an entrepreneur in face of a double sided moral hazard problem. Two moral hazard problems exist since both, the venture capitalist and the entrepreneur, supply value-adding services.[26] In this setting Fairchild shows that total welfare is only maximized when the venture capitalist possesses high value-adding capabilities and the entrepreneur has all bargaining power.

Empirically, Gompers and Lerner (1996) find that venture capital funds in the United States are able to reduce the number of restrictive covenants imposed by their investors in years with high capital supply. They interpret this observation as a result of an increased bargaining power by venture capital funds. However, Schmidt and Wahrenburg (2003) are unable to confirm this observation for European venture capital funds.

2.3 Contribution of the Current Study

From the studies presented in this chapter we have seen that informational asymmetries play an important role in a firm's financing decision and, in particular, in the firm's choices of contract type and placement mode (see Section 2.1). However, depending on the kind of informational asymmetry quite different, often even contradicting theoretical predictions about the firm's optimal contract type / placement mode choice can be derived. Additionally, as mentioned on page 10, the empirical evidence is far from conclusive, neither unambiguously supporting one nor any other theory / prediction.

Since the aim of this dissertation is to demonstrate how a firm's financing decision depends on bargaining power considerations and that bargaining power considerations help to reduce the gap between theoretical predictions in financial contracting and empirical evidence, we place our analysis in an asymmetric information setting. That bargaining power considerations have an effect on firms' financing decisions is already illustrated in Section 2.2; but their implications for a firm's contract type / placement mode choice have not been analyzed explicitly so far. Therefore, we concentrate on the latter question and study explicitly how the ex-ante distribution of bargaining power between a lender and a borrower affects the borrower's preferred contract type / placement mode choice.

Our contribution to the literature is threefold. Firstly, as already mentioned, we incorporate bargaining power considerations in a model of financial contracting under asymmetric information and examine the implications for the firm's preferred contract type / placement mode choice. Secondly, the

---

[26] Two moral hazard problems exist due to the entrepreneur's shirking and the capitalist's hold-up opportunity, i. e. the capitalist can threat to exit the contract before maturity in order to enforce contract renegotiations to his benefit.

current study is one of the first studies examining a firm's choices of contract type and placement mode jointly in order to account for all potential interdependencies among these choices. That a firm's choices of contract type and placement mode are interrelated and have to be considered jointly is empirically confirmed by Gomes and Phillips (2005).[27] Thirdly, the model we develop allows us to examine a firm's financing decision either in the presence of an interim and an ex-post informational asymmetry or in the presence of all three types of informational asymmetry (see Chapter 3 respectively Chapter 4). It is important to consider different kinds of informational asymmetry simultaneously when examining a firm's financing decision since otherwise some possible interdependencies between occurring effects would be neglected.

---

[27] Another study which also takes the interaction between a firm's contract type and its placement mode choices into account is Fulghieri and Lukin (2001). The authors show that the lender's incentives for information production in financial contracting not only depends on the firm's, i. e. the borrower's placement mode choice but also on the degree of information sensitivity of the chosen contract type.

3

A Model to Analyze Bargaining Power Effects

In this chapter an analytical principal-agent framework with a lender and a borrower is developed to examine bargaining power effects in financial contracting. The lender and the borrower bargain how to finance a risky project. Since public as well as private debt and equity are considered as potential *contract type / placement mode* (*ct/pm*)-constellations, this study focuses on borrowers with access to these markets. Public debt financing can be thought of as (corporate) bond financing, whereas private debt resembles a bank loan (cf. Carey and Rosen (2001)). Similarly, equity financing can either be obtained publicly from the stock market or privately from a single investor (cf. Cronqvist and Nilsson (2003)).

Effects on the preferred *ct/pm*-constellation caused by variations in the ex-ante distribution of bargaining power between the lender and the borrower are considered. Power effects are analyzed by a comparison of optimal contracting decisions in alternative bargaining power scenarios. In each scenario certain rights, such as the right to determine the contract conditions or the *ct/pm*-choice, are either assigned to the lender or to the borrower, both seeking to maximize their expected profits.

The chapter is divided into five sections. Section 3.1 describes the main assumptions of the principal-agent framework in which our analysis is set before the lender's and the borrower's expected profits are determined in Section 3.2. The feasible *ct/pm*-constellations which either the lender or the borrower can choose to maximize their expected profits are defined in Section 3.3. In Section 3.4 the preferred *ct/pm*-choices in the alternative bargaining scenarios are derived. Finally, in Section 3.5 bargaining power effects in financial contracting are determined by a comparison of the optimal *ct/pm*-choices in the different scenarios.

3.1 Main Assumptions

The main assumptions, forming the analytical principal-agent framework we employ to examine bargaining power effects in financial contracting, are presented and motivated in this section. Assumptions 1 to 4 specify an optimal contracting framework. Assumption 5 presents four alternative bargaining power scenarios for this setting.

As Hart (2001) points out, the theory of financial contracting is about deals between financiers and those who need financing. Therefore, we assume a standard one-period setting where an entrepreneur (the borrower) seeks to raise funds from a potential investor (the lender) in order to finance a risky project.

Assumption 1
One-Period Setting
- In $t = 0$ the lender L (principal) has funds, for example 1\$, to invest but no project to invest in. The borrower B (agent) has a one-period risky project but no funds. The lender can either hold the funds riskless in his pocket and receive no interest,[1] or lend the 1\$ in $t = 0$ to the borrower so that the latter can undertake the project. The lender and the borrower can write different payoff based financial contracts.
- The borrower's liability is limited to the project's payoff in $t = 1$.
- Both parties are completely rational, risk neutral and are trying to maximize their expected profits Pj ($j \in \{L, B\}$).
- The lender and the borrower must at least achieve a non-negative expected profit ($Pj \geq 0\ \forall j \in \{L, B\}$) in order to guarantee contract participation.[2] □

To find out whether the agents can agree on a financing contract more specifications are necessary.

---

[1] Alternatively, we could assume a positive riskless interest rate offered by the capital market, but this would unnecessarily result in more complex calculations without obtaining new insights. The only effect a positive interest rate has is that some projects financed for a riskless interest rate of zero, since their net present value exceeds the opportunity costs of zero, are not considered anymore. However, the implications of bargaining power variations for financial contracting remain unaffected. Therefore, we set the riskless rate to zero.

[2] This simplification is relaxed in Chapter 4 when an ex-ante informational asymmetry concerning the lender's financing and the borrower's profit opportunity is introduced.

Assumption 2
Project and Interim as well as Ex-post Uncertainty Specification

- Project: The return of the borrower's risky project $\tilde{y}$ is distributed equally in the range $[y_{min}, y_{max}]$, where $y_{min} = \mu(1 - \zeta) \land y_{max} = \mu(1 + \zeta)$.[3] μ stands for the expected project return and ζ ($\zeta \in [0, \zeta_{max}]$) for the relative return disparity (symmetric risk).[4] The expected project return μ and the *maximum* risk ζ_{max} ($\zeta_{max} \in [0, 1]$) are common knowledge.
- Interim uncertainty: The borrower can unobservably vary the project risk ζ with $\zeta \in [0, \zeta_{max}]$ just after the financial contract is signed.[5]
- Ex-post uncertainty: In $t = 1$ the borrower obtains private information about the actual project return realization which cannot be observed by the lender. □

Such a situation is possible in a wide range of circumstances. For example, imagine the entrepreneur plans to open a trouser shop in $t = 0$ while the demand for trousers is uncertain (risky project), and the entrepreneur can unobservably choose the range he offers, which can range from a diversified mix to a specialization in green leggings (interim uncertainty). Additionally, the potential lender may be unable to evaluate the trouser market and, in particular, the entrepreneur's success adequately enough, so that the borrower's actual revenue in $t = 1$ is private knowledge (ex-post uncertainty).

The financial contract both parties prefer in such a framework depends not only on the repayment structures as specified in Assumption 3, as one might expect, but also on the placement modes available (see Assumption 4). The placement mode choice is important as it basically determines the information transmission mechanism between the lender and the borrower, which is, in particular, essential for bargaining outcomes under informational asymmetries (see Crawford and Sobel (1982)). As the aim of this study is to

---

[3] The return distribution is chosen deliberately to obtain explicit solutions (technical assumption). However, the equal distribution possesses the property that it is possible, given the expected return μ, to determine the relative return disparity ζ at least necessary to achieve a certain project return realization $\hat{y}$ ($\zeta \geq \dfrac{\mu - \hat{y}}{\mu}$).
But, as the financial contracts we consider do not allow the fixing of contract conditions contingent on the project's proclaimed return realization $\hat{y}$ and, therefore, on the implicit risk level, it is not possible to contract upon a specific risk level.

[4] Possible reasonable symmetric risk measures are the spread of the return distribution $S(\tilde{y}) = y_{max} - y_{min} = 2\zeta\mu$ or the variance $Var(\tilde{y}) = \dfrac{\zeta^2 \mu^2}{3}$.

[5] Alternatively, we could assume that an infinite number of projects with the same expected return μ but different disparities ζ exist ($\zeta \in [0, \zeta_{max}]$), while the borrower's project choice is unobservable for the lender.

examine bargaining power effects in financial contracting, we are concerned about the effects of variations in bargaining power on the preferred ct/pm-choice. We do not want to derive optimal contracts for the described situation. Therefore, we restrict our analysis to a given set of available contracts, such as public and private debt as well as equity, with reasonable pre-determined monitoring/verification schemes.

Assumption 3
Contract Types: Debt vs. Equity
- There are two types of financial contracts, one equity-like and one debt-like.
- The EQUITY CONTRACT is characterized by a proportional return participation of the lender $q\widehat{y}$ ($q \in [0,1]$) in $t = 1$ dependent on the borrower's proclaimed project return $\widehat{y}$, and the unconditional right to monitor the borrower's behavior throughout the period which includes auditing his books and, thereby, resolving all potential interim and ex-post informational asymmetries. We assume that the lender always uses his right to monitor and has to bear the costs mc in $t = 1$.[6]
- The DEBT CONTRACT is characterized by a fixed repayment h ($h \in [0, y_{max}]$) from the borrower to the lender, and the conditional right to verify the proclaimed project return $\widehat{y}$ in the proclaimed default case ($\widehat{y} < h$).[7] In this case, the borrower has to hand over the whole project return to the lender. We assume that the lender verifies $\widehat{y}$ in $t = 1$ in every default state at costs vc with $vc > mc$. □

Assumption 3 seems reasonable, considering (i) the combination of fractional return rights and unconditional rights to monitor and intervene is consistent with equity contracts observed in practice (cf. Myers (2000) and Hvide and Leite (2002)), while (ii) the mixture of promised constant repayment over all possible states with the right to verify the stated project return and eventually seize the whole return if obligations are not fully satisfied is widely accepted in real world debt contracts (cf. Harris and Raviv (1992)).

However, while the ex-ante determined, state independent monitoring setup (à la Diamond (1984)) is typically assumed as reasonable in equity financing, the efficiency of the common ex-post determined, state dependent

---

[6] Monitoring costs mc can either occur as assumed constantly throughout the period and are due in $t = 1$, or can alternatively occur in $t = 0$ as an up-front fee, as the time structure is irrelevant in our model due to zero discounting.

[7] Of course, to be exact, a third case has to be considered, i. e. when the fixed repayment exceeds the maximum project return y_{max} ($h > y_{max}$). However, we rule out this case by $h \in [0, y_{max}]$ as the borrower always defaults and the resulting expected profits are identical to the case $h = y_{max}$. Moreover, for $h = y_{max}$ debt financing resembles equity financing, as the lender obtains the whole risky project return in all circumstances.

deterministic verification mechanism in debt financing (à la Townsend (1979) and Gale and Hellwig (1985)), also preventing the borrower from cheating concerning y, remains debatable. Difficulties due to lacking subgame completeness at the repayment stage and missing Pareto-superior randomization arise.[8] Section 5.2.3 handles these concerns.

Due to possible difficulties associated with the entrepreneur's potential adverse behavior anticipating defaulting, for example hiding funds, it is realistic to only consider cases where $vc > mc$ (cf. Hellwig (2000)).[9] This condition also prevents equity financing from being dominated by debt financing, as otherwise $(vc \leq mc)$ the arising agency costs of debt financing are bounded below the costs associated with equity financing even if the borrower defaults nearly in every possible state.[10] Additionally, due to the fixed absolute costs associated with either debt or equity financing, a financing mix is always inefficient, as costs increase without any advantages.

Of course, in the real world, firms mix debt and equity financing (cf., e. g., Harris and Raviv (1991), Bancel and Mittoo (2003) and Frank and Goyal (2003a)), while even firms operating under the Anglo-Saxon tradition (mainly US, UK and Canada) are not as leveraged as firms in other major economies (Italy, Japan and Germany).[11] However, whatever reason exists for this phenomenon,[12] it is not due to the circumstances described here where all financing mixes turn out to be inefficient.

We must note that, only simple one-period, cash flow based financial contracts between the lender and the borrower are analyzed, which do not allow the fixing of contract conditions contingent on the proclaimed project return.

---

[8] The deterministic verification mechanism lacks subgame completeness at the repayment stage as it is assumed that the lender verifies the project return, even if the potential gains are lower than the costs vc, i. e. $h - vc < \widehat{y} \leq h$. Additionally, a randomization of the verification procedure can reduce the expected costs, while still preventing the borrower from proclaiming false defaults.

[9] Moreover, as discussed in Section 2.1.1.3, Hvide and Leite (2002) argue that the monitoring costs of equity financing are lower than the verification costs of debt financing, as debtholders and equityholders possess different control rights, which implies that their incentives to invest in "cheap" monitoring technology ex-ante differ. Furthermore, Habib and Johnsen (2000) point out that equityholders care about the firm in its primary use, while debtholders are concerned about the firm's alternative use, i. e. its liquidation value.

[10] This argument is valid independent of the bargaining power distribution between the lender and the borrower. The power distribution, of course, affects the contract conditions (h, q), and therefore the arising agency costs. However, this causes for $vc \leq mc$ only a variation in the absolute difference between the agency costs of debt and equity, and *not* in their ranking.

[11] See Rajan and Zingales (1995) and Antoniou et al. (2002).

[12] See, e. g., Harris and Raviv (1991) and Hart (2001) for possible explanations.

We further abstract from other possible contract elements such as a non-monetary penalties (cf. Diamond (1984)). Therefore, the contracts offer the lender and the borrower *no* opportunity to contract upon a certain risk level.

So far, we have modeled two famous effects considered in financial contracting based on standard principal-agent considerations (cf. Campbell and Chan (1992) and Dowd (1992)). Dependent on the negotiated financial contract between the lender and the borrower, two possible *moral hazard* difficulties arise because of the borrower's informational advantage. Firstly, if the borrower has a risk-shifting opportunity, i. e. the opportunity to unobservably vary the project risk after contract agreement in a way that reduces his expected financial obligation, and therefore increases his expected profit, he will set ζ ($\zeta \leq \zeta_{max}$) to ζ_{max} to shift as much downside risk to the lender as possible.[13] Such a risk-shifting opportunity exists for debt financing. Secondly, independent of the contract type, the borrower has an incentive to understate the actual project return y in $t = 1$ ($\widehat{y} < y$) to reduce his financial obligation (cf. Hart and Holmström (1987)). These difficulties affect the common tradeoff between debt's advantage that verification is *only* necessary in the default state while having to cope with the disadvantage of higher absolute costs ($vc > mc$) (cf. Dowd (1992) and Campbell and Chan (1992)).

As mentioned earlier, similar to the repayment structure of the financial contracts, the placement modes available have to be considered.

Assumption 4
Placement Mode: Public Offering vs. Private Placement
- The lender and the borrower have either the opportunity to transact via the capital market by a public offering or to negotiate personally in a private placement.
- PRIVATE PLACEMENTS cause negotiation costs nc for both parties.[14]
- PUBLIC OFFERINGS, in comparison, do not require extra costs.[15]
  □

---

[13] Harris and Raviv (1991) refer to this as the "asset substitution effect".

[14] Realistically, negotiation costs should occur for both parties in $t = 0$, but in order to be exact, at this stage the wealthless borrower has no free funds to cover these costs. However, we imagine that the borrower can somehow manage it, e. g. by delaying his effort compensation to $t = 1$. This is not problematic due to zero discounting, while even for a positive risk adjusted interest rate the effects of compounding and discounting cancel out in the borrower's profit evaluation in $t = 0$.

[15] Alternatively, we could assume that public offerings also cause some costs, but we refrain from that as only the cost differential between public and private placement is relevant to evaluate whether or not it is worth bearing the extra costs. Therefore, we actually only assume that private placements are more expensive than public offerings.

We introduce both available placement modes for the sake of completeness, as during the remainder of this chapter the lender and the borrower bargain over four different ct/pm-constellations. However, during the entire chapter private offerings are inefficient as they cause costs nc without offering any advantage so far. But the distinction between both placement modes becomes relevant when ex-ante informational asymmetries are considered in the next chapter as the personal negotiations in private placements offer the opportunity to resolve such possible asymmetries. Public offerings are costless but will not solve these potential ex-ante informational asymmetries (see Assumption 4', Chapter 4).[16]

The importance of the placement mode choice in financial contracting is empirically confirmed by Mackie-Mason (1990).[17] Furthermore, it seems reasonable to assume that private placements are more expensive than public offerings, despite all possible transaction costs occurring in public placements, e. g. for road shows or rating agencies. Hertzel and Smith (1993), as well as Wu (2004), empirically support evidence that a private *equity* issue is more expensive for a firm than a public offering, due to investor compensation for their evaluation and screening service. Best and Zhang (1993), as well as Krishnaswami et al. (1999), obtain similar results for *debt* issues. It is also reasonable to assume that costs occur for both parties, as a private placement causes extra costs or effort for the borrower and the lender, as sensitive information has to be presented by one side and evaluated by the other. The assumption that the same negotiation costs arise for both parties can of course be relaxed but this would not provide additional insights. Therefore, we stick to this assumption. Again, due to the fixed absolute costs associated with a private placement, a placement mix is inefficient.

However, in the real world, firms do mix public and private placements (cf. Kwan and Carleton (2004), Krishnaswami et al. (1999) and Denis and Mihov (2003)), but obviously for a reason not covered in this principal-agent framework.[18]

After defining the basic principal-agent framework to examine bargaining power effects in financial contracting (see Assumptions 1 to 4) we have to

---

[16] Of course, in public offerings bond ratings, issue prospects, etc. might convey information, but they are not costless and do not resolve all ex-ante informational asymmetries. In this setting, we simply assume two extreme potential placement modes, one costly with the opportunity to resolve ex-ante informational asymmetries, while the other mode is costless and does not offer this opportunity. In reality this distinction is not as clear- cut as we assume, but resolving ex-ante informational asymmetries is always associated with costs.

[17] The author points out that ex-ante informational asymmetries are one of the significant determinants of the choice between private and public fund raising, even after controlling for the security type.

[18] Carey and Rosen (2001), for example, illustrate that a mix of public and private debt can be optimal in an incomplete contracting framework.

specify the way in which the lender and the borrower can negotiate how to finance the borrower's project.

As it is not clear which is the suitable non-cooperative bargaining game between the lender and the borrower (cf. Schmitz (2001)), we extend the common *two-scenario approach*, where simply one party is *exogenously* given absolute power or none (cf. Besanko and Thakor (1987), Schmitz (2001), Schäfer (2002) or Mukhopadhyay (2002)), to a *four-scenario approach*. This extension is motivated by the observation that capital markets are still mostly highly segmented (e. g. due to government constraints, institutional practices or investor perceptions), in the form that large institutional investors, like pension funds or banks, are basically restricted to one kind of financial contract via a predefined placement mode (cf., e. g., Allen and Jagtiani (1996)). For instance, commercial banks are mostly limited to the provision of private debt, while this limitation becomes less tense due to the trend towards universal banking. However, portfolio managers, and in particular investment funds, are often also restricted to invest in public equity, as their customers want to participate from particular stock market developments, e. g. high tech shares vs. blue chips. On the other hand, for example, firms might be restricted to raise debt instead of equity, as their controlling parties are control dilution averse (cf. Cronqvist and Nilsson (2003)) or they require private financing instead of public financing in order to shield valuable information from the firm's competitors (cf. Dhaliwal et al. (2003)). Therefore, fund suppliers as well as borrowers might be restricted to a particular capital market segment. Bolton and Freixas (2000) even show that in an asymmetric information context capital market segmentation is likely to occur. See Figure 3.1 for an illustration of a segmented capital market for the setting described so far.[19]

Fig. 3.1: Illustration of Assumed Capital Market Segmentation

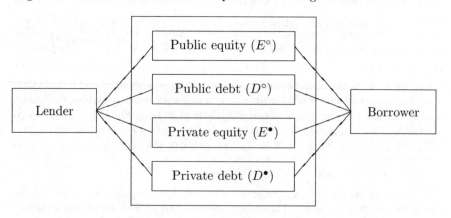

---

[19] We indicate public placements by ○ as no costs occur for both parties, while ● stands for private placements, where negotiation cost arise for both parties.

In such a segmented environment the *competition between and within* these market segments plays a significant role for a firm's financing decision, and therefore in our four bargaining power scenarios, while the traditional two-scenario approach neglects possible market segmentation.[20] For example, a high degree of competition amongst fund suppliers in the private equity market segment indicates low bargaining power of private equity finance companies and vice versa. Recently, for example, Inderst and Müller (2004) found that for parts of the private equity market segment, the competition between fund suppliers, the relationship between fund supply and demand as well as the degree of market transparency are all important aspects when venture capitalists and entrepreneurs bargain over financial claims and these aspects also influence the value of the venture.

While, of course, different (restricted) fund suppliers and borrowers exist in a segmented capital market, a firm having to choose among the different market segments also chooses among different suppliers. However, in our principal-agent framework of one lender and one borrower we abstract from this effect. Instead, we assume that capital market competition affects the distribution of bargaining power when the lender and the borrower negotiate how to finance the borrower's risky project. Moreover, we ignore that the degree of capital market competition can differ among the capital market segments, since the aim of this study is to analyze bargaining power effects in financial contracting and not the implications of capital market segmentation.

Due to the relevance of competition between and within certain market segments for a firm's financing decision, *two* aspects are important in pre-contract negotiations, which we both use in our *four-scenario approach*. Firstly, *only one* party, either the lender or the borrower, has to determine the relevant market segment, i. e. the actual ct/pm-choice. Secondly, the contract conditions have to be negotiated. Due to the fact that the first aspect involves a *discrete* choice, as either the lender or the borrower determines ct/pm, we also assume that only one party can set the contract conditions. To examine how bargaining power affects financial contracting decisions, the following four bargaining power scenarios emerge.

---

[20] For example, Mukhopadhyay (2002) also illustrates his two-scenario approach with reference to the degree of market competition either on the lender or the borrower side, but he ignores possible market segmentation.

Assumption 5
Four Power Scenarios

- LENDER SCENARIO WITH ABSOLUTE POWER (LS^a): The lender proposes a take-it-or-leave-it offer concerning placement mode, contract type *and* conditions, which the borrower can either accept or reject. If he accepts, he receives the necessary funds to finance his project and proceeds, otherwise the game ends.

- LENDER SCENARIO WITH RESTRICTED POWER (LS^r): The lender chooses the placement mode and the contract type. Conditional on this choice, the borrower makes a take-it-or-leave-it proposal concerning contract conditions, which the lender can either accept or the game ends. If the lender accepts, the borrower receives the necessary funds to finance his project and proceeds.

- BORROWER SCENARIO WITH RESTRICTED POWER (BS^r): The borrower chooses the placement mode and the contract type. Conditional on this choice, the lender makes a take-it-or-leave-it proposal concerning contract conditions, which the borrower can either accept or the game ends. If the borrower accepts, he receives the necessary funds to finance the project and proceeds.

- BORROWER SCENARIO WITH ABSOLUTE POWER (BS^a): The borrower proposes a take-it-or-leave-it offer concerning placement mode, contract type *and* conditions which the lender can either accept or reject. If the lender accepts, the borrower receives the necessary funds to finance his project and proceeds, otherwise the game ends. □

The order of the restricted power scenarios is ambiguous, as it is not clear yet whether the lender or the borrower prefers the power to determine the ct/pm-choice to the right to set the contract condition, or vice versa.[21] All four scenarios are based on ultimatum bargaining settings, but, as we will see, they are sufficient to demonstrate how financial contracts depend on the ex-ante bargaining power distribution.[22]

From a borrowing entrepreneur's perspective facing a segmented capital market, a lender scenario with absolute power (LS^a) can be thought of as if a large number of competing entrepreneurs seek financing while just a small number of potential lenders with funds available exists. The borrowers compete for these funds. A lender scenario with restricted power (LS^r) might arise when competing investors exist, who still have enough power to choose the market segment where to supply their funds, but are unable to determine the contract conditions. For example, an universal bank might refuse to grant a loan to a potential borrower but offers to supply funds by the bank's

---

[21] As we will show, the lender's and the borrower's preferences for the different power types can vary and are significantly influenced by the project type.

[22] For reviews of more advanced bargaining settings see, for example, Roth (1985b).

private equity fund. On the other hand, a borrower scenario with restricted power (BS^r) seems reasonable if funds are supplied in all different market segments so that the borrower can choose the ct/pm, but due to an excess capital demand in each market segment, the borrower has to accept all offered conditions. Finally, a borrower scenario with absolute power (BS^a) is imaginable when only a small number of entrepreneurs offers investment opportunities to a large number of competing potential investors.[23]

Figure 3.2 illustrates the structure of the assumed sequential contract negotiation game.

Fig. 3.2: Financial Contracting Game (BM)

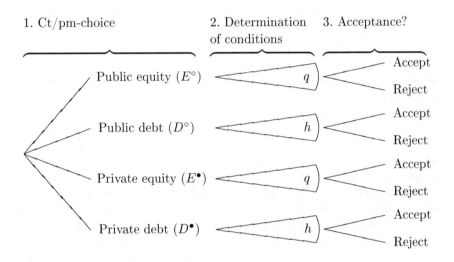

---

[23] While the presented scenario motivation hinges on general capital market characteristics, such as the number of potential investors and entrepreneurs or fund supply vs. demand, individual firm characteristics unrelated to the specific project offer an alternative scenario motivation. The characteristics have to be unrelated to the project as the lender and the borrower bargain in all four scenarios how to finance the same project. Therefore, a firm's reputation or past performance is a suitable example. For example, a firm with a bad reputation trying to raise funds will face difficulties trying to attract funds at all (LS^a), while a firm with a slightly better reputation still finds potential investors in each market segment (BS^r). On the other hand, a firm with a good reputation can choose where to borrow from (BS^a), while even for such firms investors might not be willing to supply all kinds of funds (LS^r). Individual investors can be affected by past market experiences with similar investments while institutional investors face investment constraints.

Table 3.1 summarizes the game sequence of the parties involved in the different power scenarios.

Table 3.1: Player Sequence in the Contract Negotiation Game for each Bargaining Power Scenario

Scenario	1. Ct/pm-choice	2. Determination of conditions	3. Acceptance
LS^a	Lender	Lender	Borrower
LS^r	Lender	Borrower	Lender
BS^r	Borrower	Lender	Borrower
BS^a	Borrower	Borrower	Lender

The whole outcome of the game hinges on the third stage, because only if the conditions offered are accepted, the borrower receives the necessary funds and proceeds. Otherwise the game ends and the lender as well as the borrower are left with their outside options which are equal to zero (see Assumption 1). It is important to realize that we therefore implicitly assume that the negotiation costs of a private placement only occur if the offered (binding) conditions are accepted. If they would occur otherwise, hold up problems would arise, as either the lender or the borrower has to make an up-front investment (nc) which the party determining contract conditions does not have to take into account since this investment is sunk, i. e. does not affect the opponent's (final) participation constraint. In our opinion it is reasonable to neglect potential hold up problems and their consequences in this setting because in reality the lender as well as the borrower possess ways to mitigate these difficulties, e. g. by demanding an up-front compensation/fee, while additionally the focus of this study is on general bargaining power effects on a firm's ct/pm-choice where the occurrence of nc is a minor issue.

We must note that in the commonly assumed *absolute power scenarios* (two-scenario approach) all bargaining outcomes are Pareto-efficient, as the bargaining powers to determine the ct/pm-choice and the power to set contract conditions are treated jointly. Therefore, one party can maximize its expected profit only with respect to the opponent's participation constraint, facing no other limitation. However, in the *restricted power scenarios* the power separation, enabling us to disentangle certain bargaining power effects, might cause Pareto-inefficient bargaining outcomes. The power separation might cause such inefficiencies since, apart from the contracting game, neither the lender nor the borrower can (credibly) coordinate their actions. Therefore, the lender and the borrower might end up in a Pareto-inefficient outcome since each party maximizing its own expected profit is not only confronted with the opponent's participation constraint but also with the latter's strategic behavior to optimize his profit. However, we stick to this power separation to disentangle the power effects observable in the absolute bargaining power scenarios.[24] In fact, it is a first result to realize that a power separation in such a setting might result in Pareto-inefficient bargaining outcomes.[25]

The outcome of the negotiation game under the alternative power scenarios, i. e. the ct/pm-choices preferred in the scenarios, form the basis for the power effect derivation. We only determine the preferred ct/pm for each scenario and do not derive any kind of pecking order à la Myers and Majluf (1984).[26] Finally, due to the discrete choices underlying the *four-scenario approach*, the power effects on financial contracting are determined by a pair-wise (binary) comparison of the preferred ct/pm-choices in these scenarios.[27]

3.2 Lender's and Borrower's Expected Profits

In this section the lender's as well as the borrower's expected profits in $t = 0$, PL respectively PB, for the four possible ct/pm-constellations are determined:

Public equity (E°):	$PLE^\circ(q)$ /	$PBE^\circ(q)$
Public debt (D°):	$PLD^\circ(h)$ /	$PBD^\circ(h)$
Private equity $(E^\bullet)$:	$PLE^\bullet(q)$ /	$PBE^\bullet(q)$
Private debt $(D^\bullet)$:	$PLD^\bullet(h)$ /	$PBD^\bullet(h)$

For equity financing the profits depend on the choice of the proportional return participation of the lender in $t = 1$ ($q \in [0,1]$) and for debt financing on the choice of the fixed repayment to the lender in $t = 1$ ($h \in [0, y_{max}]$).

---

[24] Section 5.2.1 discusses the arising difficulties.

[25] It is known since the famous Prisoner's Dilemma game that in the absence of binding commitments Nash-equilibria might produce Pareto-inefficient outcomes (cf. Varian (1992) or Mas-Colell et al. (1995)).

[26] Since our objective is to examine bargaining power effects in financial contracting, i. e. how power variations affect the preferred ct/pm-choice, we only derive the borrower's preferred ct/pm-choice in each power scenario. We do not derive his second or even his third best choice, i. e. a pecking order. The borrower can always realize his first best choice (which may be different in each scenario) since one requirement in the optimization is that the respective solutions are feasible.

[27] In our principal-agent model we are able to determine bargaining power effects by a pair-wise comparison of the preferred ct/pm-choices since only the bargaining power distribution between lender and borrower varies from scenario to scenario, all other parameters are constant. We perform a kind of comparative static analysis.

The consistency of the derived expected profits can be examined by the following equation

$$\mu = 1 + PL + PB + AC \tag{3.1}$$

since independent of the ct/pm-choice the whole project return is distributed either to the lender (PL) or to the borrower (PB) and has to cover the initial investment outlay and all arising agency costs (AC).

Before proceeding we collect the key variables defined so far are.

Project specification: y Project return

                        f_y Project return distribution[28]

                        μ Expected project return

                        ζ Possible return disparity (bounded below ζ_{max})

Cost structure: mc Monitoring costs (equity financing)

                        vc Verification costs (debt financing)

                        nc Negotiation costs (private placement)

In the described setting the lender's expected profit in case of public equity financing is

$$PLE^{\circ}(q) = \int_{y_{min}}^{y_{max}} qy\, f_y\, dy - mc - 1 = q\mu - mc - 1. \tag{3.2}$$

As one might have expected due to the lender's proportional return participation PLE° is independent of the (symmetric) project risk ζ. The lender with an equity stake in the project profits in an identical way from a potential symmetric risk increase as he loses. However, the equity stake's advantage of being risk insensitive is combined with the disadvantage of always requiring monitoring costs mc.

In the case of a private placement, the lender's expected profit $PLE^{\bullet}$ is additionally diminished by negotiation costs nc

$$PLE^{\bullet}(q) = \int_{y_{min}}^{y_{max}} qy\, f_y\, dy - mc - nc - 1 = q\mu - mc - nc - 1. \tag{3.3}$$

Since a default threshold, here h, exists in debt financing, the lender's expected profit for public debt PLD° of course significantly depends on the absolute height of the fixed repayment obligation h but also on the relation to

---

[28] By the assumption of equally distributed project returns, the return distribution is $f_y = \dfrac{1}{2\mu\zeta}$.

the minimum project return y_{min}.[29] While for $h \leq y_{min}$ the borrower never defaults on his obligation (riskless debt), a positive default probability exists for $h > y_{min}$ (risky debt). In the case of riskless debt $(y_{min} \geq h)$, the borrower does not have to compensate the lender for possibly arising verification costs (vc), as they do not arise, whereas they may occur for risky debt $(y_{min} < h)$.

The lender's expected profit for public debt financing is therefore

$$PLD^\circ(h) = \qquad\qquad\qquad\qquad\qquad\qquad\qquad\qquad (3.4)$$

$$\begin{cases} \int\limits_{y_{min}}^{y_{max}} h f_y dy - 1 = h - 1 & \text{for } y_{min} \geq h \\[2mm] \int\limits_{y_{min}}^{h} (y - vc) f_y dy + \int\limits_{h}^{y_{max}} h f_y dy - 1 = & \text{for } y_{min} < h. \\[2mm] \dfrac{-\mu^2(1-\zeta)^2 + 2vc(\mu(1-\zeta) - h) + 2h\mu(1+\zeta) - h^2}{4\zeta\mu} - 1 \end{cases}$$

In this case the project risk ζ plays a significant role since, given μ, ζ influences the minimum return $(y_{min} = \mu(1-\zeta))$ and therefore the debt case distinction, i. e. the lender's expected profit $PLD^\circ(h)$.[30]

For riskless debt $(y_{min} \geq h)$ the lender obviously receives h in all circumstances, i. e. $PLD^\circ(h) = h - 1$, while for risky debt $(y_{min} < h)$ his return is diminished by the arising agency costs. The lender's expected profit of risky public debt PLD° can be disentangled as follows

$$PLD^\circ(h) = \mu - \underbrace{\frac{(\mu(1+\zeta) - h)^2}{4\zeta\mu}}_{\text{Borrower's profit}} - \underbrace{vc\frac{(h - \mu(1-\zeta))}{2\zeta\mu}}_{\text{Agency costs}} - 1 \, .$$

Obviously, the lender's expected profit is equivalent to the expected project return less (i) the borrower's expected profit $PBD^\circ(h)$ as will be seen shortly, (ii) the agency costs of risky debt financing, and (iii) the initial investment. The agency costs of risk debt financing are the expected verification costs since $\dfrac{h - \mu(1-\zeta)}{2\zeta\mu}$ is the borrower's default probability given h $(DP(h))$.

For private debt a similar distinction between riskless and risky debt is made.

---

[29] According to Assumption 2, the project risk the borrower chooses (ζ) is unobservable but the maximum risk (ζ_{max}) and the expected project return (μ) are common knowledge, so that the lender knows $y_{min} = \mu(1-\zeta) \geq \mu(1-\zeta_{max})$.

[30] In case of risky debt, an increase of ζ given μ and h results in an increase of the default probability for $h < \mu$. The default probability stays constant for $h = \mu$ and decreases for $h > \mu$.

$$PLD^\bullet(h) = \tag{3.5}$$

$$\begin{cases} \int\limits_{y_{min}}^{y_{max}} hf_y dy - nc - 1 = h - nc - 1 & \text{for } y_{min} \geq h \\[2mm] \int\limits_{y_{min}}^{h}(y-vc)f_y dy + \int\limits_{h}^{y_{max}} hf_y dy - nc - 1 = & \text{for } y_{min} < h. \\[2mm] \dfrac{-\mu^2(1-\zeta)^2 + 2vc(\mu(1-\zeta)-h) + 2h\mu(1+\zeta) - h^2}{4\zeta\mu} - nc - 1 & \end{cases}$$

Of course, $PLD^\bullet$ can be disentangled like PLD° while additional negotiation costs (further agency costs) arise.

The borrower's expected profit for public equity financing PBE° is

$$PBE^\circ(q) = \int\limits_{y_{min}}^{y_{max}} (1-q)yf_y dy = (1-q)\mu \tag{3.6}$$

and for private equity

$$PBE^\bullet(q) = \int\limits_{y_{min}}^{y_{max}} (1-q)yf_y dy - nc = (1-q)\mu - nc . \tag{3.7}$$

For debt financing again it has to be distinguished between riskless and risky debt. In the case of public debt the borrower receives a profit

$$PBD^\circ(h) = \tag{3.8}$$

$$\begin{cases} \int\limits_{y_{min}}^{y_{max}}(y-h)f_y dy = \mu - h & \text{for } y_{min} \geq h \\[2mm] \int\limits_{h}^{y_{max}}(y-h)f_y dy = \dfrac{(\mu(1+\zeta)-h)^2}{4\zeta\mu} & \text{for } y_{min} < h . \end{cases}$$

While for riskless debt the borrower's expected profit is clearly the expected project return minus his repayment obligation, for risky debt financing he only benefits in the non-default state. The borrower's non-default probability for risky debt is $\dfrac{\mu(1+\zeta)-h}{2\zeta\mu}$ and in the non-default state (due to the equal distribution) he achieves an expected profit of $\dfrac{\mu(1+\zeta)-h}{2}$.[31]

For private debt again additional negotiation costs arise, which are assumed to occur for both parties so that

---

[31] The non-default probability for risky debt does *not* go to infinity when ζ decreases since risky debt requires $h > y_{min} = \mu(1-\zeta)$ which implies that $\mu(1+\zeta) - h$ is always smaller than $2\zeta\mu$. For $\zeta = 0$ risky debt becomes impossible since risky debt requires $h > y_{min}$ but $y_{min} = y_{max} = \mu$ while h is bounded by y_{max} ($h \leq y_{max}$) due to Assumption 3 (see p. 38).

$$PBD^\bullet(h) = \tag{3.9}$$

$$\begin{cases} \int\limits_{y_{min}}^{y_{max}} (y-h)f_y\,dy - nc = \mu - h - nc & \text{for } y_{min} \geq h \\[2ex] \int\limits_{h}^{y_{max}} (y-h)f_y\,dy - nc = \dfrac{(\mu(1+\zeta)-h)^2}{4\zeta\mu} - nc & \text{for } y_{min} < h. \end{cases}$$

In the following two sections the defined contract negotiation game is solved for the alternative bargaining power scenarios according to the backward induction principle. In Section 3.3 it is analyzed whether certain ct/pm-constellations are feasible and which expected profit the lender respective the borrower can maximally extract (Stage 2&3). In Section 3.4 the preferred ct/pm-choices in the alternative power scenarios are derived (Stage 1).

3.3 Feasible Contracts and Their Conditions

In this section, the feasible profit allocations between the lender and the borrower for a certain project (μ, ζ_{max}-specification) for the four available ct/pm-choices are determined (see Sections 3.3.1 to 3.3.4). Our goal is to examine which ct/pm-constellations the lender and the borrower can actually choose from when they maximize their profits in the different bargaining power scenarios. A profit allocation is feasible when both the lender's as well as the borrower's participation constraints are satisfied ($PL, PB \geq 0$).

We refrain from an explicit derivation of all feasible profit allocations for a given ct/pm-choice because in order to solve the contracting game it is convenient to handle Stages 2 and 3 jointly, i. e. to optimize the lender's respective the borrower's profit with respect to the opponent's participation constraint. Therefore, we determine for which projects (μ, ζ_{max}-specifications) particular ct/pm-constellations are feasible when either the lender or the borrower optimizes contracts' conditions. We do not obtain a whole range of possible contract conditions (q, h), instead we derive only two conditions, one from the lender and one from the borrower optimization. Obviously, the range of feasible ct/pm-choices for a particular project is unaffected by this separation since the lender's as well as the borrower's participation constraints have to be satisfied independent of the remaining profit to be distributed (see Result 1, p. 75). In the extreme case where both participation constraints are binding no extra profit remains to be distributed, hence the power to determine contracts' conditions becomes irrelevant. However, the distinction between feasibility from the lender's respective the borrower's perspective becomes particularly relevant in the extended model where the lender and the borrower possess different private information, which implies that contract feasibility can vary among the lender's and the borrower's perspective.

The feasible profit allocations for the available ct/pm-constellations of the lender optimization are relevant for the determination of the optimal ct/pm-choices in the LS^a and the BS^r scenario. The feasible profit allocations of the borrower optimization are relevant for the BS^a and the LS^r scenario. Finally, in Section 3.3.5, the feasible contracting solutions of the lender optimization are compared with the solutions of the borrower optimization to verify that they coincide.

Since ex-ante informational asymmetries about contractual partners' opportunities (outside options) are not considered in this chapter, private placements are never *Pareto-efficient*, because to pay nc in order to resolve ex-ante informational asymmetries which do not exist would simply be burning money. However, private placements themselves *cannot* be ignored in the absence of ex-ante uncertainties, as inefficient contracts are not necessarily always dominated by efficient public contracts. Efficient public contracts are not necessarily incentive-compatible for the lender and for the borrower. For example, the lender might be reduced to his outside option for all feasible ct/pm-choices when the borrower determines the contract conditions (LS^r) due to the lender's interim as well as ex-post informational disadvantage. This makes the lender indifferent between all feasible ct/pm-choices even if some of them are Pareto-inefficient. Unfortunately for the borrower, in this contract negotiation game any opportunity to make binding commitments is lacking, so the borrower can do nothing about it. Of course, the borrower could promise the lender a reward for rejecting inefficient private placements in the LS^r scenario, but the promise lacks subgame completeness. Once the lender has chosen a public placement, the borrower will squeeze him to his outside option.

To deal with the arising feasible contracting constraints FCC for each ct/pm-choice in both optimizations, we indicate their origin: BPC refers to the borrower's participation constraint, LPC to the lender's participation constraint, $BPCb/nb$ to the condition distinguishing BPC binding or not binding, CDr/nr to risky vs. riskless debt case distinction and finally NoN to non negativity. The feasible contracting constraints which are *not* explicitly stated for a ct/pm-choice are automatically satisfied when the explicitly stated feasible contracting constraints are satisfied.

3.3.1 Public Equity (E°)

In this section, the lender's as well as the borrower's optimization behaviors given public equity financing are examined. The feasible contracting solutions and the expected profits either the lender or the borrower can maximally achieve have to be determined. In Section 3.3.1.1, the power to set the contract conditions is given to the lender, while Section 3.3.1.2 focuses on the borrower.

3.3.1.1 Lender Optimization

The following optimization problem can determine for which projects (μ, ζ_{max}-specifications) *public equity* contracts are feasible, when the lender has the power to determine the contract conditions, and which expected profit the lender $PLE^{\circ}(q_L)$ can maximally extract.[32, 33]

$$\max_{q_L} PLE^{\circ}(q_L) \text{ s.t. } PBE^{\circ}(q_L) \geq 0 \land PLE^{\circ}(q_L) \geq 0$$

The choice of ζ is irrelevant, since no default threshold exists, and ζ can therefore be neglected. The optimization problem reduces due to $\dfrac{\partial PLE^{\circ}(q_L)}{\partial q_L} = \mu > 0 \ \forall q_L$ (from equation 3.2) to the task of *maximizing* q_L with respect to the imposed constraints. To solve this problem, it can be assumed that the BPC is binding, as $\dfrac{\partial PBE^{\circ}(q_L)}{\partial q_L} = -\mu < 0 \ \forall q_L$ (from 3.6). Hence, the optimal proportional return participation

$$q_L^* = 1$$

is found, while the respective expected profits are stated in Table 3.2.[34]

Table 3.2: Expected Profits and Feasible Contracting Constraint of a Public Equity Contract (BM, Lender Optimization)

PLE_L°	$\mu - 1 - mc$
PBE_L°	0
LPC	$\mu \geq 1 + mc$

The feasibility of public equity contracts thus only depends on μ and not on ζ_{max}, as the project's expected excess return $\mu - 1$ must be sufficient to compensate for the arising agency costs mc. This is reasonable since in equity financing any potential loss for the lender due to a project risk increase by the borrower is offset by the arising gain. The borrower is also invariant to variations in project risk. Due to the lacking default threshold no risk-shifting opportunity exists. Therefore, the lender's and the borrower's expected profits as well as equity's financing feasibility constraint do not depend on the project risk.

---

[32] We assume that the lender and the borrower participate even if they only receive their opportunity, i. e. 0, while being indifferent between participation and non-participation, i. e. their outside option.

[33] To avoid confusion we add a subscript to q indicating whether the lender (q_L) or the borrower (q_B) determines the contract conditions. The optimal values are marked by a star.

[34] We refer to $PLE^{\circ}(q_L^*)$ shortly as PLE_L° and to $PBE^{\circ}(q_L^*)$ as PBE_L°.

Intuitively, once all informational asymmetries between the lender and the borrower are resolved (causing agency costs mc) it is optimal for the lender to completely own the project when he has the power to determine the contract conditions.

3.3.1.2 Borrower Optimization

To find out for which projects (μ, ζ_{max}-specifications) *public equity* contracts are feasible, when the borrower determines the contract conditions and which profit the borrower $PBE°(q_B)$ can maximally extract, the optimization problem

$$\max_{q_B} PBE°(q_B) \text{ s.t. } PLE°(q_B) \geq 0 \wedge PBE°(q_B) \geq 0$$

has to be solved. The choice of ζ is again irrelevant and therefore neglected.[35] The optimization problem reduces due to $\dfrac{\partial PBE°(q_B)}{\partial q_B} = -\mu < 0\ \forall q_B$ (from equation 3.6) to the task of *minimizing* q_B with respect to the imposed constraints. To solve this problem, it is assumed that the LPC is binding, as $\dfrac{\partial PLE°(q_B)}{\partial q_B} = \mu > 0\ \forall q_B$ (from 3.2) and the constraint *cannot* be satisfied for $q_B = 0$ (no compensation for lending). Taking the LPC as binding, we obtain

$$q_B^* = \frac{1 + mc}{\mu}\ .$$

The respective expected profits are stated in Table 3.3.[36]

Table 3.3: Expected Profits and Feasible Contracting Constraint of a Public Equity Contract (BM, Borrower Optimization)

$PBE_B°$	$\mu - 1 - mc$
$PLE_B°$	0
BPC	$\mu \geq 1 + mc$

---

[35] See the discussion in the previous Section 3.3.1.1 (lender optimization).

[36] Note: The structure of Table 3.3 is identical to the structure of Table 3.2: The expected profit of the player determining contracts' conditions is always stated in the first line. The expected profit of the player who can not directly affect contracts' conditions is stated in the second line. The relevant FCCs are stated below.

In our opinion this structure is helpful since from the first line it can always be seen which expected profit the player with the right to set contracts' conditions can maximally extract for the respective ct/pm-constellation, etc.

In the borrower optimization, the lender becomes a residual claimant $q_B^* \leq 1$, he simply gets back his initial investment.[37]

3.3.2 Public Debt ($D°$)

The lender's as well as the borrower's optimization behavior given public debt financing is examined in this section. The feasible contracting solutions as well as the expected profits either the lender or the borrower can maximally extract are determined. Section 3.3.2.1 considers the case where the power to set the contract conditions is given to the lender, while in Section 3.3.2.2 the borrower possesses this power.

3.3.2.1 Lender Optimization

The optimization problem for *public debt* contracts for the case that the lender determines the contract conditions is quite similar to the one for public equity contracts (see Section 3.3.1.2). The lender's expected profit $PLD°(h_L)$ has to be maximized with respect to both participation constraints (LPC and BPC)

$$\max_{h_L} PLD°(h_L) \text{ s.t. } PBD°(h_L) \geq 0 \wedge PLD°(h_L) \geq 0.$$

However, for debt contracts the choice of ζ is *not* irrelevant because debt contracts contain a default threshold because of the fixed repayment obligation h_L. If a risk-shifting opportunity exists the borrower sets $\zeta = \zeta_{max}$ in order to shift as much downside risk as possible to the lender,[38] who is anticipating this. Therefore, we have to distinguish between two debt financing cases, one

---

[37] From the results stated so far (see Sections 3.3.1.1 and 3.3.1.2) it becomes obvious that for public equity financing always the player with the power to determine contracts' conditions can extract the whole excess return of the project (after monitoring costs), i. e. "the winner takes it all". The other player simply receives a non-negative expected profit to guarantee his contract participation.

[38] Having once agreed on h_L, the borrower will always, contrary to all previous promises, set $\zeta = \zeta_{max}$, as he possesses a "call option" on the profit realization. Hence, any risk increase is beneficial for the borrower due to an "option price" reduction. For example, the borrower is supposed to pay h_L when $y \geq h_L$ and y otherwise. A risk increase now reduces the borrower's expected (re)payment while this coincides with the lender's additional arising expected loss which increases in ζ. The lender's total expected payoff loss due to the borrower's default is $LPM_{h_L}^1(\mu, \zeta) = \int_{y_{min}}^{h_L} (h_L - y) f_y dy = \frac{(h_L - y_{min})^2}{4\zeta\mu} = \frac{(h_L - \mu(1-\zeta))^2}{4\zeta\mu}$ for $y_{min} < h_L$ with $\frac{\partial LPM_{h_L}^1(\mu,\zeta)}{\partial \zeta} = \frac{\zeta^2\mu^2 - (\mu - h_L)^2}{4\zeta^2\mu} \geq 0$. The gain of the borrower due to any risk increase is equal to the loss of the lender since the return distribution is symmetric and the lender as well as the borrower are risk neutral, i. e. they are only concerned about their expected pay-offs.

with a risk-shifting opportunity which can trigger a borrower default ($h_L > y_{min}\big|_{\zeta=\zeta_{max}}$) and one without ($h_L \leq y_{min}\big|_{\zeta=\zeta_{max}}$). While the latter results in riskless debt financing, the former results in risky debt financing. At the end, it must be examined which debt type the lender prefers if both debt types are feasible for certain project specifications.

- *Case I – Riskless debt:* $h_{L,I} \leq y_{min}\big|_{\zeta=\zeta_{max}}$

  As $\dfrac{\partial PLD^\circ(h_{L,I})}{\partial h_{L,I}} = 1 \ \forall h_{L,I}$ (from equation 3.4), the problem is a simple *maximization* problem with respect to the CDr/nr ($h_{L,I} \leq y_{min}\big|_{\zeta=\zeta_{max}}$), avoiding verification/default costs, and the BPC.[39] By presuming the CDr/nr as binding, we obtain the optimal fixed repayment

  $$h^*_{L,I} = y_{min}\big|_{\zeta=\zeta_{max}} = \mu(1 - \zeta_{max}) .$$

  Table 3.4 summarizes the results.

Table 3.4: Expected Profits and Feasible Contracting Constraint of a Public Debt Contract – Case I (BM, Lender Optimization)

$PLD^\circ_{L,I}$	$\mu(1 - \zeta_{max}) - 1$
$PBD^\circ_{L,I}$	$\zeta_{max}\mu$
LPC[40]	$\mu(1 - \zeta_{max}) \geq 1$

Riskless debt for a given expected return μ is only feasible up to a specific maximum risk level (ζ_{crit}) such that the LPC can be satisfied without causing agency costs (CDr/nr-requirement ($h^*_{L,I} \leq y_{min}\big|_{\zeta=\zeta_{max}}$)). Therefore, the feasible parameter constellations are bounded for a certain μ by $\zeta_{crit} = 1 - 1/\mu$ (a higher μ allows for higher ζ_{max} without violating the LPC). The BPC is always satisfied due to the borrower's positive informational rent ($\zeta_{max}\mu$) which he receives because of his one-sided informational advantage. A rent reduction is only possible by incorporating agency costs (CDr/nr-violation), see Case II.

---

[39] In this situation the BPC is never binding since the borrower keeps a positive informational rent ($\zeta_{max}\mu$, as can be see below) due to his one-sided informational advantage. However, if the borrower may have a positive outside option, a further distinction becomes necessary, i. e. BPC binding or not, as can be seen in Chapter 4.

[40] The conditions arise when evaluating the LPC for $h^*_{L,I}$.

- *Case II – Risky debt:* $y_{min}\big|_{\zeta=\zeta_{max}} < h_{L,II}$

Due to $\dfrac{\partial PLD^\circ(h_{L,II})}{\partial h_{L,II}} = \dfrac{y_{max} - h_{L,II} - vc}{2\zeta_{max}\mu}$ (from equation 3.4) being

negative for $h_{L,II} = y_{max}$ this time an *interior* solution might be optimal

since $\dfrac{\partial^2 PLD^\circ(h_{L,II})}{\partial h_{L,II}^2} < 0$. By setting $\dfrac{\partial PLD^\circ(h_{L,II})}{\partial h_{L,II}} = 0$ we obtain[41]

$$h_{L,II}^* = \mu(1 + \zeta_{max}) - vc \ .$$

Table 3.5: Expected Profits and Feasible Contracting Constraints of a Public Debt Contract – Case II (BM, Lender Optimization)

$PLD_{L,II}^\circ$	$\mu + \dfrac{vc^2}{4\zeta_{max}\mu} - 1 - vc$
$PBD_{L,II}^\circ$	$\dfrac{vc^2}{4\zeta_{max}\mu}$
CDr/nr[42]	$\zeta_{max} > \dfrac{vc}{2\mu}$
LPC	$\mu \geq 1 + vc \vee \zeta_{max} \leq \dfrac{vc^2}{4\mu(1 + vc - \mu)}$

The reason for this *interior* solution, where $y_{min}\big|_{\zeta=\zeta_{max}} < h_{L,II}^*$, is that the lender faces a tradeoff. He knows that the borrower will always set $\zeta = \zeta_{max}$ independent of $h_{L,II}$. The lender could now set $h_{L,II} = y_{max}\big|_{\zeta=\zeta_{max}}$ which still satisfies the BPC but increases the lender's expected verification/default costs to a maximum. A slight decrease in $h_{L,II}$ lowers not only the expected borrower repayment in a *convex* way

---

[41] See footnote 39.

[42] The CDr/nr and the LPC are obtained by a transformation of the initial CDr/nr and LPC evaluated for $h_{L,II}^*$.

but also reduces the lender's expected verification/default costs *linearly*.[43]
The interior solution balances the costs and benefits (for the lender) from
an infinitesimally small reduction of $h_{L,II}$.

According to the LPC, risky debt is always feasible if $\mu - 1 \geq vc$, i. e.
the expected excess return $(\mu - 1)$ is sufficient to compensate even for
permanent verification/default costs. However, risky debt is also possible
for $\mu - 1 < vc$ if ζ_{max} is bounded below $\dfrac{vc^2}{4\mu(1 + vc - \mu)}$ [LPC]. The available excess return $\mu - 1$ to compensate for the arising agency costs in the
default case can no longer compensate for permanent default costs, but
due to the ζ_{max} limitation, the default probability is bound, guaranteeing
sufficient compensation for the agency costs. The CDr/nr determines the
minimum spread of the project return $(2\zeta_{max}\mu)$ so that an interior solution becomes feasible $(2\zeta_{max}\mu > vc)$. If this condition is violated, i. e. if
the return spread is too small, the risky debt case collapses to the riskless
debt case $(h_{L,I} \leq y_{min}\big|_{\zeta=\zeta_{max}})$.

A better picture of the above stated feasibility constraints for public debt
and for which constellations riskless as well as risky debt are possible can be
obtained from Figure 3.3. This figure illustrates the conditions for an arbi-
trary parameter choice of $vc = 0.4$. For $vc = 0.4$, 40% of the initially invested
funds have to be spent in $t = 1$ to verify the project return.

---

[43] This can easily be seen, since a reduction in h influences the expected verifica-
tion/default costs only through the decreasing default probability (fixed absolute
costs), while the expected cash flow is influenced through a decreasing default
probability and a changing payoff structure, making larger reductions more severe. For a given project $(\mu, \zeta_{max}$-specification) the borrower's default probability
is

$$DP(h) = \begin{cases} 0 & \text{for } h \leq y_{min}\big|_{\zeta=\zeta_{max}} \\[2ex] \dfrac{h - y_{min}\big|_{\zeta=\zeta_{max}}}{y_{max}\big|_{\zeta=\zeta_{max}} - y_{min}\big|_{\zeta=\zeta_{max}}} = \dfrac{h - \mu(1 - \zeta_{max})}{2\zeta_{max}\mu} & \text{for } y_{min}\big|_{\zeta=\zeta_{max}} < h \leq y_{max}\big|_{\zeta=\zeta_{max}} \end{cases}.$$

Therefore, the expected verification/default costs are $vc \cdot DP(h)$ with $\dfrac{\partial DP(h)}{\partial h} = \dfrac{1}{2\zeta_{max}\mu}$. (Deliberately, no subscript for h is used in this derivation since this
derivation is valid when the lender or when the borrower has the power to set
contract conditions.)

Fig. 3.3: Feasible Contracting Area for Public Debt (BM, Lender Optimization, $vc = 0.4$)

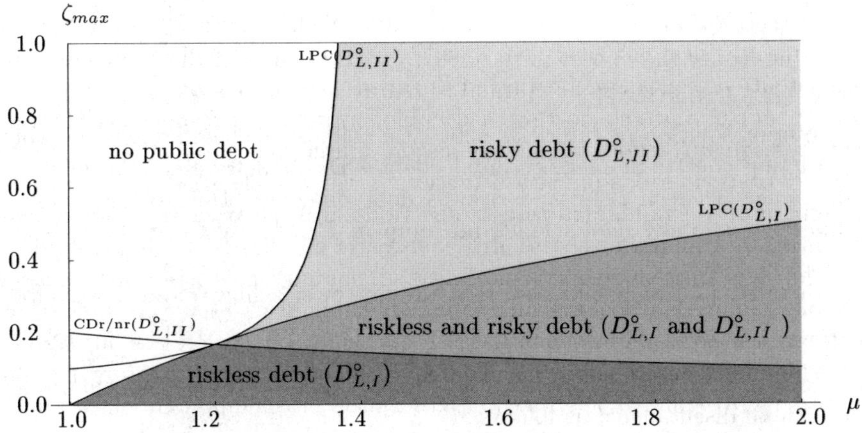

Note: The shaded areas indicate for which projects (μ, ζ_{max}-specifications) a public debt contract is feasible. While the feasible contracting area of riskless public debt ($D^\circ_{L,I}$) is restricted below the LPC($D^\circ_{L,I}$), the feasible contracting area of risky public debt ($D^\circ_{L,II}$) is bounded above the CDr/nr($D^\circ_{L,II}$) and bounded below the LPC($D^\circ_{L,II}$).[44] Therefore, four different areas occur: One where no public debt contract is feasible; one where only riskless, respectively risky, debt is feasible and one where even both are feasible.

Figure 3.3 illustrates for which project specifications public debt contracts are feasible. A contract is feasible when the explicitly stated feasible contracting constraints (FCCs), here the LPC and the CDr/nr, are satisfied. The FCCs not stated explicitly like the BPC are satisfied automatically. Risky debt contracts are feasible for high expected project returns with at least a certain return spread [CDr/nr($D^\circ_{L,II}$)]. For lower spreads the risky debt case collapses. Riskless debt is feasible also for lower expected returns since no agency costs arise, but the return volatility is bound below a critical level [LPC($D^\circ_{L,I}$)] to avoid risk-shifting opportunities. For certain project specifications even both debt types are feasible.

If only either risky or riskless debt is feasible (bright and dark grey area), the lender has no debt type choice. However, when both debt types are feasible (intersection regions of $D^\circ_{L,I}$ and $D^\circ_{L,II}$, grey area), the lender has a choice. Comparing the lender's expected profits $PLD^\circ_{L,I}$ and $PLD^\circ_{L,II}$, it

---

[44] The LPC($D^\circ_{L,II}$) has a point of discontinuity at $\mu = 1 + vc$. For $\mu \geq 1 + vc$ the LPC($D^\circ_{L,II}$) is satisfied for all possible ζ_{max} while for $\mu < 1 + vc$ ζ_{max} is bounded below $\frac{vc^2}{4\mu(1+vc-\mu)}$, see Table 3.5.

becomes obvious that risky debt $(D^\circ_{L,II})$ always dominates the riskless alternative $(D^\circ_{L,I})$ when both are feasible. Risky debt dominates riskless debt since the lender's expected profit $PLD^\circ_{L,I} = \mu(1 - \zeta_{max}) - 1$ is bound below $PLD^\circ_{L,II} = \mu + \dfrac{vc^2}{4\zeta_{max}\mu} - 1 - vc$.[45] This is not surprising since risky debt offers the lender the opportunity to extract extra parts of the borrower's return. When it appears to be optimal to reduce h such that $h^* \leq y_{min}\big|_{\zeta=\zeta_{max}}$, the interior solution of the risky debt case collapses. Therefore, the feasible contracting constraint of risky public debt CDr/nr determines the lender's indifference between risky and riskless debt financing. Due to the lender's indifference at risky debt's FCC, no jump in PLD°_L occurs at the CDr/nr $(2\zeta_{max}\mu = vc)$.[46]

Alternatively, one could note that the borrower achieves a positive informational rent from debt financing (either $\zeta_{max}\mu$ for $D^\circ_{L,I}$ or $\dfrac{vc^2}{4\zeta_{max}\mu}$ for $D^\circ_{L,II}$) due to his one-sided informational advantage. The borrower's informational rent can be disentangled as follows

$$IR(D^\circ_{L,I}) = 1 \quad \cdot \zeta_{max}\mu = \zeta_{max}\mu$$

$$IR(D^\circ_{L,II}) = \frac{vc}{2\zeta_{max}\mu} \cdot \frac{vc}{2} = \frac{vc^2}{4\zeta_{max}\mu} \quad .$$

The first component is the borrower's non-default probability, while the second stands for the borrower's expected return in the non-default state.[47] The borrower's informational rent coincides for (costless) public offerings with the borrower's expected profit $(IR(D^\circ_{L,I}) = PBD^\circ_{L,I}$ and $IR(D^\circ_{L,II}) = PBD^\circ_{L,II})$, as the remaining surplus is extracted by the lender (lender optimization).

The reason for the $D^\circ_{L,II}$ rent has been pointed out already, i. e. the costs to extract the whole surplus from the borrower are higher than its gains in the end. The same is true for the $D^\circ_{L,I}$ rent, as the CDr/nr prevents a further increase in h and gives rise to the borrower's informational rent.[48] In our setting, debt financing in general leaves the borrower with a positive informational rent

---

[45] This becomes obvious since $\dfrac{vc^2}{4\zeta_{max}\mu} - vc \geq -\zeta_{max}\mu$ is equivalent to $(2\zeta_{max}\mu - vc)^2 \geq 0$.

[46] $PLD^\circ_{L,I}$ and $PLD^\circ_{L,II}$ are equal to $\mu - vc/2 - 1$ for $2\zeta_{max}\mu = vc$.

[47] The borrower's non-default probability of risky debt results from $y_{max}\big|_{\zeta=\zeta_{max}} - h^*_{L,II} = vc$. The related default probability is $DP(h) = \dfrac{h^*_{L,II} - y_{min}\big|_{\zeta=\zeta_{max}}}{2\zeta_{max}\mu} = \dfrac{2\zeta_{max}\mu - vc}{2\zeta_{max}\mu}$.

[48] One can note that the *borrower's* informational rent increases with ζ_{max} for the riskless debt case while decreasing for risky debt (given a fixed μ). The reason is

since the resulting agency costs are related to the negotiated fixed repayment causing the described tradeoff. When both debt types are feasible risky debt is preferable for the lender, as the informational rent extractable from the borrower by a shift from riskless to risky debt overcompensates the lender's arising agency costs. Thereby, the borrower's potential informational rent is bound by $vc/2$, as he receives $\zeta_{max}\mu$ for $\zeta_{max} < \dfrac{vc}{2\mu}$ and $\dfrac{vc^2}{4\zeta_{max}\mu}$ otherwise.

3.3.2.2 Borrower Optimization

For the case of a *public debt* contract and the borrower has the power to determine the contract conditions, the optimization problem is

$$\max_{h_B} PBD^\circ(h_B) \text{ s.t. } PLD^\circ(h_B) \geq 0 \wedge PBD^\circ(h_B) \geq 0.$$

As we have shown in the lender optimization for debt type contracts (see Section 3.3.2.1), the choice of ζ is *not* irrelevant as the borrower will try to shift as much downside risk to the lender as possible. The borrower sets $\zeta = \zeta_{max}$ when a risk-shifting opportunity exists ($h_B > y_{min}\big|_{\zeta=\zeta_{max}}$), which, however, the lender anticipates. Due to the borrower's potential adverse behavior, a distinction whether a risk-shifting opportunity exists or not is necessary in the borrower's optimization. Alternatively, it could be distinguished whether risky debt financing emerges or not. Due to $\dfrac{\partial PBD^\circ(h_B)}{\partial h_B} \leq 0 \ \forall h_B$ (from equation 3.8) the borrower tries to *minimize* h_B with respect to the imposed constrains. In the end, it must be examined which public debt type the borrower will prefer if both are feasible for certain projects.

- *Case I – Riskless debt:* $h_{B,I} \leq y_{min}\big|_{\zeta=\zeta_{max}}$

  The solution for this case is easy to obtain due to $\dfrac{\partial PLD^\circ(h_{B,I})}{\partial h_{B,I}} = 1 \ \forall h_{B,I}$ (from 3.4). Since the LPC is not satisfied for $h_{B,I} = 0$, the LPC must be binding. No default can occur, verification as well as default costs can be neglected. Therefore,

---

that for riskless debt an increase in ζ_{max} decreases on the one hand $y_{min}\big|_{\zeta=\zeta_{max}}$ which requires a reduction in h_L^* in order not to violate the CDr/nr($D_{L,I}^\circ$), which leaves the borrower with a higher rent ($\zeta_{max}\mu$). On the other hand, for risky debt an increase in ζ_{max} increases $y_{max}\big|_{\zeta=\zeta_{max}}$ such that h^* can also increase (constant absolute difference $y_{max}\big|_{\zeta=\zeta_{max}} - h_L^* = vc$) which reduces the non-default probability and therefore the underlying reason for borrower's informational rent ($\dfrac{vc^2}{4\zeta_{max}\mu}$). For $\zeta_{max} = \dfrac{vc}{2\mu}$ the borrower's informational rents in riskless and risky debt are identical.

$$h^*_{B,I} = 1$$

is optimal.

Table 3.6: Expected Profits and Feasible Contracting Constraint of a Public Debt Contract – Case I (BM, Borrower Optimization)

$PBD^\circ_{B,I}$	$\mu - 1$
$PLD^\circ_{B,I}$	0
CDr/nr[49]	$\mu(1 - \zeta_{max}) \geq 1$

- *Case II – Risky debt:* $y_{min}\big|_{\zeta=\zeta_{max}} < h_{B,II}$

The risky debt case is more complicated. Not only can the borrower default on his debt obligation $h_{B,II}$, but also $\dfrac{\partial PLD^\circ(h_{B,II})}{\partial h_{B,II}} = \dfrac{y_{max} - h_{B,II} - vc}{2\zeta_{max}\mu}$

(from 3.4) is not always positive. However, since $\dfrac{\partial PLD^\circ(h_{B,II})}{\partial h_{B,II}}$ is decreasing in $h_{B,II}$ and presuming that the LPC is *not* satisfied for $h_{B,II} = y_{min}\big|_{\zeta=\zeta_{max}}$ (riskless debt), we can proceed by taking the LPC as binding. We might get two solutions where the LPC is binding due to the quadratic polynomial, from which we choose the smaller one, as the borrower (contrary to the lender) tries to minimize his repayment obligation[50]

$$h^*_{B,II} = \mu(1 + \zeta_{max}) - vc - \sqrt{4\zeta_{max}\mu(\mu - 1 - vc) + vc^2}.$$

Table 3.7: Expected Profits and Feasible Contracting Constraints of a Public Debt Contract – Case II (BM, Borrower Optimization)

$PBD^\circ_{B,II}$	$\dfrac{\left[vc + \sqrt{4\zeta_{max}\mu(\mu - 1 - vc) + vc^2}\right]^2}{4\zeta_{max}\mu}$
$PLD^\circ_{B,II}$	0
NoN	$\mu \geq 1 + vc \vee \zeta_{max} \leq \dfrac{vc^2}{4\mu(1 + vc - \mu)}$
CDr/nr	$\mu(1 - \zeta_{max}) < 1 \wedge \zeta_{max} > \dfrac{vc}{2\mu}$

---

[49] We do not need to examine the BPC since if $\mu(1 - \zeta_{max}) \geq 1$ [CDr/nr], it is guaranteed that $\mu \geq 1$.

[50] Note: The CDr/nr assures additional to $h^*_{B,II} > y_{min}\big|_{\zeta=\zeta_{max}}$ that $h^*_{B,II} > 0$. The new emerging non-negativity constraint (NoN) guarantees a real square root. Hence, the BPC does not have to be examined explicitly since if the NoN is satisfied the BPC is automatically satisfied as well.

Since both cases do not intersect due to their respective CDr/nr, $PBD^{\circ}_{B,I}$ and $PBD^{\circ}_{B,II}$ do not have to be compared. Of course, the borrower can determine h_B and, therefore, choose between riskless and risky debt financing, however, when he maximizes his profit, his debt financing choice is uniquely defined for each project (μ, ζ_{max}-specification). Since the borrower's main goal is to maximize his profit and consequently to minimize his repayment obligation h_B, he prefers riskless debt financing ($h_{B,I} \leq y_{min}\big|_{\zeta=\zeta_{max}}$) as long as possible, while risky debt financing ($h_{B,II} > y_{min}\big|_{\zeta=\zeta_{max}}$) is considered otherwise. Since the lender does not possess any informational advantage he never receives an informational rent in debt financing. Therefore, the borrower only tries to reduce the agency costs (AC) of debt financing and the agency costs are lower for riskless debt financing (AC=0) than for risky debt (AC>0).[51, 52]

Figure 3.4 illustrates this observation in accordance with Figure 3.3, i. e. $vc = 0.4$.

---

[51] Risky debt financing collapses if it is possible for the borrower to set $h_B \leq y_{min}\big|_{\zeta=\zeta_{max}}$ without violating the LPC, see the CDr/nr($D^{\circ}_{B,II}$).

[52] Contrary, in the lender optimization, the lender additionally has to examine whether risky debt is preferable despite causing higher agency costs, since offering the opportunity to reduce the borrower's informational rent which requires an extra comparison of both debt financing cases.

Fig. 3.4: Feasible Contracting Area for Public Debt (BM, Borrower Optimization, $vc = 0.4$)

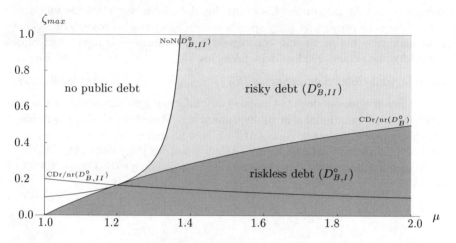

Note: The shaded areas indicate for which projects (μ, ζ_{max}-specifications) a public debt contract is feasible. While the feasible contracting area of riskless public debt ($D_{B,I}^\circ$) is restricted below the CDr/nr(D_B°), the feasible contracting area of risky public debt ($D_{B,II}^\circ$) is bounded above the CDr/nr(D_B°) and the CDr/nr($D_{B,II}^\circ$), and below the NoN($D_{B,II}^\circ$).[53] Therefore, three areas occur: One where no public debt contract is feasible; one where only riskless and one where only risky debt is feasible.

3.3.3 Private Equity ($E^\bullet$)

This section examines the lender's as well as the borrower's optimization behavior given private equity financing. In Section 3.3.3.1, the power to set the contract conditions is given to the lender, while the second Section 3.3.3.2 focuses on the borrower.

Due to the missing ex-ante informational asymmetries, the derivation of the feasible contracting solutions and the optimal contract conditions is similar for *private* placements and *public* offerings. Therefore, we mainly state the optimization problems with the obtained results. Differences in the derivations are explained. Due to so far missing ex-ante informational asymmetries, all private placements are *inefficient* due to the costs nc for both parties (Assumption 4). However, they are *not* necessarily *dominated* by efficient public offerings due to public placement's potentially lacking incentive compatibility.

---

[53] The NoN($D_{B,II}^\circ$) has a point of discontinuity at $\mu = 1 + vc$.

3.3.3.1 Lender Optimization

The optimization problem

$$\max_{q_L} PLE^\bullet(q_L) \text{ s.t. } PBE^\bullet(q_L) \geq 0 \wedge PLE^\bullet(q_L) \geq 0$$

has to be solved to find out for which project specifications *private equity* contracts are feasible and which expected profit $PLE^\bullet(q_L)$ the lender maximally can extract for the case that the lender determines the contract conditions. It turns out that the proportional return participation

$$q_L^* = 1 - \frac{nc}{\mu}$$

is optimal. Table 3.8 contains the related expected profits and feasible contracting constraints.

Table 3.8: Expected Profits and Feasible Contracting Constraint of a Private Equity Contract (BM, Lender Optimization)

PLE_L^*	$\mu - 1 - mc - 2nc$
PBE_L^*	0
LPC	$\mu \geq 1 + mc + 2nc$

As we have already observed for public equity contracts, the feasibility of private equity contracts only depends on μ and not on ζ_{max}. However, now the project's expected excess return $\mu - 1$ must be at least sufficient to cover the total arising agency costs of $mc + 2nc$. Whether private equity is dominated in the end, for example, by public equity, depends on which party has the right to determine the ct/pm-choice. If the lender has the choice, he prefers public to private equity since the profit he can extract is larger. However, if the borrower has the choice, he is indifferent between public and private equity since he is reduced to his participation constraint anyway.

3.3.3.2 Borrower Optimization

On the other hand, to determine for which project specifications private equity contracts are feasible when the borrower determines the contract conditions,

$$\max_{q_B} PBE^\bullet(q_B) \text{ s.t. } PLE^\bullet(q_B) \geq 0 \wedge PBE^\bullet(q_B) \geq 0$$

has to be solved. The optimum is

$$q_B^* = \frac{1 + mc + nc}{\mu} \ .$$

Table 3.9: Expected Profits and Feasible Contracting Constraint of a Private Equity Contract (BM, Borrower Optimization)

$PBE_B^{\bullet}$	$\mu - 1 - mc - 2nc$
$PLE_B^{\bullet}$	0
BPC	$\mu \geq 1 + mc + 2nc$

3.3.4 Private Debt $(D^{\bullet})$

Analogously to private equity financing, the lender and the borrower can negotiate personally about debt financing. The feasible contracting solutions as well as the expected profits either the lender or the borrower can maximally extract are determined. In Section 3.3.4.1, the power to set the contract conditions is given to the lender, while Section 3.3.4.2 focuses on the borrower.

Due to missing ex-ante informational asymmetries, the derivations are similar for *private* placements and for *public* offerings. Basically, only the optimization problems are stated with the obtained results. Differences in the derivations are explained.

3.3.4.1 Lender Optimization

For *private debt* contracts the lender's optimization problem is

$$\max_{h_L} PLD^{\bullet}(h_L) \text{ s.t. } PBD^{\bullet}(h_L) \geq 0 \land PLD^{\bullet}(h_L) \geq 0.$$

Since nc arises for both parties for a private placement, a further case distinction among risky and riskless debt is necessary. In particular, two new subcases arise. One of these subcases concerns where the borrower's informational rent due to his interim and ex-post informational advantage is high enough to compensate for the nc he has to bear ((a), BPC not binding) and one where it is not ((b), BPC binding). If the BPC is still not binding we proceed in the common way. If the BPC is binding, the solution is found by simply solving the BPC. The informational rent the borrower receives in debt financing due to his interim and ex-post informational advantage is independent of the placement mode as even private negotiations only offer the opportunity to overcome potential ex-ante informational asymmetries. Difficulties due to interim and ex-post uncertainty can only be mitigated by monitoring the borrower's behavior and/or verifying the proclaimed project return. However, the placement mode choice still affects the borrower's expected profit.

- *Case I – Riskless debt:* $h_{L,I} \leq y_{min}\big|_{\zeta=\zeta_{max}}$
  - *Ia – BPC not binding since* $\zeta_{max}\mu \geq nc$

  $h^*_{L,Ia} = \mu(1 - \zeta_{max})$

Table 3.10: Expected Profits and Feasible Contracting Constraints of a Private Debt Contract – Case Ia (BM, Lender Optimization)

$PLD^\bullet_{L,Ia}$	$\mu(1 - \zeta_{max}) - 1 - nc$
$PBD^\bullet_{L,Ia}$	$\zeta_{max}\mu - nc$
BPCb/nb	$\mu \geq \dfrac{nc}{\zeta_{max}}$
LPC	$\mu(1 - \zeta_{max}) \geq 1 + nc$

This case is similar to the case $D^\circ_{L,I}$. The feasible solutions are bound below the LPC. The BPCb/nb now additionally bounds the solutions, guaranteeing that the borrower's informational rent ($\zeta_{max}\mu$) is high enough to compensate for his negotiation costs nc. Since the informational rent *increases* with ζ_{max}, ζ_{max} needs to hold a critical level nc/μ. If ζ_{max} drops *below* nc/μ the BPC becomes binding (case Ib).

  - *Ib – BPC binding since* $\zeta_{max}\mu < nc$

  $h^*_{L,Ib} = \mu - nc$

Table 3.11: Expected Profits and Feasible Contracting Constraints of a Private Debt Contract – Case Ib (BM, Lender Optimization)

$PLD^\bullet_{L,Ib}$	$\mu - 1 - 2nc$
$PBD^\bullet_{L,Ib}$	0
BPCb/nb	$\mu < \dfrac{nc}{\zeta_{max}}$
LPC	$\mu \geq 1 + 2nc$

Now, since the BPC is binding, ζ_{max} is restricted below nc/μ [BPCb/nb], while $\mu - 1$ has to be large enough to compensate at least for the total agency costs $2nc$.

Therefore, the optimal repayment obligation for riskless debt is

$h^*_{L,I} = \mu - \max\{\mu\zeta_{max}, nc\}$.

- *Case II – Risky debt:* $y_{min}\big|_{\zeta=\zeta_{max}} < h_{L,II}$

  – *IIa – BPC not binding since* $\dfrac{vc^2}{4\zeta_{max}\mu} \geq nc$

  $h^*_{L,IIa} = \mu(1 + \zeta_{max}) - vc$

Table 3.12: Expected Profits and Feasible Contracting Constraints of a Private Debt Contract – Case IIa (BM, Lender Optimization)

$PLD^{\bullet}_{L,IIa}$	$\mu + \dfrac{vc^2}{4\zeta_{max}\mu} - 1 - vc - nc$
$PBD^{\bullet}_{L,IIa}$	$\dfrac{vc^2}{4\zeta_{max}\mu} - nc$
CDr/nr	$\dfrac{vc}{2\mu} < \zeta_{max}$
BPCb/nb[54]	$\zeta_{max} \leq \dfrac{vc^2}{4\mu nc}$
LPC	$\mu \geq 1 + vc + nc \vee \zeta_{max} \leq \dfrac{vc^2}{4\mu(1 + vc + nc - \mu)}$

This case is similar to the case $D^{\circ}_{L,II}$. The feasible solutions are bound by the LPC, the CDr/nr and the BPCb/nb. The BPCb/nb guarantees that the borrower's informational rent ($\dfrac{vc^2}{4\zeta_{max}\mu}$) is high enough to compensate for nc. Since the borrower's informational rent *decreases* with ζ_{max}, ζ_{max} needs to stay *below* the critical level $\dfrac{vc^2}{4\mu nc}$. If ζ_{max} increases above $\dfrac{vc^2}{4\mu nc}$, the BPC becomes binding (case IIb). Additionally, the CDr/nr makes the derived risky debt solution feasible.

---

[54] The CDr/nr and the BPCb/nb require $vc > 2nc$, otherwise case IIa is impossible.

- *IIb – BPC binding since* $\dfrac{vc^2}{4\zeta_{max}\mu} < nc$

$$h^*_{L,IIb} = \mu(1 + \zeta_{max}) - 2\sqrt{nc\zeta_{max}\mu}$$

Table 3.13: Expected Profits and Feasible Contracting Constraints of a Private Debt Contract – Case IIb (BM, Lender Optimization)

$PLD^{\bullet}_{L,IIb}$	$\mu + vc\sqrt{\dfrac{nc}{\zeta_{max}\mu}} - 1 - vc - 2nc$
$PBD^{\bullet}_{L,IIb}$	0
CDr/nr	$\zeta_{max} > \dfrac{nc}{\mu}$
BPCb/nb	$\zeta_{max} > \dfrac{vc^2}{4\mu nc}$
LPC	$\mu \geq 1 + vc + 2nc \vee \zeta_{max} \leq \dfrac{ncvc^2}{\mu(1 + vc + 2nc - \mu)^2}$

$h^*_{L,IIb} = \mu(1 + \zeta_{max}) - 2\sqrt{nc\zeta_{max}\mu}$ causes a non-default probability of $\dfrac{2\sqrt{nc\zeta_{max}\mu}}{2\zeta_{max}\mu}$ which implies that the borrower's expected surplus $(\dfrac{2\sqrt{nc\zeta_{max}\mu}}{2\zeta_{max}\mu} \cdot \sqrt{nc\zeta_{max}\mu})$ is just sufficient to compensate for his negotiation costs (nc).

The binding BPC restricts ζ_{max} by $\max\{\dfrac{vc^2}{4\mu nc}, \dfrac{nc}{\mu}\}$ [CDr/nr and BPCb/nb] while $\mu - 1$ has to be sufficiently large to compensate for the total agency costs.

Hence, the optimal repayment obligation is

$$h^*_{L,II} = \mu(1 + \zeta_{max}) - \sqrt{\max\{vc^2, 4nc\zeta_{max}\mu\}} \ .$$

Figure 3.5 illustrates the feasible contracting conditions of private debt for the parameter specification ($vc = 0.4, nc = 0.1$). $vc = 0.4$ is kept constant for our illustrative purpose (compare Figure 3.3), while we assumed lower negotiation costs $nc = 0.1$. $nc = 0.1$ indicates that the costs of a private placement amount for the lender and the borrower totals to 20% ($2nc$) of their investment.

Fig. 3.5: Feasible Contracting Area for (i) Riskless and (ii) Risky Private Debt (BM, Lender Optimization, $vc = 0.4, nc = 0.1$)

(i) Riskless debt

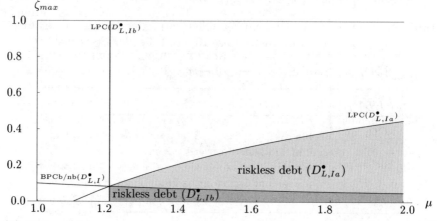

(ii) Risky debt

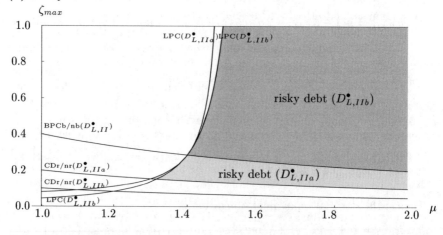

Note: The shaded areas indicate for which projects (μ, ζ_{max}-specifications) a private debt contract is feasible. While the feasible contracting area of riskless private debt with non-binding BPC, $D_{L,Ia}^{\bullet}$, is bounded below the LPC($D_{L,Ia}^{\bullet}$) and above the BPCb/nb($D_{L,I}^{\bullet}$), the area of riskless private debt with binding BPC, $D_{L,Ib}^{\bullet}$, is bound below the BPCb/nb($D_{L,I}^{\bullet}$) and right of the LPC($D_{L,Ib}^{\bullet}$). The area of risky debt with non-binding BPC, $D_{L,IIa}^{\bullet}$, is fixed between the CDr/nr($D_{L,IIa}^{\bullet}$) and the BPCb/nb($D_{L,II}^{\bullet}$) and bounded below the LPC($D_{L,IIa}^{\circ}$) with a point of discontinuity at $\mu = 1 + vc + nc$. The area of risky debt with binding BPC, $D_{L,IIb}^{\bullet}$, is restricted above the BPCb/nb($D_{L,II}^{\bullet}$) and below the LPC($D_{L,IIb}^{\bullet}$) with a point of discontinuity at $\mu = 1 + vc + 2nc$.

It becomes obvious that, similar to the public debt case, riskless private debt is feasible for expected project returns which cover at least the initial outlay and the arising costs of the private placement $2nc$ while ζ_{max} is bounded below the LPC($D^\bullet_{L,Ia}$). Since the borrower's informational rent for riskless debt ($IR(D^\bullet_{L,I}) = \zeta_{max}\mu$) increases in ζ_{max} for a fixed μ, the BPC is binding for low ζ_{max}.[55] $D^\bullet_{L,Ia}$ lies above $D^\bullet_{L,Ib}$. Risky debt again requires a higher expected return to compensate for the additional arising verification/default costs while ζ_{max} is restricted above the minimum of the CDr/nr($D^\bullet_{L,IIa}$) and the CDr/nr($D^\bullet_{L,IIb}$) to keep the interior solutions feasible. Contrary to the riskless debt case, the borrower's informational rent ($IR(D^\bullet_{L,II}) = \dfrac{vc^2}{4\zeta_{max}\mu}$) now decreases with ζ_{max} for a fixed μ so that the BPC becomes binding while ζ_{max} increases.[56] $D^\bullet_{L,IIa}$ lies below $D^\bullet_{L,IIb}$.

Additionally, the feasible regions of riskless and risky private debt with binding BPC cannot intersect, as BPCb/nb($D^\bullet_{L,I}$)=CDr/nr($D^\bullet_{L,IIb}$). The regions cannot intersect as in both cases the borrower is left with no profit and for riskless debt financing ($D^\bullet_{L,Ib}$) he receives even more than the whole return distribution spread, while for risky debt financing ($D^\bullet_{L,IIb}$) he is left with less return variation and additionally the non-default probability is lower than one. Therefore, the FCCs of $D^\bullet_{L,Ib}$ and $D^\bullet_{L,IIb}$ cannot be satisfied simultaneously (see the BPCb/nb and the LPC for $D^\bullet_{L,Ib}$ in Table 3.11 and the CDr/nr, the BPCb/nb and the LPC for $D^\bullet_{L,IIb}$ in Table 3.13).

Furthermore, it becomes clear from our discussion so far that the borrower achieves an expected profit ($PB > 0$) if his debt financing informational rent is higher than the negotiation costs arising in the private placement (case Ia and IIa). Otherwise (case Ib and IIb), the borrower is left with his outside option ($PB = 0$). In comparison to a public placement where the borrower obtains the full informational rent ($PBD^\circ_L = IR(D_L)$) the borrower's expected profit is reduced through the negotiation costs of a private placement ($PBD^\bullet_L = \max\{IR(D_L) - nc, 0\}$). This is possible since the borrower is compensated for his *own* negotiation costs nc either fully ($IR(D_L) \geq nc$) or at

---

[55] The borrower's informational rent of riskless debt ($\zeta_{max}\mu$) increases with ζ_{max} for a given μ, since an increase in ζ_{max} forces the lender to reduce the borrower's repayment obligation ($h^*_{L,Ia} = y_{min}\big|_{\zeta=\zeta_{max}}$) to keep debt financing riskless. The borrower's non-default probability remains equal to one, while the borrower's extractable surplus in the non-default state ($\zeta_{max}\mu$) increases.

[56] For risky debt the borrower's informational rent ($\dfrac{vc^2}{4\zeta_{max}\mu}$) decreases with ζ_{max} for a given μ, as when ζ_{max} increases the lender will increase his demandable repayment $h^*_{L,IIa}$ ($h^*_{L,IIa} = y_{max}\big|_{\zeta=\zeta_{max}} - vc = \mu(1+\zeta_{max}) - vc$), such that the non-default probability ($1 - DP(h_{L,IIa})$) reduces while the borrower's extractable surplus in the non-default case ($vc/2$) remains constant.

least partly $(IR(D_L) < nc)$ through his *own* uncontractable informational rent $(IR(D_L))$.[57]

However, private placements are still *inefficient* from the lender's perspective, since for (Ib and IIb) the borrower requires extra funds as compensation for his negotiation costs, while for (Ia and b) and (IIa and b) the lender's nc must be compensated anyway.

Before we can proceed, it has to be examined which debt case the lender prefers when both cases $D^\bullet_{L,I}$ and $D^\bullet_{L,II}$ are feasible. By comparing the relevant expected profits PL, it becomes obvious that from the lender's perspective

- $D^\bullet_{L,IIa}$ dominates $D^\bullet_{L,Ia}$ for the intersection region of $D^\bullet_{L,Ia}$ and $D^\bullet_{L,IIa}$ (the argument is equivalent to that of public offering's comparison on page 59).
- $D^\bullet_{L,IIa}$ dominates $D^\bullet_{L,Ib}$ for the intersection region of $D^\bullet_{L,Ib}$ and $D^\bullet_{L,IIa}$, as on the one hand no extra money has to be spent to compensate for the borrower's costs nc and on the other hand the risky debt type is chosen.[58]
- No dominance relation exists for the remaining comparison of $D^\bullet_{L,IIb}$ and $D^\bullet_{L,Ia}$. Since $D^\bullet_{L,IIb}$ has the risky debt advantage, no extra funds are required for $D^\bullet_{L,Ia}$. It turns out that $D^\bullet_{L,IIb}$ is preferable to $D^\bullet_{L,Ia}$
  $(PLD^\bullet_{L,IIb} \geq PLD^\bullet_{L,Ia})$ for $\zeta_{max} \geq \dfrac{vc + nc}{\mu}$ or $\zeta_{max} < \dfrac{vc + nc}{\mu}$ with
  $vc > 2nc$ and $\dfrac{nc}{\mu} \leq \zeta_{max} \leq \dfrac{nc + 2vc - \sqrt{nc}\sqrt{nc + 4vc}}{2\mu}$ and vice versa.[59]

Therefore, the lender's optimization behavior for private debt financing is determined. The remaining cases $D^\bullet_{L,Ia}, D^\bullet_{L,Ib}$ (see BPCb/nb($D^\bullet_{L,I}$)), $D^\bullet_{L,IIa}$, $D^\bullet_{L,IIb}$ (see BPCb/nb($D^\bullet_{L,II}$)) and $D^\bullet_{L,Ib}, D^\bullet_{L,IIb}$ (see BPCb/nb($D^\bullet_{L,I}$) and CDr/nr($D^\bullet_{L,IIb}$)) do not have to be compared as they do not intersect.

---

[57] Uncontractable means that contracting this rent is more expensive than the extractible surplus. Therefore, contracting this rent results in a net loss.

[58] A substitution of nc by $\zeta_{max}\mu$ in the comparison of $PLD^\bullet_{L,Ib}$ and $PLD^\bullet_{L,IIa}$ is helpful to see $PLD^\bullet_{L,IIa} \geq PLD^\bullet_{L,Ib}$, as $(vc - 2\zeta_{max}\mu)^2 \geq 0$.

[59] The reason behind this phenomenon is that $PLD^\bullet_{L,IIb} \geq PLD^\bullet_{L,Ia}$ does not always hold, although $h^*_{L,IIb} \geq h^*_{L,Ia}$ results from the same tradeoff, which drives the interior risky debt solution $D^\bullet_{L,IIa}$. In the case of risky debt, the borrower's informational rent might not be sufficient to cover the arising negotiation costs $(\dfrac{vc^2}{4\zeta_{max}\mu} < nc)$ such that the lender has to reduce h_L below $h^*_{L,IIa}$. This reduction might shift h_L below $\mu(1 - \zeta_{max}) + vc$. In this region $(\mu(1-\zeta_{max}) < h_L < \mu(1-\zeta_{max})+vc)$ a marginal increase in h_L increases the expected verification costs in a linear way, while the additional return (only) grows in a convex way. Therefore, a net loss occurs such that in total the possibility of $PLD^\bullet_{L,Ia} > PLD^\bullet_{L,IIb}$ arises, despite $h^*_{L,IIb} \geq h^*_{L,Ia}$.

3.3.4.2 Borrower Optimization

For private debt, the borrower's optimization problem looks familiar to the borrower's optimization problem of public debt.

$$\max_{h_B} PBD^\bullet(h_B) \text{ s.t. } PLD^\bullet(h_B) \geq 0 \wedge PBD^\bullet(h_B) \geq 0.$$

Again, a distinction between riskless (case I) and risky debt (case II) is necessary to solve the optimization problem, as the lender never achieves an informational rent. Up to now he has only an informational disadvantage. Hence, no further case distinction as in the lender optimization is required. The LPC is always binding.

- *Case I – Riskless debt:* $h_{B,I} \leq y_{min}\big|_{\zeta=\zeta_{max}}$

$$h^*_{B,I} = 1 + nc$$

Table 3.14: Expected Profits and Feasible Contracting Constraints of a Private Debt Contract – Case I (BM, Borrower Optimization)

$PBD^\bullet_{B,I}$	$\mu - 1 - 2nc$
$PLD^\bullet_{B,I}$	0
CDr/nr	$\mu(1 - \zeta_{max}) \geq 1 + nc$
BPC	$\mu \geq 1 + 2nc$

- *Case II – Risky debt:* $y_{min}\big|_{\zeta=\zeta_{max}} < h_{B,II}$

$$h^*_{B,II} = \mu(1 + \zeta_{max}) - vc - \sqrt{4\zeta_{max}\mu(\mu - 1 - nc - vc) + vc^2}$$

Table 3.15: Expected Profits and Feasible Contracting Constraints of a Private Debt Contract – Case II (BM, Borrower Optimization)

$PBD^\bullet_{B,II}$	$\dfrac{\left[vc + \sqrt{4\zeta_{max}\mu(\mu - 1 - nc - vc) + vc^2}\right]^2}{4\zeta_{max}\mu} - nc$
$PLD^\bullet_{B,II}$	0
NoN	$\mu \geq 1 + nc + vc \vee \zeta_{max} \leq \dfrac{vc^2}{4\mu(1 + vc + nc - \mu)}$
CDr/nr	$\mu(1 - \zeta_{max}) < 1 + nc \vee \zeta_{max} > \dfrac{vc}{2\mu}$
BPC[60]	$\mu \geq 1 + vc + 2nc \vee \zeta_{max} \leq \dfrac{ncvc^2}{\mu(1 + vc + 2nc - \mu)^2}$

Since both debt types do not intersect, no comparison is required.

3.3.5 Comparison of Feasible Contracting Solutions from the Lender's and from the Borrower's Perspective

As expected, a comparison of the feasible contracting solutions for the different ct/pm-constellations and optimization scenarios reveals that the range of feasible contracting solutions for a given ct/pm-choice is independent of the bargaining power distribution. This has to be the case due to our definition of feasibility and due to the absence of an ex-ante informational asymmetry between the lender and the borrower. Of course, the expected profits the lender/borrower can maximally extract vary. But the range of feasible contracting solutions determined by the relevant FCC is independent of the bargaining power distribution.

- For public equity, see Section 3.3.1, the range is restricted in the lender's as well as in the borrower's optimization scenario by the lender's as well as the borrower's simultaneously binding participation constraints. The lender and the borrower only receive their minimum for participation ($PL = PB = 0$),[61] while no further project return is left to distribute. Hence, the range where public equity financing is feasible is independent of the bargaining power distribution ($\mathrm{FCC}(E_L^\circ) = \mathrm{FCC}(E_B^\circ) = \mathrm{FCC}(E^\circ)$).
- For public debt, see Section 3.3.2, the range is restricted in the lender's as well as in the borrower's optimization scenario by the lender's binding participation constraint, while the borrower keeps his "uncontractible" informational rent (see Tables 3.4 and 3.5 for the lender optimization and Tables 3.6 and 3.7 for the borrower optimization). This informational rent is "uncontractible", as the costs to extract exceed the achievable gains, which implies that the lender refrains from extracting this rent in the lender's optimization. Similarly, when the borrower has the power to determine the contract conditions he could promise to pay an extra return to the lender in order to relax the range restriction, but the lender anticipating the borrower's adverse ex-post behavior will demand an appropriate agency cost compensation, making this behavior unprofitable for the borrower. Consequently,

---

[60] Actually, ζ_{max} is restricted by the maximum of $\left\{ \dfrac{vc^2}{2\mu(1 + vc + 2nc - \mu)}, \right.$

$\left. \dfrac{ncvc^2}{\mu(1 + vc + 2nc - \mu)^2} \right\}$. But since $\dfrac{ncvc^2}{\mu(1 + vc + 2nc - \mu)^2}$ is binding for

$\mu \geq 1 + vc$, while for $\mu < 1 + vc$ the NoN is more restrictive than the BPC, we do not state the whole condition explicitly.

[61] The project return of $\mu = 1 + mc$ is just sufficient to compensate for the initial outlay and the occurring agency costs. No surplus remains to be distributed ($\mu = 1 + PL + PB + AC$ and $\mu = 1 + AC$ require $PL + PB = 0$ and due to both participation constraints $PL = PB = 0$).

the range where public debt financing is feasible is independent of the
bargaining power distribution $(\text{FCC}(D_L^\circ) = \text{FCC}(D_B^\circ) = \text{FCC}(D^\circ))$.

- By a similar reasoning it can be shown that the range where private equity,
respectively private debt, financing is feasible (see Sections 3.3.3 and 3.3.4)
is also independent of the bargaining power distribution despite occur-
ring negotiation costs $(\text{FCC}(E_L^\bullet) = \text{FCC}(E_B^\bullet) = \text{FCC}(E^\bullet); \text{FCC}(D_L^\bullet) = \text{FCC}(D_B^\bullet) = \text{FCC}(D^\bullet))$.

The main implication of this "finding" is stated in Result 1.

Result 1
Basic Model: Feasible Contracting Solutions
Since the range of feasible contracting solutions for $E^\circ, D^\circ, E^\bullet$ and
$D^\bullet$ is independent of the bargaining power distribution, variations in
the bargaining power distribution do not necessarily have to cause
changes in the preferred ct/pm-choice. □

However, changes in the bargaining power distribution *can* influence a
firm's financing decision. The distribution of bargaining power might be
relevant for projects $(\mu, \zeta_{max}$-specifications) where more than one ct/pm-
constellation is feasible.

- Since due to the pre-defined monitoring and verification schemes of the
considered *cts* (Assumption 3), the agency costs associated with equity
financing are risk-insensitive. The respective FCC determined by the
lender's as well as the borrower's participation constraints is therefore
also risk insensitive (a vertical line in a μ, ζ_{max}-diagram).
The agency costs of riskless debt financing are also risk insensitive since
equal to zero, while the costs of risky debt can either decrease, increase or even stay constant for variations in project risk depending
on the relationship between h and μ.[62] The respective FCC is determined by the lender's binding participation constraint, the minimiza-
tion of the arising agency costs and borrower's informational rent, which
sums up to $0 + \zeta_{max}\mu = \zeta_{max}\mu$ for riskless public debt financing and to
$vc \cdot \dfrac{2\zeta_{max}\mu - vc}{2\zeta_{max}\mu} + \dfrac{vc^2}{4\zeta_{max}\mu} = \dfrac{4\zeta_{max}\mu vc - vc^2}{4\zeta_{max}\mu}$ for risky debt. Obviously,
the FCC is risk sensitive in a way that for a "low" possible return volatil-
ity, only a "low" expected project return is required to satisfy the LPC,
the BPC, and the arising agency costs, while for a "high" possible return
volatility, a "high" expected return is required (upward slopping line in a
μ, ζ_{max}-diagram).
Since additionally we only consider reasonable cases where $vc > mc$, neither debt nor equity contracts are feasible for each project where the other

---

[62] The variation in the agency costs of risky public debt financing, $vc \cdot DP(h)$, is
determined by the variation in the default probability DP. The DP decreases
with ζ_{max} for $h < \mu$, stays constant for $h = \mu$ and increases with ζ_{max} for $h > \mu$.

ct is feasible. Therefore, both FCCs divide the possible projects (μ, ζ_{max}-specifications) into four groups (the FCC($E°$) and the FCC($D°$) intersect once). Either no ct is feasible, or just one (either debt or equity), while at least for certain projects both ct are feasible.

• Private placements always cause higher total agency costs for a particular ct than public offerings without offering any advantage so far (see Assumption 4). Hence, the private placement feasible sets are subsets of the public offering feasible sets (the FCCs of debt and equity are shifted to the right in a μ, ζ_{max}-diagram).

Figure 3.6 illustrates the feasible contracting areas and, thereby, the potentially feasible ct/pm-choices for each possible project specification.

Obviously for certain project specifications, of course, depending on the underlying parameters (vc, mc and nc), the borrower is unable to acquire the required funds to finance the project, independent of the bargaining power distribution (Area I in Table 3.16). For other projects only public equity (Area II in Table 3.16) or respectively public debt (Area III in Table 3.16) financing is feasible, while an actual ct/pm-choice exists for the remaining projects, as there is more than one ct/pm-constellation feasible (Areas IV-IX in Table 3.16). Basically, we observe that the number of feasible ct/pm-constellations is positively correlated with the project's expected return given ζ_{max}, as the excess return $\mu - 1$ available to compensate for the arising agency costs increases.

Fig. 3.6: Feasible Contracting Areas and Possible Contract Type / Placement Mode-Choices, (i) for $nc = 0.05$, (ii) for $nc = 0.175$ (BM, $vc = 0.4, mc = 0.15$)

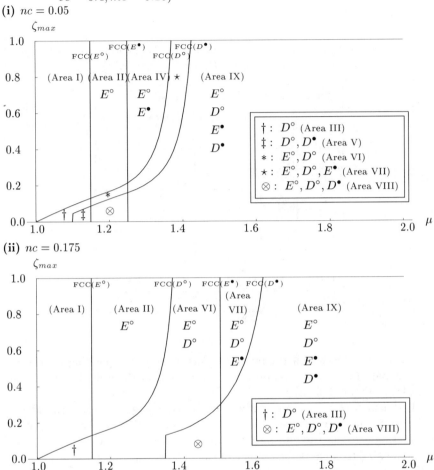

Note: The overall feasible contracting constraints (FCC) illustrated define the feasible contracting areas of $E^\circ, D^\circ, E^\bullet$ and $D^\bullet$. An actual ct/pm-choice is only possible when more than one ct/pm-constellation is feasible for this project specification. All the FCCs restrict the expected project return μ to a certain minimum level dependent on the maximum project return volatility ζ_{max}. The feasible contracting areas of private placements are subsets of the respective public offerings' contracting areas due to the additionally arising negotiation costs nc. See Table 3.16 for a summary of all possible contracting areas and their corresponding numbers.

To cope with the parameter dependency of the FCCs, the division of the feasible contracting areas is also stated depending on vc, mc and nc. Table 3.16 summarizes the possible ct/pms for the areas separated by the respective FCC.

Table 3.16: Definitions of Feasible Contracting Areas and Possible Contract Type / Placement Mode-Constellations (BM)

Area	FCC satisfied	Possible ct/pms
I	–	–
II	E°	E°
III	D°	D°
IV	$E^\circ, E^\bullet$	$E^\circ, E^\bullet$
V	$D^\circ, D^\bullet$	$D^\circ, D^\bullet$
VI	E°, D°	E°, D°
VII	$E^\circ, D^\circ, E^\bullet$	$E^\circ, D^\circ, E^\bullet$
VIII	$E^\circ, D^\circ, D^\bullet$	$E^\circ, D^\circ, D^\bullet$
IX	$E^\circ, D^\circ, E^\bullet, D^\bullet$	$E^\circ, D^\circ, E^\bullet, D^\bullet$

All of these areas can occur, but do not have to. Depending on the exogenous parameter constellation (mc, vc, nc) some of them might vanish, like Area IV in Figure 3.6 (ii).

The next section examines which ct/pm-constellation either the lender, respectively the borrower, actually prefers in each of the nine possible feasible contracting areas in the alternative bargaining power scenarios.[63]

To be able to solve the last stage of the defined contract negotiation game in the next section we not only determined the feasible ct/pms, but also derived the expected profits the lender respective the borrower can maximally extract.

---

[63] The bargaining power assigned either to the lender or to the borrower is independent of the project specification. Just one particular project is considered examining the effects of variations in the bargaining power distribution on the preferred ct/pm-choice.

3.4 Contract Type / Placement Mode-Choices in the Bargaining Power Scenarios

After analyzing which *ct/pm*-constellations are feasible for particular projects and which expected profits the lender respective the borrower can extract in Section 3.3, the *optimal ct/pm*-choices in each of the four bargaining power scenarios are derived in this section (see Sections 3.4.1 to 3.4.4). In the LS^a and the LS^r scenarios, the lender has the right to determine ct/pm, i. e. the first stage of the contracting game, while in the BS^a and the BS^r scenarios the borrower chooses. Section 3.5 then examines bargaining power effects in financial contracting by a pair-wise comparison of these optimal *ct/pm*-choices in the different bargaining power scenarios.

3.4.1 Lender Scenario with Absolute Power (LS^a)

The *lender scenario with absolute power* is characterized by the fact that the lender can propose a take-it-or-leave-it offer concerning ct/pm *and* contract conditions which the borrower either has to accept or reject. If the borrower accepts, he receives the necessary funds to finance his project and proceeds, otherwise the game ends.

The lender's expected profits PL for all feasible *ct/pm*-choices for a given project specification must be compared to determine which *ct/pm*-constellation (E°, D°, $E^\bullet$, $D^\bullet$) the lender prefers.[64] As mentioned before, inefficient private placements are not necessarily dominated by efficient public offerings in the current setting, due to potentially lacking incentive compatibility. However, in the LS^a scenario, public offerings dominate private placements, as private placements simply cause higher total agency costs without providing any advantage. The only effect of switching from a public offering to a private placement is that the lender's extractable surplus is diminished. Furthermore, due to private placement's higher total agency costs, their feasible solution sets are subsets of public offerings. Therefore, private placements can be neglected in the LS^a scenario.

To examine which *publicly* offered contract type (E°, D°) the lender prefers, the expected profits PLE_L° and PLD_L° must be compared for all projects where at least both of them are feasible.[65] The lender prefers public equity to public debt if

- for riskless public debt $\zeta_{max} > \dfrac{mc}{\mu}$ ($PLE_L^\circ \geq PLD_{L,I}^\circ$) and

---

[64] Since the lender has the power to determine the contract conditions, the comparison of the different *ct/pm*-choices is based on the results obtained in the lender optimizations of the previous Section 3.3.

[65] The respective expected profits are: $PLE_L^\circ = \mu - 1 - mc$, $PLD_{L,I}^\circ = \mu(1 - \zeta_{max}) - 1$, $PLD_{L,II}^\circ = \mu + \dfrac{vc^2}{4\zeta_{max}\mu} - 1 - vc$ (see Tables 3.2, 3.4 and 3.5)

- for risky public debt $\zeta_{max} > \dfrac{vc^2}{4\mu(vc - mc)}$ $(PLE^\circ_L \geq PLD^\circ_{L,II})$.[66]

As can be seen, the favorability of equity financing increases with ζ_{max} for projects where both types of public offerings are feasible. Intuitively, an increase in ζ_{max} for a given μ results in a higher default probability of debt financing, increasing debt financing's agency costs, while the agency costs of equity financing are risk insensitive. The favorability of equity financing also increases with μ for a given ζ_{max}. An increase in μ for a given ζ_{max} increases, of course, the extractable surplus but also widens the maximum project return spread $(2\zeta_{max}\mu)$. The spread increase causes (after an appropriate debt repayment adjustment) again the default probability and thereby debt's agency costs to increase while equity's agency costs stay constant.

Figure 3.7 illustrates for the parameter specification $vc = 0.4$ and $mc = 0.1$ for which project specifications a contract will be agreed upon at all and whether equity or debt is preferred by the lender. vc is chosen to be constant to keep Figure 3.6, illustration of all feasible contracting solutions, comparable. For the given parameter constellation, risky public debt is always dominated by public equity since the CDr/nr($D^\circ_{L,II}$) $(\zeta_{max} \leq \dfrac{vc}{2\mu})$ is stricter than the respective preference condition $(\zeta_{max} \geq \dfrac{vc^2}{4\mu(vc - mc)})$. Therefore, equity financing can only be dominated by riskless debt.[67]

---

[66] Remember that $vc > mc$ (Assumption 3).

[67] Since the indifference curves related to the lender's profit comparison never intersect, they can only change their order dependent on whether (i) $vc \geq 2mc$ or (ii) $vc < 2mc$. Therefore, the structure of the figure is independent of the exact parameter constellation. Hence, we illustrate only one case, while both cases are of course considered in Table 3.17, summarizing our findings for the LS^a scenario.

Fig. 3.7: Contract Type / Placement Mode-Choices in the Lender Scenario with Absolute Power (BM, $vc = 0.4, mc = 0.1, nc = 0.1$)

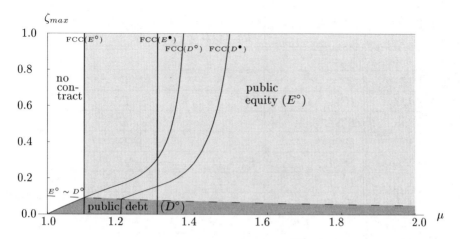

Note: The shaded areas indicate whether a contract will be agreed upon for a certain project specification and, if so, which *ct/pm*-choice the lender prefers. No filling stands for no contract agreement, the light grey filling for public equity and the grey filling for public debt. The feasible contracting constraints (FCC) are printed continuously while the indifference curve ($E^\circ \sim D^\circ$) is dotted.

As expected, the lender chooses a public equity contract when only a public equity contract is feasible and public debt when only public debt is feasible. In the case in which both types are feasible, a comparison of the lender's expected profits (PL) reveals that he favors public equity to public debt above the indifference curve $E^\circ \sim D^\circ$, and vice versa. This is due to the increasing agency costs of debt financing with the agreed repayment h, while the agency costs of equity are independent of the proportional return participation q.

Table 3.17 summarizes the optimal *ct/pm*-choices for all possible projects (μ, ζ_{max}-specifications) in the LS^a scenario. The notation of the areas is based on Table 3.16, while areas subdivided according to the lender's preferences are indicated by subscripts.

Table 3.17: Contract Type / Placement Mode-Choices in the Lender Scenario with Absolute Power (BM)

Area	FCC satisfied	Lender's preference $(E^\circ$ vs. $D^\circ)$	Ct/pm-choices
I	–		–
II	E°		E°
III	D°		D°
IV	$E^\circ, E^\bullet$		E°
V	$D^\circ, D^\bullet$		D°
VI_E	E°, D°	$E^\circ \succ D^\circ$	E°
VI_D	E°, D°	$E^\circ \prec D^\circ$	D°
VII_E	$E^\circ, D^\circ, E^\bullet$	$E^\circ \succ D^\circ$	E°
VII_D	$E^\circ, D^\circ, E^\bullet$	$E^\circ \prec D^\circ$	D°
VIII_E	$E^\circ, D^\circ, D^\bullet$	$E^\circ \succ D^\circ$	E°
VIII_D	$E^\circ, D^\circ, D^\bullet$	$E^\circ \prec D^\circ$	D°
IX_E	$E^\circ, D^\circ, E^\bullet, D^\bullet$	$E^\circ \succ D^\circ$	E°
IX_D	$E^\circ, D^\circ, E^\bullet, D^\bullet$	$E^\circ \prec D^\circ$	D°

3.4.2 Lender Scenario with Restricted Power (LS^r)

In the *lender scenario with restricted power*, the lender can still choose ct/pm, but the borrower determines the contract conditions. Finally, the lender has the choice to either accept the conditions or to reject the financing of project.

To determine the lender's preferred ct/pm-choice, the feasible choices have to be evaluated from the lender's perspective, while the borrower determines the conditions.[68] The derivation simplifies since the range of feasible contracting solutions for $E^\circ, D^\circ, E^\bullet$ and $D^\bullet$ is independent of the bargaining power distribution [Result 1]. Therefore, the lender's potential feasible ct/pm-choices are identical to the choices in the LS^a scenario, as is made obvious in Figure 3.6. Moreover, the lender's expected profits are identical for each feasible ct/pm: $PLE_B^\circ = PLD_B^\circ = PLE_B^\bullet = PLD_B^\bullet = 0$ (see Tables 3.3, 3.6, 3.7, 3.9, 3.14 and 3.15).

Due to the lender's interim and ex-post informational disadvantage, the borrower has the opportunity to squeeze the lender's expected profit to his

---

[68] Since the borrower can choose contracts' conditions, the comparison of the different ct/pm-choices (from the lender's perspective) is based on the results obtained in the borrower optimization of the previous Section 3.3.

profit reservation ($PL = 0$) for all four *ct/pm*-constellations. Hence, the lender becomes indifferent between *all* feasible *ct/pm*-choices. Inefficient private placements cannot be ruled out as constantly dominated by public offerings, despite the absence of ex-ante informational asymmetry, as private placements might be as attractive as public offerings for the party choosing the *ct/pm* due to the separation of bargaining power.

The fact that the lender is indifferent between all feasible *ct/pm*-choices is also illustrated in Figure 3.8.

Fig. 3.8: Contract Type / Placement Mode-Choices in the Lender Scenario with Restricted Power (BM, $vc = 0.4, mc = 0.1, nc = 0.1$)

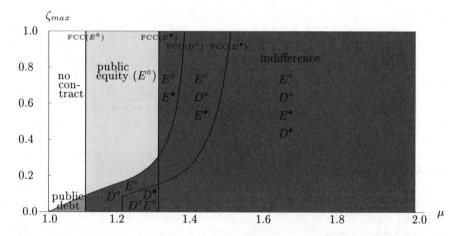

Note: The shaded areas indicate whether a contract will be agreed upon for a certain project specification and, if so, which *ct/pm* the lender chooses. No filling stands for no contract agreement, the light grey filling for public equity, the grey filling for public debt, while the dark grey filling indicates indifference between all feasible choices. The feasible contracting constraints (FCC) are printed continuously.

The lender chooses a public equity (debt) contract when only a public equity (debt) is feasible. When more than one *ct/pm*-constellation is feasible, the lender is indifferent between them, since he will be squeezed to his participation constraint anyway.

Due to the assumed contract negotiation game, see Figure 3.2, the borrower cannot influence the lender's *ct/pm*-choice, e. g. by offering not to squeeze the lender to his profit reservation when choosing an efficient public offering, as such a promise lacks subgame completeness. Therefore, the borrower might end up with a Pareto-inefficient lender's (*ct/pm* optimizer's) *ct/pm*-choice.

Table 3.18 summarizes our findings.[69]

Table 3.18: Contract Type / Placement Mode-Choices in the Lender Scenario with Restricted Power (BM)

Area	FCC satisfied	Ct/pm-choice
I	–	–
II	E°	E°
III	D°	D°
IV	$E^\circ, E^\bullet$	$E^\circ, E^\bullet$
V	$D^\circ, D^\bullet$	$D^\circ, D^\bullet$
VI	E°, D°	E°, D°
VII	$E^\circ, D^\circ, E^\bullet$	$E^\circ, D^\circ, E^\bullet$
VIII	$E^\circ, D^\circ, D^\bullet$	$E^\circ, D^\circ, D^\bullet$
IX	$E^\circ, D^\circ, E^\bullet, D^\bullet$	$E^\circ, D^\circ, E^\bullet, D^\bullet$

3.4.3 Borrower Scenario with Restricted Power (BS^r)

In the *borrower scenario with restricted power*, the borrower chooses ct/pm, while the lender determines contracts' conditions depending on the borrower's ct/pm-choice. Finally, the borrower can decide whether to accept the offered conditions and undertake the project or not.

To find out which ct/pm-choice the borrower prefers, his expected profits (PB) have to be evaluated when the lender optimizes the contract conditions.[70] The determination of the borrower's optimal ct/pm-choices simplifies again due to Result 1. The borrower's expected profits for each ct/pm-choice are: $PBE_L^\circ = 0$, $PBD_{L,I}^\circ = \zeta_{max}\mu$, $PBD_{L,II}^\circ = \dfrac{vc^2}{4\zeta_{max}\mu}$, $PBE_L^\bullet = 0$,

$PBD_{L,Ia}^\bullet = \zeta_{max} - nc$, $PBD_{L,Ib}^\bullet = 0$, $PBD_{L,IIa}^\bullet = \dfrac{vc^2}{4\zeta_{max}\mu} - nc$ and $PBD_{L,IIb}^\bullet = 0$ (see Tables 3.2, 3.4, 3.5, 3.8, 3.10, 3.11, 3.12 and 3.13).

Comparing the borrower's expected profits it becomes clear that the lender squeezes the borrower's profits for equity financing to his profit reservation, as all informational asymmetries are resolved. However, for debt financing

---

[69] The notation of the areas is based on Table 3.16.

[70] The following comparison is therefore based on the results of the lender optimization of the previous Section 3.3.

it is unprofitable for the lender to squeeze the borrower's profit to his profit reservation.[71] The borrower is left with a positive informational rent due to his interim and ex-post informational advantage. Therefore, the borrower prefers debt to equity financing. The borrower also favors public debt offerings to private placements since public offerings do not cause extra negotiation costs reducing the borrower's expected profit. Moreover, since the feasible private debt solutions are a subset of public debt solutions, private debt financing is dominated by public debt in the BS^r scenario. However, for equity financing the borrower is indifferent between public and private placement, as he is squeezed to his profit reservation anyway.[72]

The borrower's preferred ct/pm-choices are illustrated in Figure 3.9. It can be seen which projects are financed at all and, if so, which ct the borrower prefers.

Fig. 3.9: Contract Type / Placement Mode-Choices in the Borrower Scenario with Restricted Power (BM, $vc = 0.4, mc = 0.1, nc = 0.1$)

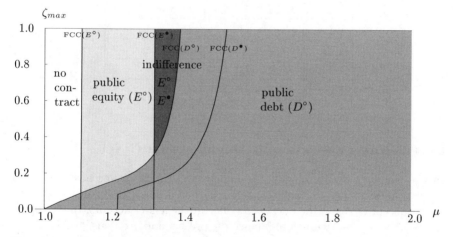

Note: The shaded areas indicate whether a contract will be agreed upon for a certain project specification and, if so, which ct/pm-choice the borrower prefers. No filling stands for no contract agreement, the light grey filling for public equity, the grey filling for public debt, while the dark grey filling indicates indifference between public and private equity. The feasible contracting constraints (FCC) are printed continuously.

---

[71] The costs to do so exceed the potential gains. Therefore, the lender refrains from extracting the whole surplus.

[72] Like in the LS^r scenario (inefficient), private placements are not generally dominated by (efficient) public offerings.

The borrower chooses a public equity contract when only public equity is feasible. When private equity is feasible in addition, he is indifferent between both. However, he prefers public debt whenever possible to secure his positive informational rent, as otherwise he is squeezed to his profit reservation.

Table 3.19 summarizes our findings.[73]

Table 3.19: Contract Type / Placement Mode-Choices in the Borrower Scenario with Restricted Power (BM)

Area	FCC satisfied	Ct/pm-choices
I	–	–
II	E°	E°
III	D°	D°
IV	$E^\circ, E^\bullet$	$E^\circ, E^\bullet$
V	$D^\circ, D^\bullet$	D°
VI	E°, D°	D°
VII	$E^\circ, D^\circ, E^\bullet$	D°
VIII	$E^\circ, D^\circ, D^\bullet$	D°
IX	$E^\circ, D^\circ, D^\bullet, E^\bullet$	D°

3.4.4 Borrower Scenario with Absolute Power (BS^a)

In the *borrower scenario with absolute power*, the borrower has the power to determine ct/pm as well as contracts' conditions, while the lender can finally accept or reject the borrower's offer. If the lender accepts, the borrower receives the necessary funds to finance the project and proceeds, otherwise the game ends.

To solve the final stage of the contract negotiation game for the BS^a scenario, the borrower's expected profits, obtained in the borrower optimization of the previous Section 3.3, have to be compared. Due to his informational advantage, the borrower is able to reduce the lender's expected profit to the latter's profit reservation for all feasible ct/pm-constellations. The borrower can extract the whole remaining surplus less the arising agency costs. Therefore, the borrower prefers the agency costs minimizing ct/pm-choice. Hence, private placements become irrelevant as in the LS^a scenario, see Section 3.4.1.

---

[73] The notation of the areas is based on Table 3.16.

To evaluate the borrower's preference among public equity and public debt, his expected profits must be compared, revealing that[74]

- public equity financing (E_B°) is always dominated by riskless public debt financing ($D_{B,I}^\circ$), and
- risky public debt financing ($D_{B,II}^\circ$) dominates public equity financing (E_B°), i. e. $PBD_{B,II}^\circ > PBE_B^\circ$, if $\zeta_{max} \leq \dfrac{vc^2}{2\mu(vc - mc)}$ for $\mu \leq 1 + \dfrac{vc + mc}{2}$

  and $\zeta_{max} < \dfrac{\mu - 1 - mc}{\mu\left(\frac{vc-mc}{vc}\right)^2} = \dfrac{1}{\left(\frac{vc-mc}{vc}\right)^2} - \dfrac{1 + mc}{\mu\left(\frac{vc-mc}{vc}\right)^2}$ for $\mu > 1 + \dfrac{vc + mc}{2}$,

  and vice versa.

Since no agency costs at all arise in riskless public debt financing, public equity is always dominated when riskless public debt is feasible. Risky debt only dominates public equity for a restricted set of projects. Similar to the LS^a scenario, an increase in ζ_{max} given μ increases the favorability of equity financing. However, contrary to the LS^a scenario, an increase in μ given ζ_{max} increases the favorability of (risky) debt financing due to a default probability reduction.[75]

The following figure illustrates for which projects which *ct* is preferred by the borrower.

---

[74] The borrower's expected profits are: $PBE_B^\circ = \mu - 1 - mc$, $PBD_{B,I}^\circ = \mu - 1$, $PBD_{B,II}^\circ = \dfrac{[vc + \sqrt{4\zeta_{max}\mu(\mu - 1 - vc) + vc^2}]^2}{4\zeta_{max}\mu}$ (see Tables 3.3, 3.6 and 3.7).

[75] The agency costs of risky debt financing ($h > y_{min}\big|_{\zeta=\zeta_{max}}$) are $vc \cdot$

$DP(h, \mu, \zeta_{max}) = vc \cdot \dfrac{h - y_{min}\big|_{\zeta=\zeta_{max}}}{2\zeta_{max}\mu} = vc \cdot \dfrac{h - \mu(1 - \zeta_{max})}{2\zeta_{max}\mu}$ with

$\dfrac{\partial DP(h, \mu, \zeta_{max})}{\partial \mu} \leq 0$.

Fig. 3.10: Contract Type / Placement Mode-Choices in the Borrower Scenario with Absolute Power (BM, $vc = 0.4, mc = 0.1, nc = 0.1$)

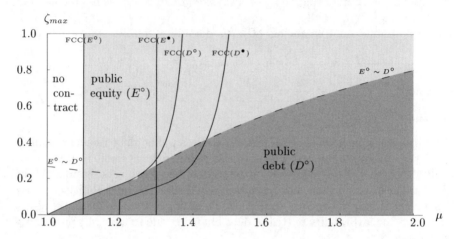

Note: The shaded areas indicate whether a contract will be agreed upon for a certain project specification and, if so, which ct/pm-choice the borrower prefers. No filling stands for no contract agreement, the light grey filling for public equity and the grey filling for public debt. Feasible contracting constraints (FCC) are printed continuously while the indifference curve with a sharp turning point at $\mu = 1 + \dfrac{vc + mc}{2}$ ($E^\circ \sim D^\circ$) is dotted.

The borrower chooses a public equity contract when only a public equity contract is feasible and public debt when only public debt is feasible. When both types are feasible, the borrower's expected profit (PB) comparison reveals that he favors public equity to public debt above the indifference curve $E^\circ \sim D^\circ$, while preferring public debt below. This is due to the varying agency costs of debt financing with the agreed repayment h, while the agency costs of equity are independent of the proportional return participation q.

Table 3.20 summarizes our findings.[76] The obtained results look similar to the results of the LS^a scenario, see Table 3.17. However, they differ substantially since the borrower's instead of the lender's preference is considered, as can be seen in the following section.

---

[76] The notation of the areas is based on Table 3.16, while areas subdivided according to the borrower's preferences are indicated by subscripts.

Table 3.20: Contract Type / Placement Mode-Choices in the Borrower Scenario with Absolute Power (BM)

Area	FCC satisfied	Borrower's preference (E° vs. D°)	Ct/pm-choices
I	–		–
II	E°		E°
III	D°		D°
IV	$E^\circ, E^\bullet$		E°
V	$D^\circ, D^\bullet$		D°
VI_E	E°, D°	$E^\circ \succ D^\circ$	E°
VI_D	E°, D°	$E^\circ \prec D^\circ$	D°
VII_E	$E^\circ, D^\circ, E^\bullet$	$E^\circ \succ D^\circ$	E°
VII_D	$E^\circ, D^\circ, E^\bullet$	$E^\circ \prec D^\circ$	D°
VIII_E	$E^\circ, D^\circ, D^\bullet$	$E^\circ \succ D^\circ$	E°
VIII_D	$E^\circ, D^\circ, D^\bullet$	$E^\circ \prec D^\circ$	D°
IX_E	$E^\circ, D^\circ, E^\bullet, D^\bullet$	$E^\circ \succ D^\circ$	E°
IX_D	$E^\circ, D^\circ, E^\bullet, D^\bullet$	$E^\circ \prec D^\circ$	D°

3.5 Bargaining Power Effects – Scenario Comparison

Finally, in this section bargaining power effects are determined by a pairwise (binary) comparison of the optimal contract choices for certain projects (fixed μ, ζ_{max}-specifications) in the analyzed bargaining power scenarios. Since bargaining power cannot vanish in our context, only the distribution among the lender and the borrower can change. The power effects are determined by a comparison of the optimal contract choices while the respective power component is redistributed.

We thereby focus on changes in the preferred ct/pm-choice since despite the (binary) comparison of both restricted power scenarios (LS^r, BS^r), the lender's and the borrower's expected profit bargaining power dependency is doubtless. The power redistributions $LS^a \to \{LS^r, BS^r, BS^a\}$ and $\{LS^r, BS^r\} \to BS^a$ are beneficial for the borrower ($\Delta PB \geq 0$)[77] but disadvantageous for the lender ($\Delta PL \leq 0$), and vice versa. Absolute bargaining power as well as each power component is desirable for the lender as well as for

---

[77] ΔPB refers to a change in the borrower's expected profit due to a redistribution of bargaining power.

the borrower as they can not lose due to an appropriate power redistribution. In the extreme case they can simply copy the predecessor's (opponent's) use of power, e. g. his ct/pm-choice and/or conditions set, and are therefore indifferent about such redistributions. However, such power redistributions usually even offer the power "recipient" the opportunity to increase his profit (share) from project financing. The "donor's" profit cannot increase due to the power redistribution if his previous use of power was optimal. In the following we mainly examine power redistributions where simply one party obtains power from the opponent and nothing more. The exception is the (binary) comparison of both restricted power scenarios where each party loses one power component but gains the other one.

Changes in the optimal ct/pm-choice due to an absolute bargaining power redistribution ($LS^a \rightleftharpoons BS^a$) are analyzed at first (see Section 3.5.1) as only agency cost aspects are relevant in the absolute power scenarios. The separation of the bargaining power components causes strategic aspects to become additionally pertinent. Section 3.5.2 deals with effects of a shift of the power to determine ct/pm ($LS^a \rightleftharpoons BS^r$, $BS^a \rightleftharpoons LS^r$), and Section 3.5.3 focuses on power effects due to a shift of the power to choose contract conditions ($LS^a \rightleftharpoons LS^r$, $BS^a \rightleftharpoons BS^r$).[78] We distinguish between the directions of potential power variations, i. e. from the lender to the borrower or vice versa, as the resulting effects can differ. Moreover, Section 3.5.4 compares the favorability of the different bargaining power components from the lender's as well as from the borrower's perspective. Consistency checks are carried out in Section 3.5.5.

The derivation of the power effects for the preferred ct/pm-choice can of course be based on a visual comparison of the optimal ct/pm-choices in the alternative scenarios, as illustrated in Figures 3.7 to 3.10, and Figure 3.11 for an alternative parameter constellation.[79]

---

[78] Changes in the optimal ct/pm-choice due to the power variation $LS^r \rightleftharpoons BS^r$ are deliberately not examined since this power variation, i. e. the power to determine ct/pm is shifted from one party to the other, while the power to set the contract conditions shifts from the latter to the former results in a mixture of several effects which are difficult to disentangle.

[79] Note: Since the figures do not illustrate the profit distribution between the lender and the borrower for the preferred ct/pm-choices, no power effects for the profit distribution can be obtained by a visual comparison. However, this is beyond the scope of this study anyway.

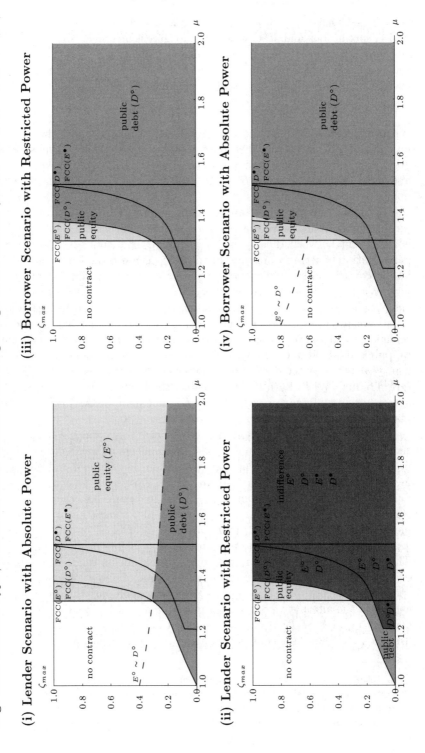

Fig. 3.11: Contract Type / Placement Mode-Choices in All Bargaining Power Scenarios (BM, $vc = 0.4, mc = 0.3, nc = 0.1$)

(i) **Lender Scenario with Absolute Power**

(ii) **Lender Scenario with Restricted Power**

(iii) **Borrower Scenario with Restricted Power**

(iv) **Borrower Scenario with Absolute Power**

Note: The shaded areas indicate whether a contract will be agreed upon for a certain project specification and, if so, which ct/pm-choice is preferred. No filling stands for no contract agreement, the light grey filling for public equity, grey for public debt, while dark grey indicates indifference among the feasible ct/pm-choices. The feasible contracting constraints (FCC) are printed continuously while the indifference curves ($E^{\circ} \sim D^{\circ}$) are dotted.

As expected for the LS^a scenario we observe that the lender chooses a public equity (debt) contract when only a public equity (debt) is feasible [light grey (grey) filling]. When both cts are feasible, the lender favors equity to debt above the indifference curve $E^{\circ} \sim D^{\circ}$, and vice versa. Private placements are dominated by public offerings. In the LS^r scenario, the lender chooses a public equity (debt) contract when only a public equity (debt) is feasible [light grey (grey) filling]. When more than one ct/pm-constellation is feasible, the lender is indifferent between them [dark grey filling]. The lender is squeezed to his participation constraint anyway. In the BS^r scenario the borrower chooses a public equity contract when only a public equity is feasible [light grey filling]. When public *and* private equity are feasible, the borrower is indifferent between them [dark grey filling] since he is squeezed to his participation constraint anyway. However, whenever public debt is feasible, the borrower prefers debt to assure a positive informational rent. Private debt is dominated by public debt. Finally, in the BS^a scenario, the borrower chooses a public equity (debt) contract when only a public equity (debt) is feasible [light grey (grey) filling]. When both cts are feasible, the borrower favors equity to debt above both indifference curves $E^{\circ} \sim D^{\circ}$ and debt below just one. Private placements are dominated by public offerings.

However, the determination of power effects for a specific project by visual inspection is quite difficult, as the power effects as well as the underlying preferred ct/pm-choices depend significantly on the assumed underlying parameter constellation of vc, mc and nc. As we have seen, for example, by the illustration of the feasible contracting areas in Figure 3.6, the areas already vary significantly depending on nc. For example, a variation in vc, mc or nc might affect the feasible contracting areas of $E^{\circ}, D^{\circ}, E^{\bullet}$ or $D^{\bullet}$ in a way that a particular ct/pm-constellation is only feasible before or after the variation. The same problem occurs for bargaining power effects related to this ct/pm occurring. Therefore, we refrain from a derivation of power effects by visual inspection of the optimal ct/pm-choice. Instead we compare the optimal ct/pm-choices in the alternative bargaining power scenarios for *each possible* feasible contracting area (I-IX) and any required subdivision, as stated in Tables 3.17 to 3.20.[80] Feasible contracting areas with divergent optimal choices are

---

[80] In total seventeen different feasible contracting areas and respective subdivisions are possible. While the feasible contracting areas differ depending on which ct/pm-constellations are feasible, the subdivisions are due to the lender's and the borrower's preference among feasible ct/pms.

subdivided according to the lender's and the borrower's preference. Table 3.21 summarizes these findings.[81] No subdivision can occur where the lender favors (public) debt while the borrower prefers (public) equity, as will be discussed in Section 3.5.1 (see Propositions 1 and 2). All potential feasible contracting areas and subdivisions are stated while the project specifications they include depend on the underlying parameter constellations of vc, mc and nc. Due to variations in the underlying parameter constellations, feasible contracting areas can vary or even vanish, but no new area can be created, as discussed in Section 3.3.5. However, for a given constellation of mc, vc and nc the feasible contracting areas and required subdivisions stay constant.

---

[81] The notation of the feasible contracting areas is based on Table 3.16, while the areas are subdivided according to the lender's as well as the borrower's preferences, as indicated by two subscripts. The first subscript refers to the lender's and the second to the borrower's preference.

Table 3.21: Summary of the Contract Type / Placement Mode-Choices in the Bargaining Power Scenarios (BM)

Area	FCCs satis-fied[82]	E° vs. D° Lender's preference	Borrower's preference	Ct/pm-choices LS^a	LS^r	BS^r	BS^a
I	–			–	–	–	–
II	E°			E°	E°	E°	E°
III	D°			D°	D°	D°	D°
IV	$E^\circ, E^\bullet$			E°	$E^\circ, E^\bullet$	$E^\circ, E^\bullet$	E°
V	$D^\circ, D^\bullet$			D°	$D^\circ, D^\bullet$	D°	D°
$VI_{E,E}$	E°, D°	$E^\circ \succ D^\circ$	$E^\circ \succ D^\circ$	E°	E°, D°	D°	E°
$VI_{E,D}$	E°, D°	$E^\circ \succ D^\circ$	$E^\circ \prec D^\circ$	E°	E°, D°	D°	D°
$VI_{D,D}$	E°, D°	$E^\circ \prec D^\circ$	$E^\circ \prec D^\circ$	D°	E°, D°	D°	D°
$VII_{E,E}$	E°, D° $E^\bullet$	$E^\circ \succ D^\circ$	$E^\circ \succ D^\circ$	E°	E°, D° $E^\bullet$	D°	E°
$VII_{E,D}$	E°, D° $E^\bullet$	$E^\circ \succ D^\circ$	$E^\circ \prec D^\circ$	E°	E°, D° $E^\bullet$	D°	D°
$VII_{D,D}$	E°, D° $E^\bullet$	$E^\circ \prec D^\circ$	$E^\circ \prec D^\circ$	D°	E°, D° $E^\bullet$	D°	D°
$VIII_{E,E}$	E°, D° $D^\bullet$	$E^\circ \succ D^\circ$	$E^\circ \succ D^\circ$	E°	E°, D° $D^\bullet$	D°	E°
$VIII_{E,D}$	E°, D° $D^\bullet$	$E^\circ \succ D^\circ$	$E^\circ \prec D^\circ$	E°	E°, D° $D^\bullet$	D°	D°
$VIII_{D,D}$	E°, D° $D^\bullet$	$E^\circ \prec D^\circ$	$E^\circ \prec D^\circ$	D°	E°, D° $D^\bullet$	D°	D°
$IX_{E,E}$	E°, D° $E^\bullet, D^\bullet$	$E^\circ \succ D^\circ$	$E^\circ \succ D^\circ$	E°	E°, D° $E^\bullet, D^\bullet$	D°	E°
$IX_{E,D}$	E°, D° $E^\bullet, D^\bullet$	$E^\circ \succ D^\circ$	$E^\circ \prec D^\circ$	E°	E°, D° $E^\bullet, D^\bullet$	D°	D°
$IX_{D,D}$	E°, D° $E^\bullet, D^\bullet$	$E^\circ \prec D^\circ$	$E^\circ \prec D^\circ$	D°	E°, D° $E^\bullet, D^\bullet$	D°	D°

---

[82] FCCs not stated explicitly are not satisfied.

3.5.1 Absolute Power Effects ($LS^a \rightleftharpoons BS^a$)

First of all, we examine the effects caused by the redistribution of the absolute bargaining power from the lender to the borrower ($LS^a \rightarrow BS^a$), before analyzing the effects of a power redistribution from the borrower to the lender ($BS^a \rightarrow LS^a$).

As stated already in the previous sections, feasible contracting areas are not only independent of the bargaining power distribution [Result 1] but (inefficient) private placements are also dominated by (efficient) public offerings in the scenarios with absolute bargaining power (LS^a, BS^a). Therefore, projects are financed independently of the bargaining power distribution by public offerings. The only effect due to a redistribution of the absolute bargaining power among lender and borrower ($LS^a \rightleftharpoons BS^a$) which might occur is a change of the preferred ct.

For the power variation $LS^a \rightarrow BS^a$, it becomes obvious by examination of the debt-equity-comparisons in the absolute power scenarios that the preferred ct in fact depends on the bargaining power distribution. While in the LS^a scenario, see Section 3.4.1, public equity (E°) is preferable to riskless public debt (D_I°) for $\zeta_{max} \geq mc/\mu$ or to risky public debt (D_{II}°) for $\zeta_{max} \geq \dfrac{vc^2}{4\mu(vc - mc)}$. Both ζ_{max}-requirements decrease with μ in a μ, ζ_{max}-diagram, i. e. the range of ζ_{max} ($\zeta_{max} \in [0,1]$) satisfying both requirements widens with μ. The requirements for the favorability of public equity (E°) in the BS^a scenario, see Section 3.4.4, are much stricter. Riskless public debt financing (D_I°) always dominates public equity financing (E°), while risky public debt (D_{II}°) is preferable to equity for $\zeta_{max} \leq \dfrac{vc^2}{2\mu(vc - mc)}$ if

$$\mu \leq 1 + \frac{vc + mc}{2} \text{ and } \zeta_{max} < \frac{\mu - 1 - mc}{\mu\left(\frac{vc - mc}{vc}\right)^2} = \frac{1}{\left(\frac{vc - mc}{vc}\right)^2} - \frac{1 + mc}{\mu\left(\frac{vc - mc}{vc}\right)^2} \text{ for}$$

$\mu > 1 + \dfrac{vc + mc}{2}$ in a μ, ζ_{max}-diagram, i. e. the range of ζ_{max} satisfying the requirement decreases. The risky debt's favorability requirement even increases in μ for $\mu \geq 1 + \dfrac{mc + vc}{2}$. Therefore, debt's favorability increases when absolute power is redistributed from the lender to the borrower.[83]

Intuitively, a redistribution of the absolute bargaining power from the lender to the borrower makes public debt contracts more favorable, as the bargaining power distribution influences the advantages, respective disadvantages, of public equity and public debt in different ways. The agency costs arising from equity financing (mc) are independent of q and, therefore, the bargaining power distribution does not matter for these costs. Contrary, the

---

[83] We should note that this is the reason why we did consider a subdivision of feasible contracting areas where the lender favors (public) debt, while the borrower prefers (public) equity.

agency costs arising from debt financing ($vc \cdot DP(h)$) depend on h and, therefore, on the bargaining power distribution.[84] For example, an increase in the borrower's bargaining power, $LS^a \to BS^a$, potentially results in a reduction of debt's repayment obligation ($h_L^* \geq h_B^*$) by which the borrower's default probability ($DP(h)$) as well as the resulting agency costs decrease. Hence, debt contracts are more often preferable to equity contracts in the BS^a scenario than in the LS^a scenario. By this reasoning we have proven Proposition 1.

Proposition 1
Basic Model: Power Effect $LS^a \to BS^a$
A redistribution of the absolute bargaining power from the lender to the borrower increases the favorability of public debt contracts in comparison to public equity when both are feasible. Some projects which were financed by public equity in the LS^a scenario, are financed by public debt in the BS^a scenario, see Areas $VI_{E,D}$, $VII_{E,D}$, $VIII_{E,D}$, $IX_{E,D}$ in Table 3.21. ($VI_{E,D}, VII_{E,D}, VIII_{E,D}, IX_{E,D} : E^\circ \Rightarrow D^\circ$)[85]
□

For the second absolute power variation $BS^a \to LS^a$, the same reasoning applies analogously. Intuitively, as the agency costs of equity financing are independent of the bargaining power distribution while the agency costs of debt financing increase with the lender's bargaining power due to $h_L^* \geq h_B^*$, the favorability of public equity contracts in comparison to debt contracts increases due to the power shift. Therefore, Proposition 2 emerges.

---

[84] Remember:

$$DP(h) = \begin{cases} 0 & \text{for } h \leq y_{min}\big|_{\zeta=\zeta_{max}} \\ \dfrac{h - \mu(1 - \zeta_{max})}{2\mu\zeta_{max}} & \text{for } y_{min}\big|_{\zeta=\zeta_{max}} < h \leq y_{max}\big|_{\zeta=\zeta_{max}} \end{cases}$$

with

$$\frac{\partial DP(h)}{\partial h} = \begin{cases} 0 & \text{for } h \leq y_{min}\big|_{\zeta=\zeta_{max}} \\ \dfrac{1}{2\mu\zeta_{max}} & \text{for } y_{min}\big|_{\zeta=\zeta_{max}} < h \leq y_{max}\big|_{\zeta=\zeta_{max}} \end{cases}.$$

[85] We should note that if a change in the preferred ct/pm-choice for project specifications in the respective areas *can occur* this is indicated by $\to$, while $\Rightarrow$ refers to changes which *have to occur*.

Proposition 2
Basic Model: Power Effect $BS^a \to LS^a$
A redistribution of the absolute bargaining power from the borrower
to the lender decreases the favorability of public debt contracts in
comparison to public equity when both are feasible. Some projects
which were financed by public debt in the BS^a scenario are financed by
public equity in the LS^a scenario, see Areas $VI_{E,D}, VII_{E,D}, VIII_{E,D},$
$IX_{E,D}$ in Table 3.21. $(VI_{E,D}, VII_{E,D}, VIII_{E,D}, IX_{E,D} : D^\circ \Rightarrow E^\circ)$
$\square$

3.5.2 Effects of the Bargaining Power to Determine Contract Type / Placement Mode ($LS^a \rightleftharpoons BS^r$, $BS^a \rightleftharpoons LS^r$)

The examination of power effects of the bargaining power to determine ct/pm
requires a distinction between two cases: The bargaining power to set contract
conditions can either belong to the lender or to the borrower. Firstly, we
analyze the case where the bargaining power to determine contract conditions
is given to the lender ($LS^a \rightleftharpoons BS^r$), before considering the case where this
power is given to the borrower ($BS^a \rightleftharpoons LS^r$).

Obviously the feasible contracting areas are still independent of the bargaining power distribution [Result 1]. However, contrary to the examination
of the absolute power effects, (inefficient) private placements can no longer be
ignored, as (efficient) public placements might lack incentive-compatibility in
the restricted power scenarios. The potentially lacking incentive-compatibility
of the efficient ct/pm-choices is due to the separation of bargaining power.
Thereby, in addition to agency cost considerations like in the absolute power
scenarios, "strategic" aspects gain importance for the party choosing ct/pm.
Hence, a redistribution of the bargaining power to determine ct/pm can affect
the preferred ct as well as pm-choice.

Firstly, a shift of the power to determine ct/pm from the lender to the
borrower is considered, while the lender keeps the power to determine contracts' conditions ($LS^a \to BS^r$). As observed in the LS^a scenario (see Section
3.4.1) the lender prefers efficient public equity or debt offerings due to his absolute power in order to maximize his expected profit with respect to the
BPC. When both cts are feasible, the lender prefers the ct which minimizes
the sum of agency costs (mc or $vc \cdot DP(h)$) and the borrower's debt financing
informational rent. Contrary, in the BS^r scenario (see Section 3.4.3) due to
the borrower's informational advantage combined with his bargaining power
limitation the borrower prefers public debt whenever feasible to secure his
positive informational rent while being indifferent between public and private
equity, otherwise just satisfying his participation constraint. Therefore, public
debt's and private equity's favorability increase when the power to determine
ct/pm is redistributed from the lender to the borrower and the lender keeps
the power to determine the contract conditions. Proposition 3 is proven.

Proposition 3
Basic Model: Power Effect $LS^a \rightarrow BS^r$
A redistribution of the bargaining power to determine ct/pm from the
lender to the borrower, while the former keeps the power to set the contract conditions increases the favorability of public debt and private
equity. Some projects which were financed by public equity in the LS^a
scenario are financed by either public debt or perhaps private equity
in the BS^r scenario, see Areas $IV, VI_{E,E}, VI_{E,D}, VII_{E,E}, VII_{E,D},$
$VIII_{E,E}, VIII_{E,D}, IX_{E,E},$ and $IX_{E,D}$ in Table 3.21. ($IV : E^\circ \rightarrow$
$E^\bullet; VI_{E,E}, VI_{E,D}, VII_{E,E}, VII_{E,D}, VIII_{E,E}, VIII_{E,D}, IX_{E,E},$
$IX_{E,D} : E^\circ \Rightarrow D^\circ$) $\square$

Examining the opposite power redistribution ($BS^r \rightarrow LS^a$) reveals that,
in contrast to the power shift to the borrower ($LS^a \rightarrow BS^r$), the favorability of
public debt as well as private equity declines. For $BS^r \rightarrow LS^a$ the favorability
of private equity decreases since projects which would be financed with private
equity in the BS^r scenario *have to be* financed by public equity in the LS^a
scenario. Therefore, Proposition 4 emerges.

Proposition 4
Basic Model: Power Effect $BS^r \rightarrow LS^a$
A redistribution of the bargaining power to determine ct/pm from
the borrower to the lender with the latter keeping the power to set
the contract conditions decreases the favorability of public debt as
well as private equity in comparison to public equity. Some projects
which were financed by public debt and all projects financed by
private equity in the BS^r scenario are financed by public equity
in the LS^a scenario, see Areas $IV, VI_{E,E}, VI_{E,D}, VII_{E,E}, VII_{E,D},$
$VIII_{E,E}, VIII_{E,D}, IX_{E,E},$ and $IX_{E,D}$ in Table 3.21. ($IV : E^\bullet \Rightarrow$
$E^\circ; VI_{E,E}, VI_{E,D}, VII_{E,E}, VII_{E,D}, VIII_{E,E}, VIII_{E,D}, IX_{E,E},$
$IX_{E,D} : D^\circ \Rightarrow E^\circ$) $\square$

Now we take a look at the effect of power shifts when the power to determine the contract conditions belongs to the borrower ($BS^a \rightleftharpoons LS^r$). Firstly,
consider a redistribution of the power to determine ct/pm from the borrower
to the lender, while the former keeps the power to determine contract conditions ($BS^a \rightarrow LS^r$). As observed in the BS^a scenario (see Section 3.4.4),
the borrower prefers efficient public equity or debt offerings due to his absolute power in order to maximize his expected profit with respect to the LPC.
When both cts are feasible, he prefers the ct minimizing agency costs. On
the contrary, in the LS^r scenario (see Section 3.4.2), due to his informational
disadvantage and the bargaining power limitation, the lender is indifferent between all feasible ct/pm-constellations, as he is squeezed to his participation
constraint anyway. Therefore, the favorability of private placements increases.
However, the favorability of public offerings does not have to increase following
the power redistribution, as the lender's actual ct/pm-choice is not uniquely

determined in the relevant contracting areas. Proposition 5 is proven by this reasoning.

Proposition 5
Basic Model: Power Effect $BS^a \rightarrow LS^r$
A redistribution of the bargaining power to determine ct/pm from the borrower to the lender, while the former keeps the power to set the contract conditions, increases the favorability of private placements, and public debt or equity might also become more preferable. Some projects which were financed by public equity or debt in the BS^a scenario are financed by either private equity or debt, or perhaps by the alternative public ct in the LS^r scenario, see Areas IV, V, VI, VII, VIII, and IX in Table 3.21. ($IV : E^\circ \rightarrow E^\bullet$; $V :$ $D^\circ \rightarrow D^\bullet$; $VI_{E,E} : E^\circ \rightarrow D^\circ$; $VI_{E,D}, VI_{D,D} : D^\circ \rightarrow E^\circ$; $VII_{E,E} :$ $E^\circ \rightarrow \{D^\circ, E^\bullet\}$; $VII_{E,D}, VII_{D,D} : D^\circ \rightarrow \{E^\circ, E^\bullet\}$; $VIII_{E,E} : E^\circ \rightarrow$ $\{D^\circ, D^\bullet\}$; $VIII_{E,D}, VIII_{D,D} : D^\circ \rightarrow \{E^\circ, D^\bullet\}$; $IX_{E,E} : E^\circ \rightarrow$ $\{D^\circ, E^\bullet, D^\bullet\}$; $IX_{E,D}, IX_{D,D} : D^\circ \rightarrow \{E^\circ, E^\bullet, D^\bullet\}$) $\square$

Focusing on the power redistribution in the opposite direction ($LS^r \rightarrow BS^a$), using the reasoning as for $BS^a \rightarrow LS^r$, it becomes obvious that the potential favorability of (inefficient) private debt and equity vanishes. The borrower with absolute power simply minimizes agency costs to maximize his expected profits. Whether (efficient) public offerings are preferred more often cannot be determined since the lender's ct/pm-choice in the LS^r scenario is not always uniquely defined. Therefore, Proposition 6 emerges.

Proposition 6
Basic Model: Power Effect $LS^r \rightarrow BS^a$
A redistribution of the power to determine ct/pm from the lender to the borrower, while the latter keeps the power to set the contract conditions, can increase the favorability of public debt and public equity. All projects which were financed by private debt or equity in the LS^r scenario are financed by public equity or debt in the BS^a scenario, while some projects financed by public debt or equity in the LS^r scenario are perhaps financed by the alternative public offering in the BS^a scenario, see Areas $IV, V, VI, VII, VIII$, and IX in Table 3.21. ($IV : E^\bullet \Rightarrow E^\circ$; $V : D^\bullet \Rightarrow D^\circ$; $VI_{E,E} : D^\circ \Rightarrow E^\circ$; $VI_{E,D}, VI_{D,D} :$ $E^\circ \Rightarrow D^\circ$; $VII_{E,E} : \{D^\circ, E^\bullet\} \Rightarrow E^\circ$; $VII_{E,D}, VII_{D,D} : \{E^\circ, E^\bullet\} \Rightarrow$ D°; $VIII_{E,E} : \{D^\circ, D^\bullet\} \Rightarrow E^\circ$; $VIII_{E,D}, VIII_{D,D} : \{E^\circ, D^\bullet\} \Rightarrow$ D°; $IX_{E,E} : \{D^\circ, E^\bullet, D^\bullet\} \Rightarrow E^\circ$; $IX_{E,D}, IX_{D,D} : \{E^\circ, E^\bullet, D^\bullet\} \Rightarrow$ D°) $\square$

3.5.3 Effects of the Bargaining Power to Determine Contracts' Conditions ($LS^a \rightleftharpoons LS^r$, $BS^a \rightleftharpoons BS^r$)

Analyzing power effects of the bargaining power to determine contract conditions requires a distinction dependent on the distribution of the bargaining power to make the ct/pm-choice. The power to determine ct/pm can either belong to the lender ($LS^a \rightleftharpoons LS^r$) or to the borrower ($BS^a \rightleftharpoons BS^r$).

The power effects analyzed in this section resemble the effects of the bargaining power to choose ct/pm since they are caused by the same forces. This similarity arises from the separation of bargaining power in the restricted power scenarios through which in addition to the pure agency cost considerations "strategic" aspects become relevant for the party choosing ct/pm. Consequently, effects of the bargaining power to determine ct/pm and effects of the bargaining power to determine contracts' conditions are both caused by agency cost and "strategic" considerations.

Examine firstly a redistribution of the bargaining power to set contract conditions from the lender to the borrower while the lender keeps the power to determine ct/pm ($LS^a \rightarrow LS^r$). As we have shown in the LS^a scenario (see Section 3.4.1), due to his absolute power the lender prefers either efficient public equity or debt offerings to maximize his expected profit. He is only concerned about arising agency cost.[86] In the LS^r scenario (see Section 3.4.2), the lender becomes indifferent between all feasible ct/pm-constellations due to his informational disadvantage and bargaining power limitation, offering him no profitable strategic opportunities. The lender is always squeezed to his participation constraint independent of his ct/pm-choice. Hence, the power redistribution increases the favorability of private placements. No clear prediction emerges for public offerings since the lender's ct/pm-choice in the LS^r scenario is not uniquely determined. Therefore, we have shown Proposition 7.

Proposition 7
Basic Model: Power Effect $LS^a \rightarrow LS^r$
A redistribution of the bargaining power to set the contract conditions from the lender to the borrower, while the lender keeps the power to determine ct/pm, increases the favorability of private placements. Some projects which were financed by public equity or debt in the LS^a scenario are financed by either private equity or debt, or perhaps by the alternative public ct in the LS^r scenario, see Areas IV, V, VI, VII, VIII, and IX in Table 3.21. ($IV : E^\circ \rightarrow E^\bullet$; $V : D^\circ \rightarrow D^\bullet$; $VI_{E,E}, VI_{E,D} : E^\circ \rightarrow D^\circ$; $VI_{D,D} : D^\circ \rightarrow E^\circ$; $VII_{E,E}, VII_{E,D} : E^\circ \rightarrow \{D^\circ, E^\bullet\}$; $VII_{D,D} : D^\circ \rightarrow \{E^\circ, E^\bullet\}$; $VIII_{E,E}, VIII_{E,D} : E^\circ \rightarrow \{D^\circ, D^\bullet\}$; $VIII_{D,D} : D^\circ \rightarrow \{E^\circ, D^\bullet\}$; $IX_{E,E}, IX_{E,D} : E^\circ \rightarrow \{D^\circ, E^\bullet, D^\bullet\}$; $IX_{D,D} : D^\circ \rightarrow \{E^\circ, E^\bullet, D^\bullet\}$) □

---

[86] This, of course, includes the borrower's potential informational rent since the agency costs to extract this rent are higher than the gains.

The opposite power redistribution ($LS^r \rightarrow LS^a$) causes the potential favorability of (inefficient) private debt and equity to vanish, as the lender with absolute power prefers efficient public placements to maximize his profit. Whether one particular (efficient) public ct is preferred more often cannot be judged since the lender's ct/pm-choice in the LS^r scenario is not uniquely defined for all possible project specifications. Proposition 8 emerges.

Proposition 8
Basic Model: Power Effect $LS^r \rightarrow LS^a$
A redistribution of the bargaining power to set the contract conditions from the borrower to the lender, while the latter keeps the power to choose ct/pm, can increase the favorability of public debt and equity. All projects which are financed by private debt or equity in the LS^r scenario are financed by public debt or equity in the LS^a scenario, while some projects financed by public debt or equity in the LS^r scenario are perhaps financed by the alternative public ct in the LS^a scenario, see Areas IV, V, VI, VII, VIII, or IX in Table 3.21. ($IV : E^\bullet \Rightarrow E^\circ$; $V : D^\bullet \Rightarrow D^\circ$; $VI_{E,E}, VI_{E,D} : D^\circ \Rightarrow E^\circ$; $VI_{D,D} : E^\circ \Rightarrow D^\circ$; $VII_{E,E}, VII_{E,D} : \{D^\circ, E^\bullet\} \Rightarrow E^\circ$; $VII_{D,D} : \{E^\circ, E^\bullet\} \Rightarrow D^\circ$; $VIII_{E,E}, VIII_{E,D} : \{D^\circ, D^\bullet\} \Rightarrow E^\circ$; $VIII_{D,D} : \{E^\circ, D^\bullet\} \Rightarrow D^\circ$; $IX_{E,E}, IX_{E,D} : \{D^\circ, E^\bullet, D^\bullet\} \Rightarrow E^\circ$; $IX_{D,D} : \{E^\circ, E^\bullet, D^\bullet\} \Rightarrow D^\circ$) $\square$

The same power redistribution can be analyzed when the bargaining power to determine the ct/pm-choice is assigned to the borrower ($BS^a \rightleftharpoons BS^r$). At first, the redistribution of bargaining power to set contracts' conditions from the borrower to the lender, while the former keeps the power to determine the ct/pm-choice, is considered ($BS^a \rightarrow BS^r$). As observed in the previous sections in the BS^a scenario (see Section 3.4.4), the borrower only chooses efficient public debt or equity offerings to maximize his expected profit. On the contrary, in the BS^r scenario (see Section 3.4.3), the borrower prefers public debt whenever feasible, due to his informational advantage combined with his bargaining power limitation. The borrower secures his positive informational rent by his public debt choice, while otherwise being indifferent between public and private equity. Therefore, public debt's as well as private equity's favorability rises when the power to set contract conditions is given to the lender, while the borrower keeps the power to determine ct/pm. Thereby, Proposition 9 is shown.

Proposition 9

Basic Model: Power Effect $BS^a \rightarrow BS^r$

A redistribution of the bargaining power to set the contract conditions from the borrower to the lender, while the former keeps the power to determine ct/pm, increases the favorability of public debt and private equity. Some projects which were financed by public equity in the BS^a scenario are financed by either public debt or eventually private equity, see Areas IV, $VI_{E,E}$, $VII_{E,E}$, $VIII_{E,E}$, and $IX_{E,E}$ in Table 3.21. ($IV : E^\circ \rightarrow E^\bullet$; $VI_{E,E}, VII_{E,E}, VIII_{E,E}, IX_{E,E} : E^\circ \Rightarrow D^\circ$)
□

For the opposite (reverse) power shift ($BS^r \rightarrow BS^a$), the potentially emerged favorability of (inefficient) private equity and public debt vanishes (see Proposition 10).

Proposition 10

Basic Model: Power Effect $BS^r \rightarrow BS^a$

A redistribution of the bargaining power to set the contract conditions from the lender to the borrower, while the latter keeps the power to determine ct/pm, decreases the favorability of public debt and private equity. All projects which were financed by private equity and some projects financed by public debt in the BS^r scenario are financed by public equity in the BS^a scenario, see Areas IV, $VI_{E,E}$, $VII_{E,E}$, $VIII_{E,E}$, or $IX_{E,E}$ in Table 3.21. ($IV : E^\bullet \Rightarrow E^\circ$; $VI_{E,E}, VII_{E,E}, VIII_{E,E}, IX_{E,E} : D^\circ \Rightarrow E^\circ$) □

3.5.4 Relevance of the Bargaining Power Components

After analyzing bargaining power effects in financial contracting, i. e. how power redistributions affect the ct/pm-choice, the relevance of each power component for the lender's as well as for the borrower's profit optimization is examined in this section. To analyze which bargaining power component the lender and the borrower prefer both restricted power scenarios are compared from the lender's as well as from the borrower's perspective. In the LS^r scenario, the lender has the power to determine ct/pm and the borrower can set the contract conditions, while in the BS^r scenario the borrower chooses ct/pm and the lender determines the conditions.

The lender facing the choice between the bargaining power to determine ct/pm (LS^r) and the power to set the contract conditions (BS^r), i. e. $BS^a \rightarrow \{LS^r, BS^r\}$, will definitely prefer to set the contract conditions. Due to the lender's informational disadvantage, the strategic opportunities offered by the bargaining power to determine ct/pm (see Section 3.4.2) are worthless to him, as he is squeezed to his participation constraint anyway ($PLE_B^\circ = PLD_B^\circ = PLE_B^\bullet = PLD_B^\bullet = 0$). However, the power to set contract conditions (see Section 3.4.3), enables the lender to extract a surplus generated from project financing even by a feasible, but from the lender's view suboptimal ct/pm-choice. In the BS^r scenario, the borrower chooses public debt whenever feasible, which for the lender implies $PLD^\circ_{L,I} = \mu(1 - \zeta_{max}) - 1$ or $PLD^\circ_{L,II} = \mu + \dfrac{vc^2}{4\zeta_{max}\mu} - 1 - vc$, while otherwise the borrower is indifferent between public and private equity ($PLE^\circ_L = \mu - 1 - mc$, $PLE^\bullet_L = \mu - 1 - mc - 2nc$).

Result 2
Basic Model: Relevance of Power Components – Lender's Perspective

The lender prefers the power to set the contract conditions to the power to determine ct/pm. $\square$

From the borrower's perspective it is more difficult to evaluate whether he prefers the bargaining power to choose ct/pm (BS^r scenario) or to set the contract conditions (LS^r scenario), i. e. $LS^a \rightarrow \{LS^r, BS^r\}$. On the one hand, due to his informational advantage the power to determine the contract conditions offers him the opportunity to extract any potential surplus from project financing for all feasible, but perhaps for the borrower suboptimal, ct/pm-choices. On the other hand, the power to determine ct/pm enables him to secure any potential (debt financing) positive informational rent. A comparison of the borrower's expected profits in the LS^r scenario $PBE^\circ_B = \mu - 1 - mc; PBD^\circ_{B,I} = \mu - 1; PBD^\circ_{B,II} = \dfrac{[vc + \sqrt{4\zeta_{max}\mu(\mu - 1 - vc) + vc^2}]^2}{4\zeta_{max}\mu}; PBE^\bullet_B = \mu - 1 - mc - 2nc; PBD^\bullet_{B,I} = \mu - 1 - 2nc; PBD^\bullet_{B,II} = \dfrac{[vc + \sqrt{4\zeta_{max}\mu(\mu - 1 - vc - nc) + vc^2}]^2}{4\zeta_{max}\mu} - nc$ and of the BS^r scenario $PBE^\circ_L = 0, PBE^\bullet_L = 0; PBD^\circ_{L,I} = \zeta_{max}\mu; PBD^\circ_{L,II} = \dfrac{vc^2}{4\zeta_{max}\mu}$ reveals that the higher the expected project return is, the more desirable the power to determine contract conditions becomes. While the expected profit the borrower can obtain with the power to chose ct/pm ($PBE^\circ_L, PBD^\circ_{L,I}, PBD^\circ_{L,II}, PBE^\bullet_L$), i. e. his informational rent, is bound below $vc/2$ (see Section 3.3.2.1),[87] the expected profit the borrower can achieve with the power to determine contract conditions ($PBE^\circ_B, PBD^\circ_{B,I}, PBD^\circ_{B,II}, PBE^\bullet_B, PBD^\bullet_{B,I}, PBD^\bullet_{B,II}$) increases (unbounded) with the expected project return μ.[88] On the contrary, the power to chose ct/pm, i. e. to secure an

---

[87] The borrower's informational rent in the BS^r scenario is bound below $vc/2$, as the lender prefers riskless debt ($D^\circ_{L,I}$) for $\zeta_{max} \leq \dfrac{vc}{2\mu}$ [CDr/nr] and risky debt ($D^\circ_{L,II}$) otherwise.

[88] Due to his informational advantage a borrower with the power to set contract conditions can extract any additional surplus due to an increase in μ and any

informational rent $(PBD^\circ_{L,I}, PBD^\circ_{L,II})$, is most valuable near public and private equity's feasible contracting constraints when public debt financing is still feasible. For such projects, the project surplus $\mu - 1$ is just sufficient to compensate for the arising agency costs of equity financing, mc respective $mc + 2nc$, while debt financing offers a positive rent. We refrain here from an exact quantification of a critical μ for a given ζ_{max}, as the lender's ct/pm-choice in the LS^r scenario is not uniquely determined given the assumptions of our framework so far. However, we must note that the borrower's preference for the power components depends on the project specification, as despite the power limitation in both cases, his informational advantage offers him the opportunity of strategical behavior in the BS^r scenario, while the resulting profit must be compared with the profit extractable otherwise by setting contract conditions (LS^r).

Result 3
Basic Model: Relevance of Power Components – Borrower's Perspective
The borrower's preference for the bargaining power components depends on the project (μ, ζ_{max}-specification). The borrower's favorability of the bargaining power to set the contract conditions in comparison to bargaining power to determine ct/pm increases with μ. $\square$

Due to the lender's informational disadvantage, the opportunity to behave strategically in the LS^r scenario is worthless to him. He is not confronted with the tradeoff the borrower is facing.

3.5.5 Consistency Checks and Concluding Remarks

This section summarizes the bargaining power effects derived in Sections 3.5.1, 3.5.2 and 3.5.3 by illustrating their consistency (see Figures 3.12 and 3.13). For each feasible contracting area, the potential changes in the preferred ct/pm-constellation due to the examined power variations are stated. The idea of this consistency check is that either the absolute power can be redistributed directly from the lender to the borrower or the power is shifted step-wise in the two alternative ways. The same is true for power shifts from the borrower to the lender.

Examining, for example, Area IV the consistency of the determined power effects is obvious, as

- $LS^a \rightarrow BS^a$ (see Figure 3.12)
  for $LS^a \rightarrow BS^a$: no change
  for $LS^a \rightarrow BS^r \rightarrow BS^a$: $E^\circ \rightarrow E^\bullet \Rightarrow E^\circ$
  for $LS^a \rightarrow LS^r \rightarrow BS^a$: $E^\circ \rightarrow E^\bullet \Rightarrow E^\circ$

surplus resulting from a debt financing agency cost reduction since $\partial DP(h)/\partial \mu \leq 0$, which implies that the agency costs of debt financing decline.

- $BS^a \rightarrow LS^a$ (see Figure 3.13)
  for $BS^a \rightarrow LS^a$: no change
  for $BS^a \rightarrow BS^r \rightarrow LS^a$: $E^\circ \rightarrow E^\bullet \Rightarrow E^\circ$
  for $BS^a \rightarrow LS^r \rightarrow LS^a$: $E^\circ \rightarrow E^\bullet \Rightarrow E^\circ$

Fig. 3.12: Power Effect's Consistency Check 1 (BM)

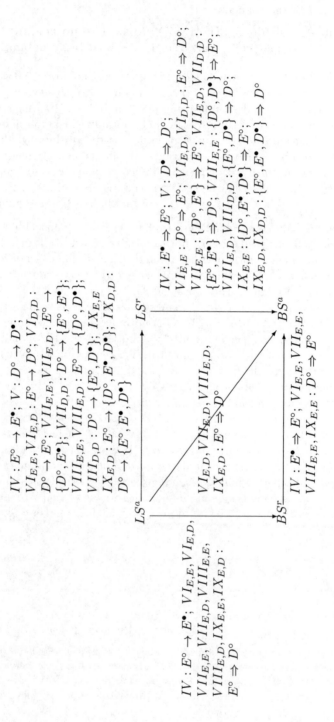

Note: This figure illustrates the consistency of the Propositions 1 ($LS^a \rightarrow BS^a$), 3 ($LS^a \rightarrow BS^r$), 6 ($LS^r \rightarrow BS^a$), 7 ($LS^a \rightarrow LS^r$) and 10 ($BS^r \rightarrow BS^a$) derived in Sections 3.5.1, 3.5.2 and 3.5.3 for each feasible contracting area. $\rightarrow$ refers to power effects which *can* occur, while $\Rightarrow$ indicates effects which *definitely* occur.

Fig. 3.13: Power Effect's Consistency Check 2 (BM)

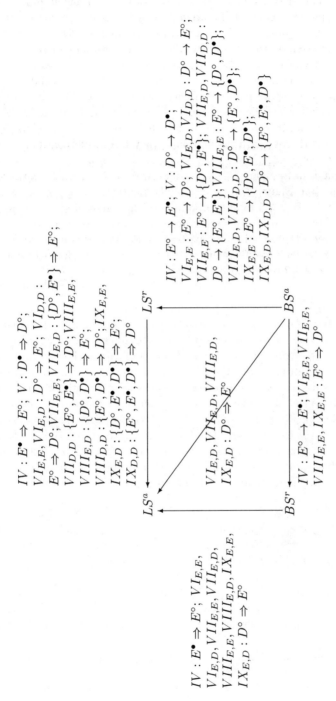

Note: This figure illustrates the consistency of the Propositions 2 ($BS^a \rightarrow LS^a$), 4 ($BS^r \rightarrow LS^r$), 5 ($BS^a \rightarrow LS^r$), 8 ($LS^r \rightarrow LS^a$) and 9 ($BS^a \rightarrow BS^r$) derived in Sections 3.5.1, 3.5.2 and 3.5.3 for each feasible contracting area. $\rightarrow$ refers to power effects which *can* occur, while $\Rightarrow$ indicates effects which *definitely* occur.

We should remember that the derivation of the bargaining power effects is based on a pair-wise (binary) comparison of the optimal ct/pm-choices in the different power scenarios. In order to cope with the parameter dependency of the underlying feasibility constraints as well as the lender's and the borrower's preferences, we divided the set of all potential projects (μ, ζ_{max}-specifications) according to general *parameter dependent* criteria into areas and analyzed each area separately (see Table 3.21). Therefore, the bargaining power effects for a certain project still significantly depend on the underlying parameter constellation. The parameters vc, mc and nc determine which criteria are satisfied for a certain project and which are not, i. e. in which feasible contracting area the project is located and, therefore, which bargaining power effects can be observed.

To conclude, in Sections 3.3 and 3.4 we solved the defined financial contracting game (see Figure 3.2), while finally the bargaining power effects are determined in Section 3.5 by a comparison of the bargaining outcomes (solutions of the game) under alternative power scenarios. We focused on effects on the preferred ct/pm-choice, an explicit profit comparison was only necessary to reveal the lender's, respectively the borrower's, preference among the different bargaining power components.

4

Extending the Model by an Ex-ante Informational Asymmetry About Contractual Partners' Opportunities

As has been shown, for example, by Wang and Williamson (1993) and Choe (1998) among others, *ex-ante* informational asymmetries play a non-negligible role in a firm's financing decision. Therefore, the basic model of Chapter 3 is extended in this chapter by ex-ante informational asymmetries about the lender's investment $(\overline{P}_L)$ respective the borrower's financing alternative $(\overline{P}_B)$. Thereby, the impact of ex-ante informational asymmetries on the bargaining power effects determined so far can be examined. In addition to the moral hazard problems of the basic principal-agent framework of Chapter 3, an adverse selection problem may now arise.

The chapter is structured as follows. In Section 4.1 ex-ante uncertainty is formally introduced (Assumption 6) and the available placement modes are revised since each *pm* offers an alternative way to cope with the arising adverse selection difficulty (Assumption 4'). Like in the basic model of Chapter 3, the three stage financial contracting game is solved by backward induction. Hence firstly, Section 4.2 determines the feasible contracting solutions and the extractable profits (Stages 2&3), while Section 4.3 derives the preferred contract choices in the four alternative bargaining power scenarios (Stage 1). Finally, in Section 4.4 bargaining power effects are analyzed by a pair-wise (binary) comparison of the preferred contract choices in the alternative power scenarios. The robustness of the derived bargaining power effects is examined in Section 4.5.

4.1 Additional and Revised Assumptions

Ex-ante informational asymmetries play a non-negligible role in a firm's financing decision. Firms, for example, acquire ratings to signal their quality to other capital market participants to obtain "better" contract conditions, e. g. cheaper funds than without rating. Sometimes the effects caused by ex-ante informational asymmetries in a firm's financing decision even dominate the effects by interim and ex-post uncertainty (cf. Wang and Williamson (1993) and Choe (1998)). Therefore, an ex-ante informational asymmetry about the lender's alternative investment $(\overline{P}_L)$ and the borrower's financing opportunity $(\overline{P}_B)$ is introduced in our principal-agent setting.[1]

Assumption 6

Ex-ante Uncertainty Specification

- The contracting parties know whether they have an alternative business opportunity (positive outside option $\overline{P}_j = P_j^+$) or not $(\overline{P}_j = 0)$; but they are unaware of the contractual partner's opportunity.

- The parties' outside options exist during the whole contract negotiation game. In any case of no contract agreement the lender and the borrower are left with their outside option $(P_j^+$ or $0)$.

- The distributions of the lender's investment and the borrower's financing alternative, $\overline{P}_L \in \{0, P_L^+\}$ with $P_L^+ > 0 \wedge Prob(\overline{P}_L = P_L^+) = \lambda_L$ $(\lambda_L \in (0,1))$ and $\overline{P}_B \in \{0, P_B^+\}$ with $P_B^+ > 0 \wedge Prob(\overline{P}_B = P_B^+) = \lambda_B$ $(\lambda_B \in (0,1))$, are independent and common knowledge.[2] □

Typically in financial contracting studies, ex-ante uncertainty is assumed about a project's or a firm's expected return or market value with the firm being better informed than potential fund suppliers (see, for example, Hertzel and Smith (1993) and Heinkel and Schwartz (1986)). However, we assume (ex-ante) uncertainty about the borrower's financing respective the lender's investment alternative (outside option) since the introduction of an ex-ante uncertainty e. g. concerning the expected project return simply results in another (additional) informational advantage of the borrower with similar implications as the advantages examined so far, see Chapter 3. Moreover, the introduction of an ex-ante uncertainty concerning the lender's and the borrower's outside option limits *both* parties' potential bargaining power. For example, a borrower with absolute bargaining power is now confronted with a tradeoff. He can set contracts' conditions in a way that the lender does not receive any additional surplus, i. e. $PL = 0$, but thereby the borrower incorporates the risk that the lender might not cooperate anymore (accept the borrower's offer) due to the lender's investment alternative $(P_L^+ > 0)$. Since

---

[1] Maskin and Tirole (1990, 1992) refer to this ex-ante uncertainty modeling as the case of private values in principal-agent frameworks with two-sided uncertainty since the agent's payoff is not "directly" affected by the principal's information. Only the principal's behavior, e. g. to participate or not, affects the agent's payoff "indirectly". The authors refer to the alternative, i. e. the principal's information directly affects the agent's payoff, as the case of common values. An example of this alternative ex-ante uncertainty modeling is the Spencian labor market (cf. Spence (1973)) with a principal / employee of either high or low productivity.

[2] $Prob(X = x)$ denotes the probability that the random variable X takes the value x.

our aim is to examine bargaining power effects in financial contracting, such possible power limitations are important and have to be taken into account.

We refer to the lender respective the borrower with a positive outside option $(\overline{P}_L = P_L^+, \overline{P}_B = P_B^+)$ as the A-type while the N-type does not possess an investment or a financing alternative $(\overline{P}_L = 0, \overline{P}_B = 0)$. Due to Assumption 6, four independent combinations of lender and borrower type are possible when negotiating about financing, as illustrated in Figure 4.1. Each party is aware of the own type but unaware of the contractual partner's type.

Therefore, the determination of contracts' conditions (Stage 2 of the contracting game) is now affected by the condition optimizer's expectation about the contractual partner's type.[3] Furthermore, due to the backward induction solution procedure (anticipation of the successor's optimal behavior) the condition optimizer's expectation(s) about the contractual partner's type and the ct/pm optimizer's expectation(s) about the condition optimizer's expectation(s) influence the preferred ct/pm-choice (Stage 1 of the contracting game).[4] Of course, for absolute power scenarios the condition optimizer and the ct/pm optimizer are identical but for restricted power scenarios they differ.

---

[3] We refer to the party with the right to determine contracts' conditions as the condition optimizer.

[4] The ct/pm optimizer is the party with the bargaining power to choose ct and pm.

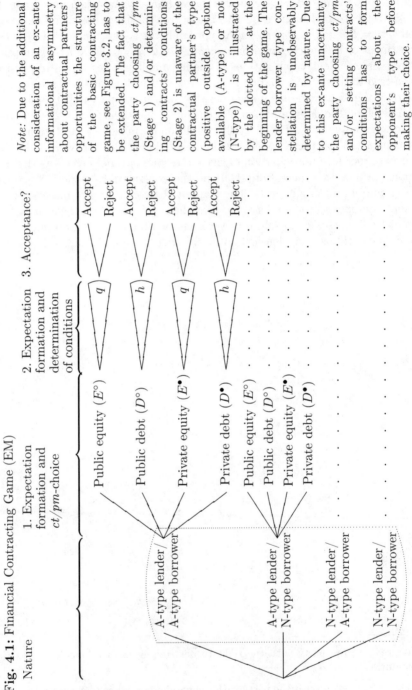

Fig. 4.1: Financial Contracting Game (EM)

Note: Due to the additional consideration of an ex-ante informational asymmetry about contractual partners' opportunities the structure of the basic contracting game, see Figure 3.2, has to be extended. The fact that the party choosing ct/pm (Stage 1) and/or determining contracts' conditions (Stage 2) is unaware of the contractual partner's type (positive outside option available (A-type) or not (N-type)) is illustrated by the dotted box at the beginning of the game. The lender/borrower type constellation is unobservably determined by nature. Due to this ex-ante uncertainty the party choosing ct/pm and/or setting contracts' conditions has to form expectations about the opponent's type before making their choice.

The extension of the basic principal-agent framework by ex-ante informational asymmetries requires a revision of Assumption 4 (public offering vs. private placement) since the available pms determine the ways to cope with this uncertainty.[5]

Assumption 4'
Placement Mode: Public Offering vs. Private Placement
- The lender and the borrower can either transact via the capital market by an anonymous public offering or negotiate personally in a private placement.
- PRIVATE PLACEMENTS solve possible ex-ante informational asymmetries about the lender's alternative investment as well as the borrower's financing opportunity, but cause negotiation costs of nc for both parties.[6]
- PUBLIC OFFERINGS do not resolve any possible ex-ante uncertainties but are costless. □

As discussed in Section 3.1, Heinkel and Schwartz (1986), Hertzel and Smith (1993) and Krishnaswami et al. (1999) empirically confirm that the possible pms are essential for financial contracting under asymmetric information. Mackie-Mason (1990) even points out that ex-ante informational asymmetries are one significant determinant of the choice between private and public fund raising, even after controlling for the security type.

The defined (extended) contract negotiation game is solved in the following sections according to the backward induction principle. While Section 4.2 defines which ct/pm-constellations are feasible and which expected profit the lender respective the borrower can maximally extract (Stages 2&3), Section 4.3 copes with the determination of the preferred ct/pm-choices in the alternative power scenarios (Stage 1).

---

[5] Assumption 4' is based on Assumption 4. Private placement's opportunity to overcome ex-ante informational asymmetries is added.

[6] In accordance with the basic framework negotiation costs only occur when both contractual partners, the lender and the borrower, agree on a private placement. Otherwise no (extra) costs occur. A private placement is only possible if, of course, the ct/pm optimizer chooses a private placement and additionally the condition optimizer is willing to offer binding contract conditions *conditioned* on each potential contractual partner's type which are finally accepted by the partner. The costly evaluation of the lender's respective the borrower's type is delayed after contract agreement to avoid otherwise occurring hold up problems. Negotiation costs occurring prior to contract agreement are sunk when negotiating about contracts' conditions since they do not affect participation constraints anymore. In our context we abstract from such potential hold up problems since the focus of this study is to examine bargaining power effects in financial contracting and not to analyze private placements in detail. Moreover, it is reasonable to assume that the lender respective the borrower find ways to mitigate such potential hold up problems, e. g. by charging an up front fee for making a binding offer.

4.2 Feasible Contracts and Their Conditions

In this section we determine the feasible contracting profit allocations between the lender and the borrower, see Sections 4.2.1 to 4.2.4, to examine among which ct/pm-constellations the lender, respectively the borrower, can choose to maximize expected profit. A profit allocation is feasible when both, the lender's as well as the borrower's, participation constraints are satisfied ($PL \geq \overline{P_L} \wedge PB \geq \overline{P_B}$).

Like in the basic model of Chapter 3, we refrain from an explicit derivation of all feasible profit allocations for a given ct/pm-choice since to solve the contracting game it is convenient to handle Stages 2 and 3 jointly, i. e. to maximize the lender's respective the borrower's profit with respect to the contractual partner's participation constraint. Hence, we (only) determine which ct/pm-constellations are feasible when either the lender or the borrower optimizes contracts' conditions. For each ct/pm-choice we obtain at most two feasible contract conditions (q, h), one from the lender and one from the borrower optimization, instead of a whole range of possible conditions. Contrary to the basic model of Chapter 3, in this extended model the range of feasible profit allocations can differ between the lender's and the borrower's optimization since both possess different (private) information. The lender and the borrower know whether they have a positive business opportunity (A-type) or not (N-type), but they are unaware of the partner's type. Therefore, we have to distinguish feasible contracting solutions given the lender's respective the borrower's private information. In Section 4.2.5 the feasible contracting solutions given the lender's respective the borrower's private information are compared.

The feasible contracting solutions of the lender optimization, i. e. given the lender's private information, are relevant for the determination of the optimal ct/pm-choices in the LS^a and the BS^r scenario (Stage 1 of the contracting game), see Sections 4.3.1 and 4.3.3. The feasible ct/pm-constellations of the borrower optimization, i. e. given the borrower's private information, are relevant for the BS^a and the LS^r scenario, see Sections 4.3.2 and 4.3.4. This does not mean that the lender's, respectively the borrower's, private information is irrelevant for the game's outcome in the other two power scenarios. Obviously, e. g. the lender's private information affects the optimal contracting solution in the LS^r scenario since the lender's (ct/pm optimizer's) participation constraint as well as his ct/pm-choice depend on the lender's type which the borrower tries to anticipate. However, since the contract negotiation game is solved by backward induction the lender's private information becomes in the BS^a and the LS^r scenario relevant at a "later" stage than in the LS^a and the BS^r scenario. Equivalently, the borrower's private information affects the game outcome of the LS^a and the BS^r scenario as well.

Since in this extended setting the contracting parties possess private information about their own opportunity $\overline{P_j}$ but are unaware of the partner's type, the pm-choice (public offering vs. private placement) gains importance because each *pm* offers a different way to cope with the ex-ante informational asymmetry. In a costless *public offering*, for example, no ex-ante uncertainty is (directly) resolved; hence, the condition optimizer (the party with the power to set contracts' conditions) is confronted with a tradeoff in his profit maximization since the surplus left to the contractual partner, PL respective PB, is positively correlated with the contract acceptance probability of the contractual partner. Therefore, the condition optimizer can choose among the following strategies to cope with the ex-ante uncertainty:[7, 8]

- STRATEGY S1: Leave a surplus to the contractual partner (L or B) which is just sufficient to satisfy his participation constraint assuming that the latter has a positive outside option ($PL \geq P_L^+$ or $PB \geq P_B^+$).[9] Hence, the condition optimizer's offer (q,h) is always accepted by the contractual partner.[10]
- STRATEGY S2: Leave a surplus to the contractual partner that only satisfies the participation constraint of a contractual partner with no positive outside option (N-type) ($P_L^+ > PL \geq 0$ or $P_B^+ > PB \geq 0$).[11] Hence, this strategy incorporates a placement risk since an S2 offer is not accepted by all opponents.

---

[7] Whether the condition optimizer actually has the choice between the following strategies is not sure. To have the choice implicitly requires that all the strategies are feasible which is not necessarily the case since the profit the lender can extract is strategy-dependent. However, we initially assume that the condition optimizer has the choice among all strategies and then examines the feasibility of these strategies.

[8] Two possible strategies exist for each public offering since only two scenarios for the contractual partner's business opportunity are considered (see Assumption 6).

[9] Since it is not clear whether the A-type's contractual partner's participation constraint is binding $PL = P_L^+$, respectively $PB = P_B^+$, or not $PL > P_L^+$ ($PB > P_B^+$), S1 requires $PL \geq P_L^+$ ($PB \geq P_B^+$). The A-type's participation constraint might be already satisfied due to the A-type's informational rent which implies that the condition optimizer does not have to take the A-type's participation constraint into account.

[10] The potential N-type partner receives a positive informational rent due to his ex-post advantage.

[11] This strategy tries to avoid S1's disadvantage of a potential informational rent payment to the N-type partner because of his ex-ante informational advantage. However, since it is still not clear whether the contractual partner's participation constraint is binding $PL = 0$, respectively $PB = 0$, or not $P_L^+ > PL > 0$ ($P_B^+ > PB > 0$), S2 requires $P_L^+ > PL \geq 0$ ($P_B^+ > PB \geq 0$). The contractual partner's participation constraint might be already satisfied due to his informational rent because of his interim and ex-post informational advantage. This implies that the condition optimizer does not always have to take the participation constraint into account.

On the other hand, in a costly *private placement* all potential ex-ante informational asymmetries are resolved. The condition optimizer knows the contractual partner's type and can maximize his expected profit with respect to the contractual partner's participation constraint, incorporating no placement risk. Contract conditions depend on the contractual partner's type.

To deal with the feasible contracting constraints *FCC* arising for each *ct/pm*-choice in both optimizations we indicate their origin, in accordance with the basic model of Chapter 3: *BPC* refers to the borrower's participation constraint, *LPC* to the lender's participation constraint, *BPCb/nb* to the condition distinguishing BPC binding or not binding, *CDr/nr* to risky vs. riskless debt case distinction, *NoN* to the non negativity of any potential square root, and finally *LCOC(AB)s/ns* respectively *BCOC(AL)s/ns* to the lender's (the borrower's) contract offer condition for the A-type borrower (lender). The mathematical solution techniques are similar to those in the basic model and are therefore not described in detail again.

4.2.1 Public Equity (E°)

In this section the lender's as well as the borrower's optimization behavior given public equity financing are examined. The feasible contracting solutions as well as the expected profits the lender, respectively the borrower, can maximally extract are determined. In Section 4.2.1.1 the power to set contract conditions is given to the lender, while Section 4.2.1.2 focuses on the case in which the borrower possesses this right.

4.2.1.1 Lender Optimization

Given public equity financing the lender as the condition optimizer can choose among the following strategies to cope with the ex-ante uncertainty about the borrower's financing alternative. Firstly, the lender can set the contract's proportional return participation q_L so that the contract is accepted for sure by the borrower, i. e. by both borrower types (strategy S1). Therefore, S1 provides the N-type borrower with a positive informational rent due to his ex-ante informational advantage.[12] Alternatively, the lender can increase q_L to reduce the borrower's potential informational rent but, thereby, he incorporates a positive probability that the contract is not accepted by the borrower, i. e. by the A-type (strategy S2). If the lender's offer is not accepted by the borrower, the lender and the borrower are left with their outside option (see Assumption 6).

To determine for which project specifications public equity contracts are feasible when conditions are set according to strategy S1 or S2 and which

---

[12] If the borrower achieves an informational rent in equity financing this rent can only result from his ex-ante informational advantage since interim and ex-post uncertainty are resolved.

expected profit $PLE°$ the lender can maximally extract, the optimization problem[13]

$$\max_{q_L} PLE°(q_L) \text{ s.t. } PBE°(q_L) \geq 0 \wedge PLE°(q_L) \geq \overline{P}_L$$

with

$$PLE°(q_L) = \begin{cases} \widetilde{PLE}°(q_L) & \text{for S1}: \ PBE°(q_L) \geq P_B^+ \\ (1 - \widehat{\lambda}_B)\widetilde{PLE}°(q_L) + \widehat{\lambda}_B\overline{P}_L & \text{for S2}: \ P_B^+ > PBE°(q_L) \end{cases}$$

has to be solved. Hence, the optimal strategy-dependent proportional return participations are

$$\text{S1}: \ q_L^* = 1 - \frac{P_B^+}{\mu}; \quad \text{S2}: \ q_L^* = 1 ,$$

guaranteeing the borrower the respective expected profit. Figure 4.2 illustrates the lender's strategy choice and the resulting borrower type-dependent expected profits.

Fig. 4.2: Lender's Strategy Choice Given Public Equity Financing (Stages 2&3 of the Financial Contracting Game, EM)

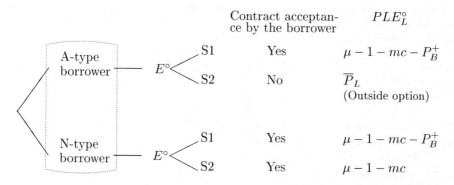

Note: The dotted box illustrates that the lender is unaware of the borrower's type when choosing between S1 and S2. The lender's strategy and borrower type-dependent expected profits are stated. In case of no contract agreement the lender and the borrower are left with their outside option.

---

[13] Due to the uncertainty about the borrower's type, the lender's expected profit depends on the lender's expectation about the borrower's type, $\widehat{\lambda}_B$ respectively $(1 - \widehat{\lambda}_B)$. Furthermore, since a feasible contracting solution does not anymore automatically imply that a contract is actually agreed and the project is financed, we have to distinguish between the lender's overall expected profit $PLE°$ and the lender's expected profit when the project is actually financed $\widetilde{PLE}°$. This distinction is only necessary for public offerings and hence neglected for private placements.

The following table summarizes the expected profits and related feasible contracting constraints.

Table 4.1: Strategy Dependent Expected Profits and Feasible Contracting Constraint of a Public Equity Contract (EM, Lender Optimization)

	PLE_L°	$\mu - 1 - mc - P_B^+$
S1	PBE_L°	$\begin{cases} P_B^+\vert(1-\widehat{\lambda}_L)P_B^+ & \text{if S1 only feasible for N-type lender} \\ P_B^+\vert P_B^+ & \text{if S1 feasible for A\&N-type lender} \end{cases}$
	LPC	$\mu \geq 1 + P_B^+ + mc + \overline{P}_L$
	PLE_L°	$(1-\widehat{\lambda}_B)(\mu - 1 - mc) + \widehat{\lambda}_B \overline{P}_L$
S2	PBE_L°	$\begin{cases} P_B^+\vert 0 & \text{if S2 only feasible for N-type lender} \\ P_B^+\vert 0 & \text{if S2 feasible for A\&N-type lender} \end{cases}$
	LPC	$\mu \geq 1 + mc + \overline{P}_L$

Note: The lender's as well as the borrower's expected profits are stated given public equity financing. The lender has the power to determine the contract's proportional return participation (q). The lender's expected profit PLE_L° and the respective LPC are stated strategy-dependent (upper vs. lower half). The borrower's expected profit PBE_L° is additionally conditioned on the lender type for which S1 respective S2 is feasible. Finally, since the BPC is borrower type-dependent, PBE_L° is stated borrower type-dependent as well (PBE_L° for A-type borrower | PBE_L° for N-type borrower).[14, 15]

All the stated results are important since they are relevant for the determination of the optimal ct/pm-choices in the LS^a and the BS^r scenario. In the LS^a scenario the lender and in the BS^r scenario the borrower makes the ct/pm-choice, hence PLE_L° respectively PBE_L° is of interest, as illustrated in the following paragraphs.

If, for example, in the LS^a scenario the lender chooses public equity financing he as the condition optimizer (also) has to choose among S1 and S2 to cope with the uncertainty about the borrower's financing alternative. Hence, to be able to evaluate both strategies from the lender's perspective

---

[14] The BPC is not explicitly stated in Table 4.1 since the BPC of the A-type (N-type) is automatically satisfied due to S1 (S2), see the lender's optimization problem on page 117.

[15] The borrower's expected profits are stated conditional on the lender's optimal strategy choice. Hence, the borrower's profits depend not only on the borrower's expectation about the lender's type but also on the lender's expectation about the borrower's type since the latter expectation affect the lender's strategy choice.

the lender's expected profit and related LPC are stated strategy-dependent. The lender is unaware of the borrower's type. He expects to be confronted with the A-type borrower with probability $\widehat{\lambda}_B$ and with the N-type with probability $(1 - \widehat{\lambda}_B)$. After the lender's optimal strategy choice for public equity financing is defined, public equity financing can be compared with other feasible ct/pm-constellations to determine the lender's preferred ct/pm-choice in the LS^a scenario.

Alternatively, if in the BS^r scenario the borrower chooses public equity financing the lender faces again both strategies to cope with the uncertainty about the borrower's financing alternative. To determine the lender's preferred strategy his strategy-dependent expected profits PLE_L° must be compared. Hence, the borrower's profits are (lender) strategy-dependent as well. Additionally, to evaluate public equity financing from the borrower's perspective his expected profits PBE_L° are contingent on the lender type for which S1 respective S2 is feasible (conditional expected profits). Since the borrower does not know the lender's type, his expected profits depend on his expectation concerning the lender's type $\widehat{\lambda}_L$ respectively $(1 - \widehat{\lambda}_L)$. Finally, the borrower's expected profits are stated borrower type-dependent, indicated by the vertical line (PBE_L° for A-type borrower$|PBE_L^\circ$ for N-type borrower), since the BPC is borrower type-dependent ($PB \geq P_B^+$ or $PB \geq 0$). E. g. both borrower types accept an S1 offer, while an S2 offer is only accepted by the N-type. If the lender and the borrower can not agree on a contract, both are left with their outside option $\overline{P}_L$ respectively $\overline{P}_B$. Unfortunately, we can not determine the borrower's expected profit more precise without analyzing the lender's optimization behavior, e. g. participation and strategy choice, at first. But as soon as the lender's optimization behavior is defined, the borrower's (borrower) type-dependent expected profits are uniquely determined.

The lender's as well as the borrower's expected profits and feasible contracting constraints depend on *expectations* about the contractual partner's business opportunity since these expectations affect contracting parties' optimization behavior (e. g. contract acceptance and lender's strategy choice). These expectations are represented by $\widehat{\lambda}_L, (1 - \widehat{\lambda}_L)$ or $\widehat{\lambda}_B, (1 - \widehat{\lambda}_B)$ respectively. The expectations *do not* necessarily coincide with the actual probabilities $\lambda_L, (1 - \lambda_L)$ and $\lambda_B, (1 - \lambda_B)$. $\widehat{\lambda}_j$ and λ_j can differ due to the agents' expectation formation as described in detail in Section 4.3 below. The main reason for this potential discrepancy is that the lender's expectation about the borrower's type ($\widehat{\lambda}_B$) is based on the lender's anticipation of the borrower's optimization behavior and vice versa. In the BS^r scenario, for example, the lender does not simply assume that both borrower types behave identically which would imply $\widehat{\lambda}_B = \lambda_B$. Instead the lender conditions his expectation about the borrower type on the borrower's ct/pm-choice ($\widehat{\lambda}_B|E^\circ, \widehat{\lambda}_B|D^\circ, \widehat{\lambda}_B|E^\bullet, \widehat{\lambda}_B|D^\bullet$) to anticipate the borrower's behavior. These conditional expectations do not necessarily coincide. Due to this potential discrepancy we stick in this section to the lender's respective the borrower's expectations of the contractual partner's type and explicitly consider the formation of these expectations in the next section when they are relevant to determine the lender's respective the borrower's optimal ct/pm-choices.

After realizing the necessity to derive all these results, we discuss them in more detail. Since the borrower's expected profit of public equity financing, relevant for the ct/pm-comparison in the BS^r scenario, cannot exactly be determined without analyzing the lender's optimization behavior (strategy choice, etc.), the N- and A-type lender's optimization behavior is examined at first. In order to analyze the lender's optimization behavior, two steps are important: Firstly, the feasible contracting areas of the alternative strategies must be determined. Secondly, a comparison of the strategy-dependent expected profits is (only) necessary for areas where both strategies are feasible.

From the *N-type lender's perspective* $(\overline{P}_L = 0)$ S2 is feasible for projects with a lower expected return than S1 due to the N-type borrower's (ex-ante) informational rent *and* the A-type borrower's profit requirement saving.[16] However, the surplus $PLE_L^\circ|S2$ extractable with S2 increases only less than proportional with μ due to the constant placement risk of S2 $((1 - \widehat{\lambda}_B) \leq 1)$, while $PLE_L^\circ|S1$ increases proportionally (no placement risk of S1). Once both strategies are feasible, the lender prefers S1 to S2 if $\mu > 1 + mc + \dfrac{P_B^+}{\widehat{\lambda}_B}$, willing to offer the N-type borrower an informational rent (expected rent $(1 - \widehat{\lambda}_B)P_B^+$) to guarantee the A-type's participation, i. e. to avoid S2's placement risk (expected gain due to A-type's participation $\widehat{\lambda}_B(\mu - 1 - mc - P_B^+)$). The N-type lender chooses strategy S2 $(E_L^\circ|S2)$ for $1 + mc \leq \mu \leq 1 + mc + \dfrac{P_B^+}{\widehat{\lambda}_B}$ and S1 $(E_L^\circ|S1)$ for $\mu > 1 + mc + \dfrac{P_B^+}{\widehat{\lambda}_B}$.[17]

The *A-type lender's perspective* $(\overline{P}_L = P_L^+)$ resembles the N-type's with the distinction of the lender's positive outside option. Again S2 is feasible for projects with a lower expected return than S1 due to analogous arguments as described above. Moreover, once both strategies are feasible the lender prefers S1 to S2 if $\mu > 1 + mc + \dfrac{P_B^+}{\widehat{\lambda}_B} + P_L^+$. The A-type lender chooses S2 $(E_L^\circ|S2)$ for $1 + mc + P_L^+ \leq \mu \leq 1 + mc + \dfrac{P_B^+}{\widehat{\lambda}_B} + P_L^+$ and S1 $(E_L^\circ|S1)$ for

---

[16] The N-type borrower's expected informational rent $((1 - \widehat{\lambda}_B)P_B^+)$ and the A-type borrower's expected additional profit requirement $(\widehat{\lambda}_B P_B^+)$ saving add up to P_B^+ which is equal to the difference of S1's and S2's FCC $((1 + mc + P_B^+) - (1 + mc) = P_B^+)$.

[17] Due to $\dfrac{P_B^+}{\widehat{\lambda}_B} \geq P_B^+ \; \forall \widehat{\lambda}_B \in [0, 1]$ $E_L^\circ|S1$'s feasibility requirement $\mu \geq 1 + mc + P_B^+$ is satisfied when S1 becomes preferable.

$\mu > 1 + mc + \dfrac{P_B^+}{\widehat{\lambda}_B} + P_L^+$.[18] The lender's positive outside option affects all conditions in the same way, i. e. all profit requirements increase by P_L^+. This guarantees that the gains from project financing can compensate the lender for his positive outside option. The tradeoff underlying the strategy comparison is unaffected by P_L^+ since once S2 is feasible S1 becomes preferable for an increase in μ of $\dfrac{P_B^+}{\widehat{\lambda}_B}$. Hence, the absolute proportion of projects financed by S2 is independent of the lender type. However, the (relative) proportion in relation to all projects which can be financed by E° varies. Since the A-type lender finances less projects in total, he prefers the risky public placement strategy (S2) more often (in relative terms) than the N-type lender.

From the *borrower's perspective* (relevant for the BS^r scenario) due to anticipation of the lender's optimization behavior, identical feasible contracting solutions and strategy choices contingent on the lender's private information arise. However, since the borrower is unaware of the lender's type, the solutions are based on the borrower's expectations of the lender's expectations about the borrower's type ($\widehat{\lambda}_B$). The borrower's optimization behavior (Stage 1 of the contracting game, ct/pm-choice) is, therefore, significantly influenced by his expectations about the lender's expectations. Moreover, the borrower's expectation about the lender's type ($\widehat{\lambda}_L$) is relevant since the lender's optimization behavior (at Stage 2 of the contracting game, participation and strategy choice) is lender type-dependent. Therefore, the borrower's expected profits for public equity financing depend on the borrower's expectation about the lender's type, the borrower's expectation about the lender's expectation about the borrower's type, and the lender's resulting (anticipated) optimization behavior (Stage 2 of the contracting game).[19]

At this stage of the contracting game only the feasibility of public equity financing given the lender's private information is considered since in the lender optimization the lender makes a *binding* offer to the borrower and thereby loses his opportunity to end the game. The lender anticipates this loss. Hence, he only offers contract conditions which satisfy $PL \geq \overline{P}_L$. The feasibility of public equity given the borrower's private information is not considered yet since in the lender optimization the borrower keeps the opportunity to end the game until Stage 3 and therefore makes *no pre-commitment* even by his ct/pm-choice in the BS^r scenario.

---

[18] Once S1 becomes preferable S1 is feasible due to $\dfrac{P_B^+}{\widehat{\lambda}_B} \geq P_B^+ \; \forall \widehat{\lambda}_B \in [0,1]$.

[19] This (complex) interdependency of the lender's and the borrower's optimization behavior is the reason why the borrower's expected profit given public equity financing (lender optimization) is conditioned on the lender's participation and strategy choice, see Table 4.1.

4.2.1.2 Borrower Optimization

Analogously to the lender optimization, in the borrower optimization the borrower (condition optimizer) can choose among two strategies to cope with the ex-ante uncertainty about the lender's financing alternative. The borrower can either set the contract's proportional return participation q_B to a level which both lender types accept (strategy S1) or reduce q_B which removes the N-type lender's potential informational rent, but incorporates a positive probability that the lender, i. e. the A-type, is not willing to accept the contract (strategy S2). The lender might achieve an informational rent due to his ex-ante informational advantage.[20]

To determine for which project specifications public equity contracts placed either according to strategy S1 or S2 are feasible and which expected profit $PBE°$ the borrower can maximally extract, the following optimization problem

$$\max_{q_B} PBE°(q_B) \text{ s.t. } PLE°(q_B) \geq 0 \wedge PBE°(q_B) \geq \overline{P}_B$$

with

$$PBE°(q_B) = \begin{cases} \widetilde{PBE}°(q_B) & \text{for S1}: \ PLE°(q_B) \geq P_L^+ \\ (1 - \widehat{\lambda}_L)\widetilde{PBE}°(q_B) + \widehat{\lambda}_L\overline{P}_B & \text{for S2}: \ P_L^+ > PLE°(q_B) \end{cases}$$

has to be solved. The optimal strategy-dependent proportional return participations are

$$\text{S1}: \quad q_B^* = \frac{1 + P_L^+ + mc}{\mu}; \quad \text{S2}: \quad q_B^* = \frac{1 + mc}{\mu}$$

with the respective expected profits and conditions.

---

[20] So far the lender never received an informational rent since he possesses an interim and ex-post (informational) disadvantage.

Table 4.2: Strategy Dependent Expected Profits and Feasible Contracting Constraint of a Public Equity Contract (EM, Borrower Optimization)

S1	PBE_B°	$\mu - 1 - mc - P_L^+$
	PLE_B°	$\begin{cases} P_L^+\vert(1 - \widehat{\lambda}_B)P_L^+ & \text{if S1 only feasible for N-type borrower} \\ P_L^+\vert P_L^+ & \text{if S1 feasible for A\&N-type borrower} \end{cases}$
	BPC	$\mu \geq 1 + mc + P_L^+ + \overline{P}_B$
S2	PBE_B°	$(1 - \widehat{\lambda}_L)(\mu - 1 - mc) + \widehat{\lambda}_L \overline{P}_B$
	PLE_B°	$\begin{cases} P_L^+\vert 0 & \text{if S2 only feasible for N-type borrower} \\ P_L^+\vert 0 & \text{if S2 feasible for A\&N-type borrower} \end{cases}$
	BPC	$\mu \geq 1 + mc + \overline{P}_B$

Note: The borrower's as well as the lender's expected profits are stated given public equity financing. The borrower has the power to determine the contract's proportional return participation (q). The borrower's expected profit PBE_B° and respective BPC are strategy-dependent. The lender's expected profit PLE_B° is additionally conditioned on the borrower type for which S1 respective S2 is feasible. Finally, since the LPC is lender type-dependent, PLE_B° is stated lender type-dependent (PLE_B° for A-type lender | PLE_B° for N-type lender).

From the *N-type borrower's perspective* ($\overline{P}_B = 0$), S2 is feasible for projects with a lower expected return than S1 as the borrower prevents with S2 any (ex-ante) informational rent payment to the N-type lender as well as the satisfaction of the A-type lender's additional profit requirement (in sum P_L^+). The N-type borrower prefers S2 as long as this amount (P_L^+) exceeds the loss associated with S2's placement risk ($\widehat{\lambda}_L(\mu - 1 - mc - P_L^+)$). Since the surplus extractable with S2 increases only less than proportionally with μ due to the constant placement risk of S2 while the surplus achievable with S1 increases proportionally, the borrower prefers S1 to S2 when both are feasible and $\mu > 1 + mc + \dfrac{P_L^+}{\widehat{\lambda}_L}$. For $\mu > 1 + mc + \dfrac{P_L^+}{\widehat{\lambda}_L}$ the N-type borrower is willing to pay the N-type lender an (ex-ante) informational rent to guarantee the A-type lender's participation. Therefore, the N-type borrower chooses strategy S2 ($E_B^\circ\vert S2$) for $1 + mc \leq \mu \leq 1 + mc + \dfrac{P_L^+}{\widehat{\lambda}_L}$ and S1 ($E_B^\circ\vert S1$) for $\mu > 1 + mc + \dfrac{P_L^+}{\widehat{\lambda}_L}$.

Quite similar is the *A-type borrower's perspective* ($\overline{P}_B = P_B^+$). The A-type borrower prefers S2 ($E_B^\circ\vert S2$) for $1 + mc + P_B^+ \leq \mu \leq 1 + mc + \dfrac{P_L^+}{\widehat{\lambda}_L} + P_B^+$ and

S1 $(E_B^\circ | S1)$ for $\mu > 1 + mc + \dfrac{P_L^+}{\widehat{\lambda}_L} + P_B^+$. The borrower's outside option affects all conditions in the same way, i. e. all profit requirements increase by P_B^+.

From the *lender's perspective* (relevant for the LS^r scenario) the same feasible contracting solutions and strategy choices contingent on the borrower's private information arise, of course dependent on the lender's expectation about the borrower's expectation about the lender's type. Moreover, the lender's expectation about the borrower's type $(\widehat{\lambda}_B)$ is relevant since the borrower's optimization behavior (at Stage 2 of the contracting game, participation and strategy choice) is borrower type-dependent. Hence, the lender's expected profit of public equity financing (borrower optimization) depends on the lender's expectation about the borrower's type, the lender's expectation about the borrower's expectation about the lender type, and the borrower's resulting optimization behavior (at Stage 2 of the contract negotiation game).

4.2.2 Public Debt (D°)

This section examines the lender's as well as the borrower's optimization behavior given public debt financing. The feasible contracting solutions and the expected profits the lender, respectively the borrower, can maximally extract are determined. In Section 4.2.2.1 the power to set contract conditions is given to the lender (lender optimization), while Section 4.2.2.2 focuses on the borrower optimization.

We refrain from an explicit consideration of each possible condition optimizer type since the lender's respective the borrower's outside option affects all derived conditions (expected profit requirements) in the same way as seen for public equity.

4.2.2.1 Lender Optimization

Parallel to public equity financing, public debt financing offers the lender (condition optimizer) two strategies to cope with the ex-ante uncertainty about the borrower's financing alternative. The lender can set the borrower's repayment obligation h_L so that the borrower participates for sure (strategy S1) or that only the N-type borrower accepts the offer (strategy S2). In addition to the borrower's informational rent due to his interim and ex-post advantage he might receive an extra rent in this extended setting because of his added ex-ante advantage.[21]

---

[21] Of course, our ex-ante uncertainty modelling is two-sided but when the lender determines contracts' conditions (lender optimization) only the uncertainty about the borrower's financing alternative is relevant. The lender knows whether he has a profit opportunity or not, but he is unaware about the borrower's type.

For which project specifications public debt contracts are feasible and which expected profit the lender can extract can be determined by the optimization problem

$$\max_{h_L} PLD^\circ(h_L) \text{ s.t. } PBD^\circ(h_L) \geq 0 \wedge PLD^\circ(h_L) \geq \overline{P}_L$$

with

$$PLD^\circ(h_L) = \begin{cases} \widetilde{PLD}^\circ(h_L) & \text{for S1}: \ PBD^\circ(h_L) \geq P_B^+ \\ (1 - \widehat{\lambda}_B)\widehat{PLD}^\circ(h_L) + \widehat{\lambda}_B\overline{P}_L & \text{for S2}: \ P_B^+ > PBD^\circ(h_L). \end{cases}$$

Like in the basic model without ex-ante uncertainty two debt financing cases have to be distinguished: Riskless (I) and risky debt (II). Due to the additional uncertainty about the borrower's financing alternative a further distinction is required: Participation constraint of the A-type borrower ($PBD^\circ \geq P_B^+$) *not* binding due to the borrower's informational rent (a) and the A-type's BPC binding despite this rent (b).[22] If the BPC is not binding (a), we can proceed as in the basic model since no strategy choice emerges (see Section 3.3.2). No strategy choice emerges since the participation constraints of both borrower types are automatically satisfied which implies that the lender can simply optimize the borrower's repayment obligation, of course, restricted by the CDr/nr and the BPCb/nb but without considering any placement risk. If the BPC is binding (b), the solution can be found by transforming the BPC of the A-type.

---

[22] The participation constraint of the N-type borrower ($PBD^\circ \geq 0$) is never binding since his informational rent is always positive.

- *Case I – Riskless debt:* $h_L \leq y_{min}\big|_{\zeta=\zeta_{max}}$

  - *Ia – A-type borrower's BPC not binding:* $\zeta_{max}\mu \geq P_B^+$

    $$h_{L,Ia}^* = \mu(1 - \zeta_{max})$$

Table 4.3: Expected Profits and Feasible Contracting Constraints of a Public Debt Contract – Case Ia (EM, Lender Optimization)

$PLD_{L,Ia}^\circ$	$\mu(1 - \zeta_{max}) - 1$
$PBD_{L,Ia}^\circ$	$\begin{cases} (1 - \widehat{\lambda}_L)\zeta_{max}\mu + \widehat{\lambda}_L P_B^+ \| & \text{if } D_{L,Ia}^\circ \text{ only feasible} \\ (1 - \widehat{\lambda}_L)\zeta_{max}\mu & \text{for N-type lender} \\ \\ \zeta_{max}\mu \| \zeta_{max}\mu & \text{if } D_{L,Ia}^\circ \text{ feasible for} \\ & \text{A\&N-type lender} \end{cases}$
BPCb/nb	$\zeta_{max} \geq \dfrac{P_B^+}{\mu}$
LPC	$\mu(1 - \zeta_{max}) \geq 1 + \overline{P}_L$

Note: The lender's as well as the borrower's expected profits are stated given public debt financing. The lender possesses the right to determine the contract's repayment obligation (h). The borrower's expected profit $PBD_{L,Ia}^\circ$ is conditioned on the lender type for which $D_{L,Ia}^\circ$ is feasible. Since the BPC is borrower type-dependent, $PBD_{L,Ia}^\circ$ is stated borrower type dependent ($PBD_{L,Ia}^\circ$ for A-type borrower | $PBD_{L,Ia}^\circ$ for N-type borrower).

Equivalent to private debt financing without ex-ante informational asymmetry (case Ia) the feasible contracting solutions of riskless public debt with non-binding BPC are bounded by the type-dependent LPC and the type-*in*dependent BPCb/nb.[23] The BPCb/nb condition determines the minimum return spread required so that the borrower's informational rent ($\zeta_{max}\mu$) is sufficient to compensate for his possible positive outside option ($\zeta_{max}\mu > P_B^+$). Since the borrower's (uncontractible) informational rent in case of contract agreement ($\zeta_{max}\mu$) already satisfies both borrower types' participation constraints ($PBD_{L,Ia}^\circ \geq P_B^+$), *no* difficulties for contracting are caused by the ex-ante informational asymmetry.[24] No strategy choice is required. The lender's expectation about the borrower's type is irrelevant. Hence no distinction between the lender's and the borrower's perspective is needed. The only expectation relevant in this debt financing case is

---

[23] We refer to the BPCb/nb ($\zeta_{max} \geq P_B^+/\mu$) as type-independent since the BPCb/nb requirement for $D_{L,Ia}^\circ$ is unaffected by variations in the lender's or the borrower's type.

[24] Of course, if no contract is agreed because e. g. the A-type lender's participation constraint is not satisfied the lender as well as the borrower are left with the outside options, 0 respective P_j^+.

the borrower's expectation about the lender's type since it affects the borrower's expected profit due to the lender's type-dependent LPC.

- *Ib – A-type borrower's BPC binding:* $\zeta_{max}\mu < P_B^+$

Assuming that the A-type's BPC is not automatically satisfied from the borrower's debt financing informational rent alone, the lender is again confronted with the strategy choice to cope with his ex-ante uncertainty about the borrower's financing alternative. The optimal strategy-dependent demanded repayment obligations are

$$S1: \quad h_{L,Ib}^* = \mu - P_B^+; \quad S2: \quad h_{L,Ib}^* = \mu(1 - \zeta_{max})$$

resulting in the following expected profits and constraints.

Table 4.4: Strategy Dependent Expected Profits and Feasible Contracting Constraints of a Public Debt Contract – Case Ib (EM, Lender Optimization)

S1	$PLD_{L,Ib}^\circ$	$\mu - 1 - P_B^+$
	$PBD_{L,Ib}^\circ$	$\begin{cases} P_B^+\|(1 - \widehat{\lambda}_L)P_B^+ & \text{if S1 only feasible for N-type lender} \\ P_B^+\|P_B^+ & \text{if S1 feasible for A\&N-type lender} \end{cases}$
	BPCb/nb	$\zeta_{max} < \dfrac{P_B^+}{\mu}$
	LPC	$\mu \geq 1 + P_B^+ + \overline{P}_L$
S2	$PLD_{L,Ib}^\circ$	$(1 - \widehat{\lambda}_B)[\mu(1 - \zeta_{max}) - 1] + \widehat{\lambda}_B \overline{P}_L$
	$PBD_{L,Ib}^\circ$	$\begin{cases} P_B^+\|(1 - \widehat{\lambda}_L)\zeta_{max}\mu & \text{if S2 only feasible for N-type lender} \\ P_B^+\|\zeta_{max}\mu & \text{if S2 feasible for A\&N-type lender} \end{cases}$
	BPCb/nb	$\zeta_{max} < \dfrac{P_B^+}{\mu}$
	LPC	$\mu(1 - \zeta_{max}) \geq 1 + \overline{P}_L$

Note: The lender's as well as the borrower's expected profits are stated given public debt financing. The lender optimizes the contract's repayment obligation (h). The lender's expected profit $PLD_{L,Ib}^\circ$ and respective LPC are stated strategy-dependent. The borrower's expected profit $PBD_{L,Ib}^\circ$ is also strategy-dependent and additionally conditioned on the lender type for which $D_{L,Ib}^\circ$ is feasible. Since the BPC is borrower type-dependent, $PBD_{L,Ib}^\circ$ is stated borrower type-dependent ($PBD_{L,Ib}^\circ$ for A-type borrower | $PBD_{L,Ib}^\circ$ for N-type borrower).

As observed for public equity financing (Section 4.2.1.1) the expected profits and feasibility constraints of riskless public debt with binding BPC are strategy dependent. Furthermore, from the *lender's perspective* S2 is feasible for projects with a lower expected return than S1. This is due to S2's "saving" of the N-type borrower's informational rent and the A-type's minimum profit requirement. S1 becomes more attractive by an increase in μ given ζ_{max} since the placement risk of S2 becomes more severe. For $\mu(1-\zeta_{max}) > 1+\dfrac{P_B^+ - \zeta_{max}\mu}{\widehat{\lambda}_B}+\overline{P}_L$ the lender favors S1 to S2 since S1's disadvantage of paying the N-type borrower an (additional) ex-ante informational rent $(P_B^+ - \zeta_{max}\mu)$[25] is lower than the loss incurred due to S2's placement risk $(\widehat{\lambda}_B(\mu(1-\zeta_{max})-1-\overline{P}_L))$. As pointed out for public equity, S1 becomes more attractive due to an increase in μ since the extractable profit increases making the placement risk of S2 more severe. However, in debt financing S1 also gains attractiveness by an increase in μ, given ζ_{max}, or in ζ_{max}, given μ, due to an increase in the borrower's uncontractible informational rent because of the latter's interim and ex-post informational advantage $(\zeta_{max}\mu)$. This informational rent reduces the *additional* payoff required to satisfy the A-type's BPC $(P_B^+ - \zeta_{max}\mu)$. The disadvantage of S1 to pay an additional informational rent to the N-type borrower declines with an increase in μ or ζ_{max} since parts of the borrower's informational rent are paid anyway for reasons other than ex-ante uncertainty. Only the additional rent payment because of the ex-ante informational asymmetry about the borrower's financing opportunity affects the lender's strategy comparison. The borrower's rent due to interim and ex-post uncertainty is strategy independent.

From the *borrower's perspective* the same feasible contracting solutions and strategy choices conditioned on the lender's private information emerge, dependent on his expectation about the lender's expectation. The borrower's expectation about the lender's type $(\widehat{\lambda}_L)$ is relevant since the lender's optimization behavior (Stage 2 of the contract negotiation game) is lender type-dependent.

---

[25] In riskless debt financing the borrower receives an informational rent of $\zeta_{max}\mu$ anyway due to his interim and ex-post advantage.

- *Case II – Risky debt:* $y_{min}\big|_{\zeta=\zeta_{max}} < h_L$

    - *IIa – A-type borrower's BPC not binding:* $\dfrac{vc^2}{4\zeta_{max}\mu} \geq P_B^+$

    $h_{L,IIa}^* = \mu(1 + \zeta_{max}) - vc$

Table 4.5: Expected Profits and Feasible Contracting Constraints of a Public Debt Contract – Case IIa (EM, Lender Optimization)

$PLD_{L,IIa}^\circ$	$\mu + \dfrac{vc^2}{4\zeta_{max}\mu} - 1 - vc$	
$PBD_{L,IIa}^\circ$	$\begin{cases} (1-\widehat{\lambda}_L)\dfrac{vc^2}{4\zeta_{max}\mu} + \widehat{\lambda}_L P_B^+\vert \\[2mm] (1-\widehat{\lambda}_L)\dfrac{vc^2}{4\zeta_{max}\mu} \end{cases}$	if $D_{L,IIa}^\circ$ only feasible for N-type lender
	$\dfrac{vc^2}{4\zeta_{max}\mu}\bigg\vert\dfrac{vc^2}{4\zeta_{max}\mu}$	if $D_{L,IIa}^\circ$ feasible for A&N-type lender
CDr/nr	$\zeta_{max} > \dfrac{vc}{2\mu}$	
BPCb/nb	$\zeta_{max} \leq \dfrac{vc^2}{4\mu P_B^+}$	
LPC[26]	$\mu \geq 1 + vc + \overline{P}_L \vee \zeta_{max} \leq \dfrac{vc^2}{4\mu(1 + vc + \overline{P}_L - \mu)}$	

Note: The structure of the table is equivalent to the structure of Table 4.3.

As described for riskless debt financing with non binding BPC (case Ia), in the case of risky debt financing with non binding BPC ex-ante uncertainty about the borrower's financing alternative causes no further difficulties since the A- and the N-type's BPCs are automatically satisfied due to the borrower's debt financing informational rent.[27] No strategy choice to cope with the ex-ante uncertainty is required.

---

[26] At first sight one might expect that $D_{L,IIa}^\circ$ is feasible even for $\mu \leq 1$, but this expectation is false. Of course, the LPC for risky debt financing with non binding BPC can be satisfied for $\mu = 1$ while $\zeta_{max} \leq \dfrac{vc^2}{4(vc + \overline{P}_L)}$; however, such constellations are not feasible since the LPC, the BPCb/nb and the CDr/nr have to be satisfied simultaneously. (For $\mu = 1$ e. g. the CDr/nr requires $\zeta_{max} > vc/2$ which is not compatible with $\zeta_{max} \leq \dfrac{vc^2}{4(vc + \overline{P}_L)}$.) The same reasoning prevents the feasibility of $\mu \in [0,1)$.

[27] In risky debt financing the borrower receives an informational rent of $\dfrac{vc^2}{4\zeta_{max}\mu}$ already from his interim and ex-post advantage.

From the *lender's perspective* the feasible contracting solutions are strategy and hence expectations independent. They are restricted by the CDr/nr, the BPCb/nb, as well as by the type-dependent LPC. Since the feasible contracting solutions and expected profits are independent of private information and expectations about this information, no *borrower perspective* has to be considered explicitly. Only relevant for the borrower's expected profit is the borrower's expectation about the lender's type since the lender's LPC is type-dependent.

-   *IIb – A-type borrower's BPC binding:* $\dfrac{vc^2}{4\zeta_{max}\mu} < P_B^+$

    If the A-type's BPC is not automatically satisfied, the lender can choose among both strategies to cope with the ex-ante uncertainty. The optimal strategy-dependent repayments are

    $$S1: \quad h_{L,IIb}^* = \mu(1 + \zeta_{max}) - 2\sqrt{P_B^+ \zeta_{max}\mu}$$
    $$S2: \quad h_{L,IIb}^* = \mu(1 + \zeta_{max}) - vc ,$$

    resulting in the following expected profits and constraints.

Table 4.6: Strategy Dependent Expected Profits and Feasible Contracting Constraints of a Public Debt Contract – Case IIb (EM, Lender Optimization)

	$PLD_{L,IIb}^\circ$	$\mu + vc\sqrt{\dfrac{P_B^+}{\zeta_{max}\mu}} - vc - P_B^+ - 1$
S1	$PBD_{L,IIb}^\circ$	$\begin{cases} P_B^+\mid(1-\widehat{\lambda}_L)P_B^+ & \text{if S1 only feasible for N-type lender} \\ P_B^+\mid P_B^+ & \text{if S1 feasible for A\&N-type lender} \end{cases}$
	CDr/nr	$\zeta_{max} > \dfrac{P_B^+}{\mu}$
	BPCb/nb	$\zeta_{max} > \dfrac{vc^2}{4\mu P_B^+}$
	LPC[28]	$\mu \geq 1 + vc + P_B^+ + \overline{P}_L \vee$ $\zeta_{max} \leq \dfrac{vc^2 P_B^+}{(1 + vc + P_B^+ + \overline{P}_L - \mu)^2 \mu}$

	table continued on next page

		table continued from previous page
	$PLD^{\circ}_{L,IIb}$	$(1 - \widehat{\lambda}_B)\left[\mu + \dfrac{vc^2}{4\zeta_{max}\mu} - 1 - vc\right] + \widehat{\lambda}_B \overline{P}_L$
S2	$PBD^{\circ}_{L,IIb}$	$\begin{cases} P^+_B\vert(1 - \widehat{\lambda}_L)\dfrac{vc^2}{4\zeta_{max}\mu} & \text{if S2 only feasible} \\ & \text{for N-type lender} \\ P^+_B\vert\dfrac{vc^2}{4\zeta_{max}\mu} & \text{if S2 feasible for} \\ & \text{A\&N-type lender} \end{cases}$
	CDr/nr	$\zeta_{max} > \dfrac{vc}{2\mu}$
	BPCb/nb	$\zeta_{max} > \dfrac{vc^2}{4\mu P^+_B}$
	LPC[29]	$\mu \geq 1 + vc + \overline{P}_L \vee \zeta_{max} \leq \dfrac{vc^2}{4\mu\left(1 + vc + \overline{P}_L - \mu\right)}$

Note: The structure of the table is equivalent to the structure of Table 4.4.

From the *lender's perspective* the feasible contracting solutions of S1 are a subset of the solutions obtained for S2. If both strategies are feasible the lender favors S1 to S2 for $\widehat{\lambda}_B(\mu + \dfrac{vc^2}{4\zeta_{max}\mu} - 1 - vc - \overline{P}_L) >$ $P^+_B - \left(vc\sqrt{\dfrac{P^+_B}{\zeta_{max}\mu}} - \dfrac{vc^2}{4\zeta_{max}\mu}\right)$. This preference condition is due to two opposing effects. Firstly, S1's favorability increases with μ since the surplus extractable with S1 increases stronger in μ than the surplus extractable with S2 due to S2's placement risk. Secondly, an increase in μ also reduces S1's favorability since an increase in μ causes the "switching" costs from S2 to S1 ($P^+_B - \dfrac{vc^2}{4\zeta_{max}\mu}$) to increase due to a decrease in the borrower's informational rent from interim and ex-post uncertainty ($\dfrac{vc^2}{4\zeta_{max}\mu}$).

---

[28] Due to the LPC possible contracting solutions for $\mu = 1$ ($\zeta_{max} \leq \dfrac{vc^2 P^+_B}{(vc + P^+_B + \overline{P}_L)}$) are prevented by the CDr/nr ($\zeta_{max} > P^+_B$).

[29] Equivalent to risky debt financing with non binding BPC, solutions for $\mu = 1$ are not feasible since the LPC and the CDr/nr have to be satisfied simultaneously.

From the *borrower's perspective* (relevant for the BS^r scenario) due to anticipation of the lender's optimization behavior, identical feasible contracting solutions and strategy choices arise. However, since the borrower is unaware of the lender's type and expectations, the solutions are based on the borrower's expectations of the lender's expectations about the borrower type $(\widehat{\lambda}_B)$.

The inter-case comparison of debt financing is delayed to Section 4.3 to cope simultaneously with the case comparison and the evaluation of the feasible ct/pm-constellations.

4.2.2.2 Borrower Optimization

In public debt financing the borrower can choose among two strategies to cope with the ex-ante uncertainty about the lender's investment alternative. The borrower can either set the contract's repayment obligation h so that both lender types accept the offered term (S1) or only the N-type participates (S2).

The contracting problem is therefore

$$\max_{h_B} PBD^\circ(h_B) \text{ s.t. } PLD^\circ(h_B) \geq 0 \wedge PBD^\circ(h_B) \geq \overline{P}_B$$

with

$$PBD^\circ(h_B)=\begin{cases} \widetilde{PBD}^\circ(h_B) & \text{for S1}: \ PLD^\circ(h_B) \geq P_L^+ \\ (1-\widehat{\lambda}_L)\widetilde{PBD}^\circ(h_B) + \widehat{\lambda}_L\overline{P}_B & \text{for S2}: \ P_L^+ > PLD^\circ(h_B). \end{cases}$$

Again two debt financing cases have to be considered: Riskless (I) and risky debt (II). Since the lender does never receive an informational rent from interim and ex-post uncertainty, the A-type's LPC is in no circumstances automatically satisfied. Therefore, no further case distinction is required.

- *Case I – Riskless debt:* $h_B \leq y_{min}\big|_{\zeta=\zeta_{max}}$

The optimal strategy-dependent repayments turn out to be

$$\text{S1}: \quad h_{B,I}^* = 1 + P_L^+; \quad \text{S2}: \quad h_{B,I}^* = 1,$$

resulting in the following expected profits and constraints.

Table 4.7: Strategy Dependent Expected Profits and Feasible Contracting Constraints of a Public Debt Contract – Case I (EM, Borrower Optimization)

S1	$PBD^\circ_{B,I}$	$\mu - 1 - P^+_L$
	$PLD^\circ_{B,I}$	$\begin{cases} P^+_L \mid (1 - \widehat{\lambda}_B)P^+_L & \text{if S1 only feasible for N-type borrower} \\ P^+_L \mid P^+_L & \text{if S1 feasible for A\&N-type borrower} \end{cases}$
	CDr/nr	$\mu(1 - \zeta_{max}) \geq 1 + P^+_L$
	BPC	$\mu \geq 1 + P^+_L + \overline{P}_B$
S2	$PBD^\circ_{B,I}$	$(1 - \widehat{\lambda}_L)(\mu - 1) + \widehat{\lambda}_L \overline{P}_B$
	$PLD^\circ_{B,I}$	$\begin{cases} P^+_L \mid 0 & \text{if S2 only feasible for N-type borrower} \\ P^+_L \mid 0 & \text{if S2 feasible for A\&N-type borrower} \end{cases}$
	CDr/nr	$\mu(1 - \zeta_{max}) \geq 1$
	BPC	$\mu \geq 1 + \overline{P}_B$

Note: The borrower's as well as the lender's expected profits are stated given public debt financing. The borrower has the right to determine the contract's repayment obligation (h). The borrower's expected profit $PBD^\circ_{B,I}$ and respective BPC and CDr/nr are stated strategy-dependent. The lender's expected profit $PLD^\circ_{B,I}$ is additionally conditioned on the borrower type for which S1 respective S2 is feasible. Since the LPC is lender type-dependent, PLD°_B is stated lender type-dependent $(PLD^\circ_{B,I}$ for A-type lender $\mid PLD^\circ_{B,I}$ for N-type lender).[30]

As previously observed, from the *borrower's perspective* the feasible solutions of S1 are a subset of S2. The borrower favors S1 to S2 if both are feasible and $\mu > 1 + \dfrac{P^+_L}{\widehat{\lambda}_L} + \overline{P}_B$.

From the *lender's perspective* the same results are obtained, dependent on the lender's expectation about the borrower's expectation about the lender's type.

- *Case II – Risky debt:* $y_{min}\big|_{\zeta=\zeta_{max}} < h_B$

  For risky debt financing the following strategy-dependent repayments are optimal

  $$\text{S1}: \quad h^*_{B,II} = \mu(1 + \zeta_{max}) - vc - \sqrt{4\zeta_{max}\mu(\mu - 1 - P^+_L - vc) + vc^2}$$

  $$\text{S2}: \quad h^*_{B,II} = \mu(1 + \zeta_{max}) - vc - \sqrt{4\zeta_{max}\mu(\mu - 1 - vc) + vc^2}.$$

---

[30] The LPC is not explicitly stated in Table 4.7 since the LPC of the A-type (N-type) is automatically satisfied due to S1 (S2).

Table 4.8: Strategy Dependent Expected Profits and Feasible Contracting Constraints of a Public Debt Contract – Case II (EM, Borrower Optimization)

S1	$PBD^{\circ}_{B,II}$	$$\dfrac{\left[vc + \sqrt{4\zeta_{max}\mu(\mu - 1 - P_L^+ - vc) + vc^2}\,\right]^2}{4\zeta_{max}\mu}$$
	$PLD^{\circ}_{B,II}$	$\begin{cases} P_L^+ \mid (1 - \widehat{\lambda}_B)P_L^+ & \text{if S1 only feasible for N-type borrower} \\ P_L^+ \mid P_L^+ & \text{if S1 feasible for A\&N-type borrower} \end{cases}$
	NoN	$\mu \geq 1 + P_L^+ + vc \vee \zeta_{max} \leq \dfrac{vc^2}{4\mu(1 + P_L^+ + vc - \mu)}$
	CDr/nr	$\mu(1 - \zeta_{max}) < 1 + P_L^+ \wedge \zeta_{max} \geq \dfrac{vc}{2\mu}$
	BPC	$\mu \geq 1 + P_L^+ + vc + \overline{P}_B \vee$ $\zeta_{max} \leq \max\left\{ \dfrac{vc^2}{2\mu(1 + P_L^+ + vc + \overline{P}_B - \mu)}; \right.$ $\left. \dfrac{vc^2 \overline{P}_B}{\mu(1 + P_L^+ + vc + \overline{P}_B - \mu)^2} \right\}$
S2	$PBD^{\circ}_{B,II}$	$(1 - \widehat{\lambda}_L) \cdot \dfrac{\left[vc + \sqrt{4\zeta_{max}\mu(\mu - 1 - vc) + vc^2}\,\right]^2}{4\zeta_{max}\mu} + \widehat{\lambda}_L \overline{P}_B$
	$PLD^{\circ}_{B,II}$	$\begin{cases} P_L^+ \mid 0 & \text{if S2 only feasible for N-type borrower} \\ P_L^+ \mid 0 & \text{if S2 feasible for A\&N-type borrower} \end{cases}$
	NoN	$\mu \geq 1 + vc \vee \zeta_{max} \leq \dfrac{vc^2}{4\mu(1 + vc - \mu)}$
	CDr/nr	$\mu(1 - \zeta_{max}) < 1 \wedge \zeta_{max} \geq \dfrac{vc}{2\mu}$
	BPC	$\mu \geq 1 + vc + \overline{P}_B \vee \zeta_{max} \leq \max\left\{ \dfrac{vc^2}{2\mu(1 + vc + \overline{P}_B - \mu)}; \right.$ $\left. \dfrac{vc^2 \overline{P}_B}{\mu(1 + vc + \overline{P}_B - \mu)^2} \right\}$

Note: The structure of the table is equivalent to the structure of Table 4.7.

The feasible contracting solutions of S1 are a subset of S2 from the *borrower's perspective*. When both are feasible S1 is preferable if

$$\frac{\left[vc+\sqrt{4\zeta_{max}\mu(\mu-1-P_L^+-vc)+vc^2}\right]^2}{4\zeta_{max}\mu} > (1-\widehat{\lambda}_L) \cdot \frac{\left[vc+\sqrt{4\zeta_{max}\mu(\mu-1-vc)+vc^2}\right]^2}{4\zeta_{max}\mu} +$$

$\widehat{\lambda}_L \overline{P}_B$. As expected, for large μ S1 is preferable since no placement risk disturbs the borrower's profit maximization. Moreover, for small μ and small ζ_{max} S1 is also preferable since for small μ the extra profit the borrower might achieve with S1 in comparison to S2 decreases in ζ_{max}. So ζ_{max} has to stay below a certain level to make it worth paying the N-type lender's informational rent in S1. Since the lender never receives an informational rent due to interim and ex-post uncertainty, the borrower's strategy comparison is always affected by the full informational rent payment due to ex-ante uncertainty (P_L^+).

From the *lender's perspective* the same feasible contracting solutions and strategy preferences based on the borrower's private information occur but as usually they depend on the lender's expectations of the borrower's expectations.

The between-case comparison of $D_{B,I}^\circ|S2$ and $D_{B,II}^\circ|S1$ is postponed to Section 4.3 to cope with the comparison simultaneously to the *ct/pm*-evaluation.

4.2.3 Private Equity ($E^\bullet$)

Alternatively to a public placement the lender and the borrower can negotiate privately about contracts' conditions in a private placement. Private placements are costly but resolve all ex-ante informational asymmetries (see Assumption 4'). To avoid possible hold up problems when one party has to make an up-front investment, we assume that private placement's negotiation costs (nc) only occur if the contract is accepted by both parties. Hence, the condition optimizer makes an initial conditional condition offer for each possible contractual partner while the actual partner's type is revealed after contract agreement and the respective contract conditions are fixed. We assume that the condition optimizer makes two initial conditional contract offers and the contractual partner chooses the offer particular designed for him. This self-selection mechanism works because after contract choice and agreement the informational asymmetry about the partners' types is revealed and only if the "correct" contract has been chosen the contract is finally settled. When the "wrong" contract has been chosen the negotiation ends and both parties are left with their outside option less sunk negotiation costs. Consequently, both parties are worse off than in the case of no contract negotiation / agreement at all and therefore this simple self-selection mechanism works.[31]

---

[31] However, no self-selection or screening mechanism in the spirit of Rothschild and Stiglitz (1976) and Wilson (1977) is possible since only one contract component is variable. For a classical self-selection mechanism at least two variable contract components are necessary to separate / to distinguish the alternative contractual partners.

Whether the condition optimizer actually offers acceptable (conditional) conditions to both possible contractual partners (A or N-type) depends on his expected profit from contract agreement. The condition optimizer offers acceptable conditions to both potential partners when each individual contract agreement is despite occurring negotiation costs profitable for him. Otherwise, the optimizer only offers (ex-ante) one contractual partner type an acceptable contract or refuses to participate at all (end of negotiation game).

Private equity placements are considered in this section. The feasible contracting solutions as well as the expected profits the lender, respectively the borrower, can maximally extract are determined. The power to set contract conditions is given to the lender in Section 4.2.3.1, while Section 4.2.3.2 focuses on the borrower as the condition optimizer.

4.2.3.1 Lender Optimization

Given privately placed equity, the lender can set his proportional return participation (q) borrower type-dependent, while nc arises for both parties. Due to the A-type borrower's positive outside option, the optimal (initial) conditions for each possible borrower type ($q^*_{L,\text{A-type borrower}}$, $q^*_{L,\text{N-type borrower}}$) will satisfy

$q^*_{L,\text{A-type borrower}} \leq q^*_{L,\text{N-type borrower}}$ due to the respective borrower type-dependent BPC.

To determine for which project specifications private equity contracts are feasible and which expected profit the lender can maximally extract, the optimization problem[32]

$$\max_{\substack{q_{L,\text{A-type borrower}} \\ q_{L,\text{N-type borrower}}}} PLE^\bullet(q) = \widehat{\lambda}_B PLE^\bullet(q_{L,\text{A-type borrower}})$$
$$+(1 - \widehat{\lambda}_B)PLE^\bullet(q_{L,\text{N-type borrower}})$$

$$s.t. \ PBE^\bullet(q_{L,\text{A-type borrower}}) \geq P_B^+, \ PBE^\bullet(q_{L,\text{N-type borrower}}) \geq 0$$
$$\wedge \ PLE^\bullet(q_{L,\text{A-type borrower}}) \geq \overline{P}_L, \ PLE^\bullet(q_{L,\text{N-type borrower}}) \geq \overline{P}_L$$

has to be solved. Since the lender is unaware of the borrower's type, his expected profit $PLE^\bullet$ depends ex-ante on his expectations about the partner's (the borrower's) type, $\widehat{\lambda}_B$ respectively $(1 - \widehat{\lambda}_B)$. The borrower's type-dependent profit requirements, $PBE^\bullet(q_{L,\text{A-type borrower}})$ for the A-type and $PBE^\bullet(q_{L,\text{N-type borrower}})$ for the N-type, guarantee that the lender's conditional offer is accepted by the respective borrower type. Finally, the lender's

---

[32] Since the lender is unaware of the borrower's type when he has to decide whether to make a binding offer or not, the lender's decision depends on the latter's (ex-ante) expected profit $PLE^\bullet(q) = \widehat{\lambda}_B PLE^\bullet(q_{L,\text{A-type borrower}}) + (1 - \widehat{\lambda}_B)PLE^\bullet(q_{L,\text{N-type borrower}})$. However, the feasible contracting constraints do not depend on the lender's expectation about the borrower's type $\widehat{\lambda}_B$ since the borrower's type is revealed after the lender's binding offer is accepted by the borrower.

profit requirements ($PLE^\bullet(q_{L,\text{A-type borrower}})$ and $PLE^\bullet(q_{L,\text{N-type borrower}}) \geq \overline{P_L}$) guarantee that each individual contract is profitable for the lender.[33] If each individual contract is profitable for the lender, the lender's ex-ante participation constraint $PLE^\bullet \geq \overline{P_L}$ is always satisfied.

If the lender offers both borrower types a contract, the optimal borrower type-dependent proportional return participations are

$$q^*_{L,\text{A-type borrower}} = 1 - \frac{P_B^+ + nc}{\mu}; \quad q^*_{L,\text{N-type borrower}} = 1 - \frac{nc}{\mu}.$$

Therefore, the lender can extract $PLE^\bullet(q^*_{L,\text{A-type borrower}}) = \mu - 1 - P_B^+ - mc - 2nc$ ($PLE^\bullet(q^*_{L,\text{N-type borrower}}) = \mu - 1 - mc - 2nc$) when contracting with the A-type (N-type) borrower. However, if the lender's profit extractable from financing the A-type borrower does not satisfy the lender's respective contract offer condition (LCOC(AB)):[34] $PLE^\bullet(q_{L,\text{A-type borrower}}) \geq \overline{P_L}$) the lender at most offers an acceptable contract to the N-type borrower[35]

$$q^*_{L,\text{A-type borrower}} = -; \quad q^*_{L,\text{N-type borrower}} = 1 - \frac{nc}{\mu}.$$

$q^*_{L,\text{A-type borrower}} = -$ indicates that the lender is not willing to offer the A-type borrower a proportional return participation which is acceptable for the latter, i. e. satisfies the A-type borrower's participation constraint. Hence, the A-type borrower rejects such a conditional offer since otherwise negotiation costs would occur for both parties, reducing the lender's and the borrower's outside option due to no contract agreement anyway. However, despite no ex-ante informational asymmetry in case of contract agreement, i. e. only the N-type borrower accepts the offered conditions, negotiation costs occur since the lender and the borrower validate the other's type (see Assumption 4').

If additionally the lender's profit extractable from financing the N-type borrower does not satisfy the respective contract offer condition (LCOC(NB)):[36]

---

[33] Whether a contract is profitable for the lender depends on the relation between the lender's extractable profit $PLE^\bullet$ and his outside option. When the extractable profit exceeds, despite private placement's negotiation costs, the lender's outside option then the contract is profitable. Otherwise the lender is better off offering unacceptable contract conditions which the borrower rejects and both are left with their outside option (backward induction). Negotiation costs neither arise for the lender nor for the borrower.

[34] LCOC(AB) stands for the lender's contract offer condition to the A-type borrower.

[35] Since the A-type borrower possesses a positive outside option, the profit the lender can extract from project financing is larger when contracting with the N instead of the A-type borrower.

[36] LCOC(NB) refers to the lender's contract offer condition when negotiating with the N-type borrower.

$PLE^{\bullet}(q_{L,\text{N-type borrower}}) \geq \overline{P}_L)$, the lender refuses to offer the A and the N-type borrower acceptable conditions. Private equity financing is always unfeasible. Therefore, the only distinction which emerges for private equity financing is whether the LCOC(AB) is satisfied or not (LCOC(AB)s/ns).

Table 4.9: Offer Dependent Expected Profits and Feasible Contracting Constraints of a Private Equity Contract (EM, Lender Optimization)

• Lender offers both borrower types a contract $(E^{\bullet(2)})$	
$PLE_L^{\bullet(2)}$	$\mu - 1 - \widehat{\lambda}_B P_B^+ - mc - 2nc$
$PBE_L^{\bullet(2)}$	$\begin{cases} P_B^+\|0 \text{ if } E^{\bullet(2)} \text{ only feasible for N-type lender} \\ P_B^+\|0 \text{ if } E^{\bullet(2)} \text{ feasible for A\&N-type lender} \end{cases}$
LCOC(AB)s/ns[37]	$\mu \geq 1 + P_B^+ + mc + 2nc + \overline{P}_L$
• Lender only offers N-type borrower a contract $(E^{\bullet(1)})$	
$PLE_L^{\bullet(1)}$	$(1 - \widehat{\lambda}_B)(\mu - 1 - mc - 2nc) + \widehat{\lambda}_B \overline{P}_L$
$PBE_L^{\bullet(1)}$	$\begin{cases} P_B^+\|0 \text{ if } E^{\bullet(1)} \text{ only feasible for N-type lender} \\ P_B^+\|0 \text{ if } E^{\bullet(1)} \text{ feasible for A\&N-type lender} \end{cases}$
LCOC(AB)s/ns	$\mu < 1 + P_B^+ + mc + 2nc + \overline{P}_L$
LCOC(NB)s[38]	$\mu \geq 1 + mc + 2nc + \overline{P}_L$

Note: The lender's as well as the borrower's expected profits are stated given private equity financing. The lender has the power to determine the contract's proportional return participation (q). The lender's expected profit $PLE_L^{\bullet}$ and the respective conditions depend on whether the lender just offers the N-type borrower a contract or both borrower types. The borrower's expected profit $PBE_L^{\bullet}$ is additionally conditioned on the lender type for which $E_L^{\bullet}$ is feasible. Since the BPC is borrower type-dependent, $PBE_L^{\bullet}$ is also stated borrower type-dependent ($PBE_L^{\bullet}$ for A-type borrower| $PBE_L^{\bullet}$ for N-type borrower).

Equivalent to a public equity offering, see Section 4.2.1, in a private placement the lender's expectation about the borrower's type, $\widehat{\lambda}_B$, or respectively

---

[37] If the LCOC(AB) is satisfied, the profit extractable from financing the A-type borrower exceeds the lender's outside option which implies that the lender's participation constraint LPC is automatically satisfied.

[38] The LCOC(NB)s indicates the feasible contracting requirement which guarantees that contracting with the N-type borrower is profitable for the lender. If the LCOC(NB) is satisfied (LCOC(NB)s) the lender's LPC for $E_L^{\bullet(1)}$ is automatically satisfied.

the borrower's expectation about the lender's expectation is highly significant to evaluate this ct/pm-constellation. Whether the lender offers both borrower types a contract in private equity financing is indicated by an additional superscript. We refer to private equity financing when the lender offers both borrower types acceptable contract conditions as $E^{\bullet(2)}$. $E^{\bullet(1)}$ stands for private equity financing where a contract is only offered to the N-type borrower.

4.2.3.2 Borrower Optimization

Given private equity financing, the borrower as the condition optimizer can condition the lender's proportional return participation (q) on the lender's, i. e. the contractual partner's, type ($q^*_{B,\text{A-type lender}}, q^*_{B,\text{N-type lender}}$). The lender's type-dependent conditions will satisfy $q^*_{B,\text{A-type lender}} \geq q^*_{B,\text{N-type lender}}$ due to the lender's type-dependent outside option (LPC: $PLE^{\bullet} \geq \overline{P}_L$).

The borrower's optimization problem for private equity financing is

$$\max_{\substack{q_{B,\text{A-type lender}} \\ q_{B,\text{N-type lender}}}} PBE^{\bullet}(q) = \widehat{\lambda}_L PBE^{\bullet}(q_{B,\text{A-type lender}})$$
$$+(1 - \widehat{\lambda}_L)PBE^{\bullet}(q_{B,\text{N-type lender}})$$

$$s.t. \ PLE^{\bullet}(q_{B,\text{A-type lender}}) \geq P^+_L, \ PLE^{\bullet}(q_{B,\text{N-type lender}}) \geq 0$$
$$\wedge \ PBE^{\bullet}(q_{B,\text{A-type lender}}) \geq \overline{P}_B, \ PBE^{\bullet}(q_{B,\text{N-type lender}}) \geq \overline{P}_B$$

and implies the following lender type-dependent optimal proportional return participations

$$q^*_{B,\text{A-type lender}} = \frac{1 + mc + nc + P^+_L}{\mu}; \quad q^*_{B,\text{N-type lender}} = \frac{1 + mc + nc}{\mu}$$

when both BCOC are satisfied.[39] If the BCOC(AL): $PBE^{\bullet}(q^*_{B,\text{A-type lender}}) \geq \overline{P}_B$ is not satisfied,[40] the borrower at most offers a contract to the N-type lender

$$q^*_{B,\text{A-type lender}} = -; \quad q^*_{B,\text{N-type lender}} = \frac{1 + mc + nc}{\mu} \ .$$

The respective expected profits and feasible contracting constraints are as follows:

---

[39] From the A-type (N-type) lender the borrower can extract $PBE^{\circ}(q_{B,\text{A-type lender}}) = \mu - 1 - mc - 2nc - P^+_L$ ($PBE^{\circ}(q_{B,\text{N-type lender}}) = \mu - 1 - mc - 2nc$).

[40] The BCOC(AL) refers to the borrower's contract offer condition when negotiating with the A-type lender.

Table 4.10: Offer Dependent Expected Profits and Feasible Contracting Constraints of a Private Equity Contract (EM, Borrower Optimization)

• Borrower offers both lender types a contract ($E^{\bullet(2)}$)	
$PBE_B^{\bullet(2)}$	$\mu - 1 - \widehat{\lambda}_L P_L^+ - mc - 2nc$
$PLE_B^{\bullet(2)}$	$\begin{cases} P_L^+ \mid 0 \text{ if } E^{\bullet(2)} \text{ only feasible for N-type borrower} \\ P_L^+ \mid 0 \text{ if } E^{\bullet(2)} \text{ feasible for A\&N-type borrower} \end{cases}$
BCOC(AL)s/ns[41]	$\mu \geq 1 + P_L^+ + mc + 2nc + \overline{P}_B$
• Borrower only offers N-type lender a contract ($E^{\bullet(1)}$)	
$PBE_B^{\bullet(1)}$	$(1 - \widehat{\lambda}_L)(\mu - 1 - mc - 2nc) + \widehat{\lambda}_L \overline{P}_B$
$PLE_B^{\bullet(1)}$	$\begin{cases} P_L^+ \mid 0 \text{ if } E^{\bullet(1)} \text{ only feasible for N-type borrower} \\ P_L^+ \mid 0 \text{ if } E^{\bullet(1)} \text{ feasible for A\&N-type borrower} \end{cases}$
BCOC(AL)s/ns	$\mu < 1 + P_L^+ + mc + 2nc + \overline{P}_B$
BCOC(NL)s	$\mu \geq 1 + mc + 2nc + \overline{P}_B$

Note: The borrower's as well as the lender's expected profits are stated given private equity financing. The borrower possesses the right to set the contract's proportional return participation (q). The borrower's expected profit $PBE_B^{\bullet}$ and the respective conditions depend on whether the borrower just offers the N-type lender a contract or both lender types. The lender's expected profit $PLE_B^{\bullet}$ is additionally conditioned on the borrower type for which $E_B^{\bullet}$ is feasible. Since the LPC is lender type-dependent, $PLE_B^{\bullet}$ is stated lender type-dependent ($PLE_B^{\bullet}$ for A-type lender| $PLE_B^{\bullet}$ for N-type lender).

4.2.4 Private Debt ($D^{\bullet}$)

In this section the lender's as well as the borrower's optimization behavior given private debt financing is examined. At first, in Section 4.2.4.1 the power to set the contract conditions is given to the lender, while in Section 4.2.4.1 the borrower possesses this right.

4.2.4.1 Lender Optimization

Given a private debt placement, the lender (condition optimizer) can set contract conditions without incorporating any placement risk.[42] The lender can

---

[41] Since the BCOC(AL) guarantees that contracting with the N and even with the A-type lender is profitable for the borrower, the BPC is automatically satisfied.

[42] No placement risk arises since the lender knows the borrower's type and hence his outside option. Therefore, the lender can choose contracts' conditions which just

set contract conditions borrower type-dependent, while nc arises for both sides. Due to the positive outside option of the A-type borrower, the optimal conditions must satisfy $h^*_{L,\text{A-type borrower}} \le h^*_{L,\text{N-type borrower}}$.

The lender's optimization problem for private debt is

$$\max_{\substack{h_{L,\text{A-type borrower}} \\ h_{L,\text{N-type borrower}}}} PLD^\bullet(h) = \widehat{\lambda}_B PLD^\bullet(h_{L,\text{A-type borrower}})$$
$$+(1 - \widehat{\lambda}_B)PLD^\bullet(h_{L,\text{N-type borrower}})$$

$$s.t.\ PBD^\bullet(h_{L,\text{A-type borrower}}) \ge P_B^+,\ \ PBD^\bullet(h_{L,\text{N-type borrower}}) \ge 0$$
$$\wedge\ PLD^\bullet(h_{L,\text{A-type borrower}}) \ge \overline{P}_L,\ \ PLD^\bullet(h_{L,\text{N-type borrower}}) \ge \overline{P}_L\ .$$

Since in comparison to public offerings private debt offerings have the advantage of borrower type-dependent repayment obligations, three (instead of two) debt financing cases have to be distinguished:[43] Both repayment obligations $(h_{L,\text{A-type borrower}}, h_{L,\text{N-type borrower}})$ result in riskless debt financing, i. e. do not offer a (harmful) risk-shifting opportunity to the borrower (Case I);[44] $h_{L,\text{N-type borrower}}$ results in risky and $h_{L,\text{A-type borrower}}$ in riskless debt financing (Case II); while finally $h_{L,\text{A-type borrower}}$ and $h_{L,\text{N-type borrower}}$ result in risky debt (Case III). Now either three (a-c) or even four (a-d) subcases occur depending on whether the borrower's positive informational rent due to interim and ex-post uncertainty is sufficient to compensate for his profit alternative $\overline{P}_B$ as well as for nc or not. Figure 4.3 illustrates all potential cases.

---

satisfy the borrower's participation constraint. Of course, the lender can refrain from doing so and hence his offer is not accepted by the contractual partner, but in this case the lender deliberately chooses this option to prevent contract agreement. The lender faces no placement risk since the outcome of his offer (contract agreement or not) is always uniquely determined.

[43] Three and not four cases have to be distinguished since, as mentioned above, the repayment obligations have to satisfy $h^*_{L,\text{A-type borrower}} \le h^*_{L,\text{N-type borrower}}$ which rules out the possibility that the N-type borrower receives a riskless and the A-type a risky debt offer.

[44] A risk-shifting opportunity is economically harmful if the borrower can vary the project risk in a way that affects the borrower's default risk on his obligation to the lender since each default is associated with agency costs.

Fig. 4.3: Subcases in Private Debt Financing (EM, Lender Optimization)

CDr/nr	Requirements for the borrower's repayment obligation	Borrower's informational rent from interim and ex-post uncertainty	BPCb/nb	Requirements for the borrower's participation constraint	Case
I	$h_{L,\text{A-type borrower}} \leq y_{min}$	$IR_{\text{A-type borrower}} = \zeta_{max}\mu$	a	$\zeta_{max}\mu \geq P_B^+ + nc$	Case Ia
	$h_{L,\text{N-type borrower}} \leq y_{min}$	$IR_{\text{N-type borrower}} = \zeta_{max}\mu$	b	$nc \leq \zeta_{max}\mu < P_B^+ + nc$	Case Ib
			c	$\zeta_{max}\mu < nc$	Case Ic
II	$h_{L,\text{A-type borrower}} \leq y_{min}$	$IR_{\text{A-type borrower}} = \zeta_{max}\mu$	a	$\zeta_{max}\mu \geq P_B^+ + nc \wedge \dfrac{vc^2}{4\zeta_{max}\mu} \geq nc$	Case IIa
	$h_{L,\text{N-type borrower}} > y_{min}$	$IR_{\text{N-type borrower}} = \dfrac{vc^2}{4\zeta_{max}\mu}$	b	$\zeta_{max}\mu < P_B^+ + nc \wedge \dfrac{vc^2}{4\zeta_{max}\mu} \geq nc$	Case IIb
			c	$\zeta_{max}\mu \geq P_B^+ + nc \wedge \dfrac{vc^2}{4\zeta_{max}\mu} < nc$	Case IIc
			d	$\zeta_{max}\mu < P_B^+ + nc \wedge \dfrac{vc^2}{4\zeta_{max}\mu} < nc$	Case IId
III	$h_{L,\text{A-type borrower}} > y_{min}$	$IR_{\text{A-type borrower}} = \dfrac{vc^2}{4\zeta_{max}\mu}$	a	$\dfrac{vc^2}{4\zeta_{max}\mu} \geq P_B^+ + nc$	Case IIIa
	$h_{L,\text{N-type borrower}} > y_{min}$	$IR_{\text{N-type borrower}} = \dfrac{vc^2}{4\zeta_{max}\mu}$	b	$nc \leq \dfrac{vc^2}{4\zeta_{max}\mu} < P_B^+ + nc$	Case IIIb
			c	$\dfrac{vc^2}{4\zeta_{max}\mu} < nc$	Case IIIc

Hence, the lender might not always have to bear the full additional agency costs related to a private placement ($2nc$) since at least parts of the negotiation costs occurring for the borrower are covered by the latter's informational rent.[45]

- *Case I – Riskless debt:* $h_{L,\text{A-type borrower}}, h_{L,\text{N-type borrower}} \leq y_{min}\big|_{\zeta=\zeta_{max}}$

  When both borrower types are financed with riskless debt three subcases have to be distinguished since both borrower types receive the same informational rent due to interim and ex-post uncertainty ($IR(D_I^\circ) = \zeta_{max}\mu$).

  - *Ia – No BPC binding:* $\zeta_{max}\mu \geq P_B^+ + nc$

    If the BPC of both borrower types is not binding since both are automatically satisfied due to the borrower's informational rent from interim and ex-post uncertainty, no difficulty in the lender's contract optimization occurs because of the ex-ante uncertainty. Hence, the optimal borrower type-dependent repayment obligations are identical:

    $$h^*_{L,Ia,\text{A-type borrower}} = y_{min}\big|_{\zeta=\zeta_{max}} = \mu(1 - \zeta_{max})$$

    $$h^*_{L,Ia,\text{N-type borrower}} = y_{min}\big|_{\zeta=\zeta_{max}} = \mu(1 - \zeta_{max}) \ .$$

    Moreover, since the repayment obligations are identical, the expected profits the lender can extract from the A- respective N-type borrower coincide which implies that the lender either offers both borrower types to finance their project or non.[46] The lender's and the borrower's expected profits are as follows:

---

[45] The borrower's informational rent might even be sufficient to compensate for the borrower's total negotiation costs. Hence, the borrower does not need compensation for his private placement's negotiation costs. Unfortunately for the lender, any remaining rent is again uncontractible since the costs to extract this rent from the borrower exceed the gains. Therefore, the lender has to bear his negotiation costs by himself and can not pass the costs on to the borrower.

[46] The case distinction that the lender only finances the N-type borrower's project and not the A-type's vanishes since the A-type borrower's positive outside option does not affect the negotiated contract conditions in this circumstances.

Table 4.11: Expected Profits and Feasible Contracting Constraints of a Private Debt Contract – Case Ia (EM, Lender Optimization)

• Lender offers both borrower types a contract $(D^{\bullet(2)})$	
$PLD_{L,Ia}^{\bullet(2)}$	$\mu(1 - \zeta_{max}) - 1 - nc$
$PBD_{L,Ia}^{\bullet(2)}$	$\begin{cases} (1 - \hat{\lambda}_L)(\zeta_{max}\mu - nc) + \hat{\lambda}_L P_B^+ \mid & \text{if } D^{\bullet(2)} \text{ only feasible} \\ (1 - \hat{\lambda}_L)(\zeta_{max}\mu - nc) & \text{for N-type lender} \\ \\ \zeta_{max}\mu - nc \mid \zeta_{max}\mu - nc & \text{if } D^{\bullet(2)} \text{ feasible for} \\ & \text{A\&N-type lender} \end{cases}$
BPCb/nb	$\zeta_{max} \geq \dfrac{P_B^+ + nc}{\mu}$
LCOC(AB)s/ns	$\mu(1 - \zeta_{max}) \geq 1 + \overline{P}_L + nc$

Note: The lender's as well as the borrower's expected profits are stated given private debt financing. The lender has the power to determine the contract's repayment obligation (h). The borrower's expected profit $PBD_{L,Ia}^{\bullet}$ is conditioned on the lender type for which $D_{L,Ia}^{\bullet}$ is feasible. Since the BPC is borrower type-dependent, $PBD_{L,Ia}^{\bullet}$ is also stated borrower type-dependent ($PBD_{L,Ia}^{\bullet}$ for A-type borrower| $PBD_{L,Ia}^{\bullet}$ for N-type borrower).

- *Ib – A-type borrower's BPC binding, N-type's not:* $nc \leq \zeta_{max}\mu < P_B^+ + nc$

    Assuming alternatively that the A-type borrower's participation constraint $(PB \geq P_B^+)$ is not automatically satisfied ($\zeta_{max}\mu < P_B^+ + nc$) (A-type's BPC binding) while the N-type's participation constraint $(PB \geq 0)$ remains satisfied ($\zeta_{max}\mu \geq nc$) (N-type's BPC not binding), the optimal borrower type-dependent repayment obligations differ:

    $$h_{L,Ib,\text{A-type borrower}}^* = \mu - (P_B^+ + nc)$$

    $$h_{L,Ib,\text{N-type borrower}}^* = y_{min}\big|_{\zeta=\zeta_{max}} = \mu(1 - \zeta_{max})$$

    The lender cannot demand the repayment obligation of the N-type borrower $(h_{L,Ib,\text{N-type borrower}}^*)$ from the A-type $(h_{L,Ib,\text{A-type borrower}}^*)$ since the latter's participation constraint would not be satisfied. The lender has to leave an extra surplus to the A-type borrower to satisfy his participation constraint, which implies $h_{L,Ib,\text{N-type borrower}}^* > h_{L,Ib,\text{A-type borrower}}^*$. Since the optimal repayments differ, the extractable profits differ too, requiring a distinction whether the lender is willing to offer both borrower types a contract or only the N-type.[47]

---

[47] From a contract with the A-type (N-type) borrower the lender can extract $PLD^{\bullet}(h_{L,Ib,\text{A-type borrower}}^*) = \mu - 1 - P_B^+ - 2nc$ ($PLD^{\bullet}(h_{L,Ib,\text{N-type borrower}}^*) = \mu(1 - \zeta_{max}) - 1 - nc$).

Table 4.12: Offer Dependent Expected Profits and Feasible Contracting Constraints of a Private Debt Contract – Case Ib (EM, Lender Optimization)

• Lender offers both borrower types a contract $(D^{\bullet(2)})$	
$PLD_{L,Ib}^{\bullet(2)}$	$\mu - 1 - nc - \widehat{\lambda}_B(P_B^+ + nc) - (1 - \widehat{\lambda}_B)\mu\zeta_{max}$
$PBD_{L,Ib}^{\bullet(2)}$	$\begin{cases} P_B^+ \mid (1 - \widehat{\lambda}_L)(\zeta_{max}\mu - nc) & \text{if } D^{\bullet(2)} \text{ only feasible for} \\ & \text{N-type lender} \\ P_B^+ \mid \zeta_{max}\mu - nc & \text{if } D^{\bullet(2)} \text{ feasible for} \\ & \text{A\&N-type lender} \end{cases}$
BPCb/nb	$\dfrac{nc}{\mu} \leq \zeta_{max} < \dfrac{P_B^+ + nc}{\mu}$
LCOC(AB)s/ns[48]	$\mu \geq 1 + P_B^+ + 2nc + \overline{P}_L$
• Lender only offers N-type borrower a contract $(D^{\bullet(1)})$	
$PLD_{L,Ib}^{\bullet(1)}$	$(1 - \widehat{\lambda}_B)(\mu(1 - \zeta_{max}) - 1 - nc) + \widehat{\lambda}_B\overline{P}_L$
$PBD_{L,Ib}^{\bullet(1)}$	$\begin{cases} P_B^+ \mid (1 - \widehat{\lambda}_L)(\zeta_{max}\mu - nc) & \text{if } D^{\bullet(1)} \text{ only feasible for} \\ & \text{N-type lender} \\ P_B^+ \mid \zeta_{max}\mu - nc & \text{if } D^{\bullet(1)} \text{ feasible for} \\ & \text{A\&N-type lender} \end{cases}$
BPCb/nb	$\dfrac{nc}{\mu} \leq \zeta_{max} < \dfrac{P_B^+ + nc}{\mu}$
LCOC(AB)s/ns	$\mu < 1 + P_B^+ + 2nc + \overline{P}_L$
LCOC(NB)s	$\mu(1 - \zeta_{max}) \geq 1 + nc + \overline{P}_L$

Note: The lender's as well as the borrower's expected profits are stated given private debt financing. The lender has the power to determine the contract's repayment obligation (h). The lender's expected profit $PLD_{L,Ia}^{\bullet}$ and the respective conditions depend on whether the lender just offers the N-type borrower a contract or both borrower types. The borrower's expected profit $PBD_{L,Ia}^{\bullet}$ is additionally conditioned on the lender type for which $D_{L,Ia}^{\bullet}$ is feasible. Since the BPC is borrower type-dependent, $PBD_{L,Ia}^{\bullet}$ is also stated borrower type-dependent ($PBD_{L,Ia}^{\bullet}$ for A-type borrower| $PBD_{L,Ia}^{\bullet}$ for N-type borrower).

---

[48] Due to the LCOC(AB), the LPC is automatically satisfied.

- *Ic – A&N-type borrowers' BPC binding:* $\zeta_{max}\mu < nc$

Finally, when both borrower participation constraints are binding for riskless debt financing the optimal repayment obligations are

$$h^*_{L,Ic,\text{A-type borrower}} = \mu - (P_B^+ + nc)$$

$$h^*_{L,Ic,\text{N-type borrower}} = \mu - nc$$

when both LCOC are satisfied. The repayment obligations differ due to the borrower's type-dependent outside option. Hence, the lender's extractable profits differ which implies that the lender either offers both borrower types a contract (LCOC(AB) satisfied) or only the N-type (LCOC(AB) not satisfied).[49] The respective expected profits are stated in the following table.

Table 4.13: Offer Dependent Expected Profits and Feasible Contracting Constraints of a Private Debt Contract – Case Ic (EM, Lender Optimization)

• Lender offers both borrower types a contract ($D^{\bullet(2)}$)	
$PLD^{\bullet(2)}_{L,Ic}$	$\mu - 1 - \widehat{\lambda}_B P_B^+ - 2nc$
$PBD^{\bullet(2)}_{L,Ic}$	$\begin{cases} P_B^+\vert 0 \text{ if } D^{\bullet(2)} \text{ only feasible for N-type lender} \\ P_B^+\vert 0 \text{ if } D^{\bullet(2)} \text{ feasible for A&N-type lender} \end{cases}$
BPCb/nb	$\zeta_{max} < \dfrac{nc}{\mu}$
LCOC(AB)s/ns[50]	$\mu \geq 1 + P_B^+ + 2nc + \overline{P}_L$
• Lender only offers N-type borrower a contract ($D^{\bullet(1)}$)	
$PLD^{\bullet(1)}_{L,Ic}$	$(1 - \widehat{\lambda}_B)(\mu - 1 - 2nc) + \widehat{\lambda}_B \overline{P}_L$
$PBD^{\bullet(1)}_{L,Ic}$	$\begin{cases} P_B^+\vert 0 \text{ if } D^{\bullet(1)} \text{ only feasible for N-type lender} \\ P_B^+\vert 0 \text{ if } D^{\bullet(1)} \text{ feasible for A&N-type lender} \end{cases}$
BPCb/nb	$\zeta_{max} < \dfrac{nc}{\mu}$
LCOC(AB)s/ns	$\mu < 1 + P_B^+ + 2nc + \overline{P}_L$
LCOC(NB)s	$\mu \geq 1 + 2nc + \overline{P}_L$

Note: The structure of the table is equivalent to the structure of Table 4.12.

---

[49] The lender can extract $PLD^\bullet(h^*_{L,Ic,\text{A-type borrower}}) = \mu - 1 - P_B^+ - 2nc$ when contracting with the A-type borrower and $PLD^\bullet(h^*_{L,Ic,\text{N-type borrower}}) = \mu - 1 - 2nc$ from contract agreement with the N-type.

[50] The LCOC(AB)s/ns guarantees that the LPC is satisfied.

- *Case II – Riskless and risky debt:* $h_{L,\text{A-type borrower}} \leq y_{min}\big|_{\zeta=\zeta_{max}} < h_{L,\text{N-type borrower}}$

Alternatively to Case I where both borrower types receive riskless debt offers, it is possible that the N-type borrower receives a risky debt offer ($h > y_{min}\big|_{\zeta=\zeta_{max}}$) while the A-type still obtains riskless debt ($h \leq y_{min}\big|_{\zeta=\zeta_{max}}$) (Case II). Private debt financing Case II is feasible in the lender optimization due to two observations. Firstly, the lender tries to maximize his expected profit by optimizing the borrower's repayment obligation which can imply $h > y_{min}\big|_{\zeta=\zeta_{max}}$ (risky debt). Secondly, the lender's optimal repayment obligation for the A-type borrower ($h_{L,\text{A-type borrower}}^*$) can not be larger than the N-type's obligation ($h_{L,\text{N-type borrower}}^*$) due to the former's positive outside option. For the borrower's expected profit only the absolute hight of the repayment obligation (h) is relevant. A higher repayment obligation causes a lower profit and vice versa. However, this is not true for the lender's profit since the lender has to bear the agency costs in the default case. The lender's expected profit is not a monotonically increasing function of the borrower's repayment obligation. The lender's expected profit from financing the A-type borrower (with riskless debt) might even exceed the lender's profit from financing the N-type (with risky debt) despite $h_{L,\text{A-type borrower}}^* < h_{L,\text{N-type borrower}}^*$. The reason for this possibility is that risky debt causes agency costs while riskless debt does not. Therefore, in Case II in comparison to Case I, we additionally have to investigate whether it is optimal for the lender to offer a contract only to the A-type borrower and not to the N-type.[51] Fortunately, due to Case I we know that if it is profitable for the lender to finance the A-type borrower's project with riskless debt it is also always profitable to finance the N-type with riskless debt too. Hence, we can exclude conceivable constellations where the lender prefers to finance only the A-type borrower with riskless debt and not the N-type with risky debt since from Case I we know that it is always profitable to offer the N-type borrower (at least) a riskless debt contract when the A-type receives one. In Case II again, we only have to examine two possible scenarios: Contract offer to both borrower types ($D^{\bullet(2)}$) or only to the N-type ($D^{\bullet(1)}$).

---

[51] In Case I we only distinguish two cases: Contract offer to both borrower types ($D^{\bullet(2)}$), or only to the N-type ($D^{\bullet(1)}$). No other case distinction is required since both borrower types receive a riskless debt offer. Hence, agency costs neither arise when contracting with the A-type nor with the N-type borrower. Therefore, the lender's expected profit is a monotonically increasing function of the borrower's repayment obligation which implies together with $h_{L,\text{A-type borrower}}^* < h_{L,\text{N-type borrower}}^*$ that the lender's profit from financing the N-type borrower in Case I always exceeds the profit extractable from financing the A-type. Consequently, if the lender refuses to finance a borrower's project, he refuses the A-type borrower's and not the N-type's.

Since the borrower's "old" informational rent differs depending on the debt case ($IR(D_I^\circ) = \zeta_{max}\mu$ for riskless debt and $IR(D_{II}^\circ) = \dfrac{vc^2}{4\zeta_{max}\mu}$ for risky debt), four subcases have to be distinguished:

The A-type borrower receives a riskless debt offer ($h_{L,\text{A-type borrower}} \leq y_{min}$) while a risky contract ($h_{L,\text{N-type borrower}} > y_{min}$) is offered to the N-type borrower and

– Case IIa
  the corresponding informational rents already satisfy the A and the N-type borrower's participation constraint.

– Case IIb
  the A-type borrower's informational rent does not jet satisfy his participation constraint, but the N-type borrower's informational rent is sufficient to satisfy the latter's constraint.

– Case IIc
  the A-type borrower's informational rent does satisfy his participation constraint already while the N-type borrower's rent is insufficient to satisfy the latter's constraint.

– Case IId
  the A and the N-type borrowers' informational rents are insufficient to satisfy their participation constraints automatically.

– *IIa – No BPC binding:* $\zeta_{max}\mu \geq P_B^+ + nc \,\wedge\, \dfrac{vc^2}{4\zeta_{max}\mu} \geq nc$

In case both borrower participation constraints are automatically satisfied, the optimal borrower type-dependent repayment obligations are

$$h_{L,IIa,\text{A-type borrower}}^* = y_{min}\Big|_{\zeta=\zeta_{max}} = \mu(1 - \zeta_{max})$$

$$h_{L,IIa,\text{N-type borrower}}^* = \mu(1 + \zeta_{max}) - vc \,.$$

The N-type borrower receives a risky debt offer while the A-type borrower can finance the project with riskless debt.[52] Hence, the lender's and the borrower's expected profits are:[53]

---

[52] The lender's expected profit when contracting with the A-type borrower is $PLD^\bullet(h_{L,IIa,\text{A-type borrower}}^*) = \mu(1 - \zeta_{max}) - 1 - nc$ (riskless debt). On the other hand contracting with the N-type results in $PLD^\bullet(h_{L,IIa,\text{N-type borrower}}^*) = \mu + \dfrac{vc^2}{4\zeta_{max}\mu} - vc - 1 - nc$ (risky debt).

[53] In case that the lender only offers the N-type borrower a contract, the A-type's BPCb/nb requirement ($\zeta_{max}\mu \geq P_B^+ + nc$ vs. $\zeta_{max}\mu < P_B^+ + nc$) becomes irrelevant for the lender's expected profits and hence the lender's expected profit in Case IIa and IIb, respectively IIc and IId, coincide. However, the distinction according to the A-type's BPC is still relevant to determine the feasible contracting areas for these cases and hence we stick to the following structure of Case II.

Table 4.14: Offer Dependent Expected Profits and Feasible Contracting Constraints of a Private Debt Contract – Case IIa (EM, Lender Optimization)

• Lender offers both borrower types a contract ($D^{\bullet(2)}$)	
$PLD_{L,IIa}^{\bullet(2)}$	$\widehat{\lambda}_B \mu (1 - \zeta_{max}) + (1 - \widehat{\lambda}_B)[\mu + \dfrac{vc^2}{4\zeta_{max}\mu} - vc] - 1 - nc$
$PBD_{L,IIa}^{\bullet(2)}$	$\begin{cases} (1 - \widehat{\lambda}_L)(\zeta_{max}\mu - nc) + \widehat{\lambda}_L P_B^+\| & \text{if } D^{\bullet(2)} \text{ only} \\ & \text{feasible for} \\ (1 - \widehat{\lambda}_L)(\dfrac{vc^2}{4\zeta_{max}\mu} - nc) & \text{N-type lender} \\ \zeta_{max}\mu - nc\|\dfrac{vc^2}{4\zeta_{max}\mu} - nc & \text{if } D^{\bullet(2)} \text{ feasible for} \\ & \text{A\&N-type lender} \end{cases}$
CDr/nr[54]	$\zeta_{max} > \dfrac{vc}{2\mu}$
BPCb/nb	$\dfrac{P_B^+ + nc}{\mu} \le \zeta_{max} \le \dfrac{vc^2}{4\mu nc}$
LCOC(AB)s/ns[55]	$\mu(1 - \zeta_{max}) \ge 1 + nc + \overline{P}_L$
• Lender only offers N-type borrower a contract ($D^{\bullet(1)}$)	
$PLD_{L,IIa}^{\bullet(1)}$	$(1 - \widehat{\lambda}_B)(\mu + \dfrac{vc^2}{4\zeta_{max}\mu} - vc - 1 - nc) + \widehat{\lambda}_B \overline{P}_L$
$PBD_{L,IIa}^{\bullet(1)}$	$\begin{cases} P_B^+\|(1 - \widehat{\lambda}_L)(\dfrac{vc^2}{4\zeta_{max}\mu} - nc) & \text{if } D^{\bullet(1)} \text{ only feasible} \\ & \text{for N-type lender} \\ P_B^+\|(\dfrac{vc^2}{4\zeta_{max}\mu} - nc) & \text{if } D^{\bullet(1)} \text{ feasible for} \\ & \text{A\&N-type lender} \end{cases}$
CDr/nr	$\zeta_{max} > \dfrac{vc}{2\mu}$
BPCb/nb	$\dfrac{P_B^+ + nc}{\mu} \le \zeta_{max} \le \dfrac{vc^2}{4\mu nc}$
LCOC(AB)s/ns	$\mu(1 - \zeta_{max}) < 1 + nc + \overline{P}_L$
LCOC(NB)s	$\mu \ge 1 + vc + nc + \overline{P}_L \vee$ $\zeta_{max} \le \dfrac{vc^2}{4\mu(1 + vc + nc + \overline{P}_L - \mu)}$

Note: The structure of the table is equivalent to the structure of Table 4.12.

- *IIb – Only A-type borrower's BPC binding, N-type's not:* $\zeta_{max}\mu < P_B^+ + nc \wedge \dfrac{vc^2}{4\zeta_{max}\mu} \ge nc$

In this case the optimal required repayments are

---

[54] The CDr/nr guarantees that the N-type borrower receives a risky debt offer.

[55] The LCOC(AB)s/ns guarantees that the LPC is satisfied.

$$h^*_{L,IIb,\text{A-type borrower}} = \mu - (P^+_B + nc)$$
$$h^*_{L,IIb,\text{N-type borrower}} = \mu(1 + \zeta_{max}) - vc$$

with expected profits $PLD^\bullet(h^*_{L,IIb,\text{A-type borrower}}) = \mu - 1 - P^+_B - 2nc$

and $PLD^\bullet(h^*_{L,IIb,\text{N-type borrower}}) = \mu + \dfrac{vc^2}{4\zeta_{max}\mu} - vc - 1 - nc$. The

lender's respective unconditional profits are:

Table 4.15: Offer Dependent Expected Profits and Feasible Contracting Constraints of a Private Debt Contract – Case IIb (EM, Lender Optimization)

• Lender offers both borrower types a contract $(D^{\bullet(2)})$	
$PLD^{\bullet(2)}_{L,IIb}$	$\mu - 1 - nc - \widehat{\lambda}_B(P^+_B + nc) - (1 - \widehat{\lambda}_B)(vc - \dfrac{vc^2}{4\zeta_{max}\mu})$
$PBD^{\bullet(2)}_{L,IIb}$	$\begin{cases} P^+_B \mid (1 - \widehat{\lambda}_L)(\dfrac{vc^2}{4\zeta_{max}\mu} - nc) & \text{if } D^{\bullet(2)} \text{ only feasible} \\ & \text{for N-type lender} \\ P^+_B \mid \dfrac{vc^2}{4\zeta_{max}\mu} - nc & \text{if } D^{\bullet(2)} \text{ feasible for} \\ & \text{A\&N-type lender} \end{cases}$
CDr/nr	$\dfrac{vc}{2\mu} < \zeta_{max}$
BPCb/nb	$\zeta_{max} < \dfrac{P^+_B + nc}{\mu} \wedge \zeta_{max} \leq \dfrac{vc^2}{4\mu nc}$
LCOC(AB)s/ns	$\mu \geq 1 + P^+_B + 2nc + \overline{P}_L$
• Lender only offers N-type borrower a contract $(D^{\bullet(1)})$	
$PLD^{\bullet(1)}_{L,IIb}$	$(1 - \widehat{\lambda}_B)(\mu + \dfrac{vc^2}{4\zeta_{max}\mu} - vc - nc - 1) + \widehat{\lambda}_B\overline{P}_L$
$PBD^{\bullet(1)}_{L,IIb}$	$\begin{cases} P^+_B \mid (1 - \widehat{\lambda}_L)(\dfrac{vc^2}{4\zeta_{max}\mu} - nc) & \text{if } D^{\bullet(1)} \text{ only feasible for} \\ & \text{N-type lender} \\ P^+_B \mid (\dfrac{vc^2}{4\zeta_{max}\mu} - nc) & \text{if } D^{\bullet(1)} \text{ feasible for} \\ & \text{A\&N-type lender} \end{cases}$
CDr/nr	$\dfrac{vc}{2\mu} < \zeta_{max}$
BPCb/nb	$\zeta_{max} < \dfrac{P^+_B + nc}{\mu} \wedge \zeta_{max} \leq \dfrac{vc^2}{4\mu nc}$
LCOC(AB)s/ns	$\mu < 1 + P^+_B + 2nc + \overline{P}_L$
LCOC(NB)s	$\mu \geq 1 + nc + \overline{P}_L \vee \zeta_{max} \leq \dfrac{vc^2}{4\mu(1 + vc + nc + \overline{P}_L - \mu)}$

Note: The structure of the table is equivalent to the structure of Table 4.12.

- IIc – *Only N-type borrower's BPC binding, A-type's not:* $\zeta_{max}\mu \geq P_B^+ + nc \wedge \dfrac{vc^2}{4\zeta_{max}\mu} < nc$

For Case IIc the optimal required repayments are

$$h_{L,IIc,\text{A-type borrower}}^* = y_{min}\big|_{\zeta=\zeta_{max}} = \mu(1 - \zeta_{max})$$

$$h_{L,IIc,\text{N-type borrower}}^* = \mu(1 + \zeta_{max}) - 2\sqrt{nc\zeta_{max}\mu}$$

when both LCOCs are satisfied. The lender can extract $PLD^\bullet(h_{L,IIc,\text{A-type borrower}}^*) = \mu(1 - \zeta_{max}) - 1 - nc$ or

$$PLD^\bullet(h_{L,IIc,\text{N-type borrower}}^*) = \mu + vc\sqrt{\dfrac{nc}{\zeta_{max}\mu}} - vc - 2nc - 1 \text{ when}$$

contracting with the A respective N-type borrower. If the LCOC(AB) is violated the lender only offers to finance the N-type borrower.

Table 4.16: Offer Dependent Expected Profits and Feasible Contracting Constraints of a Private Debt Contract – Case IIc (EM, Lender Optimization)

• Lender offers both borrower types a contract $(D^{\bullet(2)})$	
$PLD_{L,IIc}^{\bullet(2)}$	$\widehat{\lambda}_B\mu(1 - \zeta_{max}) + (1 - \widehat{\lambda}_B)[\mu + vc\sqrt{\dfrac{nc}{\zeta_{max}\mu}} - vc - nc]$ $-1 - nc$
$PBD_{L,IIc}^{\bullet(2)}$	$\begin{cases} (1 - \widehat{\lambda}_L)(\zeta_{max}\mu - nc) + \widehat{\lambda}_L P_B^+\|0 & \text{if } D^{\bullet(2)} \text{ only feasible} \\ & \text{for N-type lender} \\ \zeta_{max}\mu - nc\|0 & \text{if } D^{\bullet(2)} \text{ feasible for} \\ & \text{A\&N-type lender} \end{cases}$
CDr/nr	$\zeta_{max} > \dfrac{nc}{\mu}$
BPCb/nb	$\zeta_{max} > \dfrac{vc^2}{4\mu nc} \wedge \zeta_{max} \geq \dfrac{P_B^+ + nc}{\mu}$
LCOC(AB)s/ns	$\mu(1 - \zeta_{max}) \geq 1 + nc + \overline{P}_L$
	table continued on next page

table continued from previous page	
• Lender only offers N-type borrower a contract $(D^{\bullet(1)})$	
$PLD_{L,IIc}^{\bullet(1)}$	$(1-\widehat{\lambda}_B)(\mu+vc\sqrt{\dfrac{nc}{\zeta_{max}\mu}}-vc-2nc-1)+\widehat{\lambda}_B\overline{P}_L$
$PBD_{L,IIc}^{\bullet(1)}$	$\begin{cases} P_B^+\vert 0 \text{ if } D^{\bullet(1)} \text{ only feasible for N-type lender} \\ P_B^+\vert 0 \text{ if } D^{\bullet(1)} \text{ feasible for A\&N-type lender} \end{cases}$
CDr/nr	$\zeta_{max} > \dfrac{nc}{\mu}$
BPCb/nb	$\zeta_{max} > \dfrac{vc^2}{4\mu nc} \wedge \zeta_{max} \geq \dfrac{P_B^+ + nc}{\mu}$
LCOC(AB)s/ns	$\mu(1-\zeta_{max}) < 1 + nc + \overline{P}_L$
LCOC(NB)s	$\mu \geq 1 + vc + 2nc + \overline{P}_L \vee$
	$\zeta_{max} \leq \dfrac{nc}{\mu(1+vc+2nc+\overline{P}_L-\mu)^2}$

Note: The structure of the table is equivalent to the structure of Table 4.12.

- *IId – A&N-type borrowers' BPC binding:* $\zeta_{max}\mu < P_B^+ + nc \wedge \dfrac{vc^2}{4\zeta_{max}\mu} < nc$

  $h_{L,IId,\text{A-type borrower}}^* = \mu - (P_B^+ + nc)$

  $h_{L,IId,\text{N-type borrower}}^* = \mu(1+\zeta_{max}) - 2\sqrt{nc\zeta_{max}\mu}$

Table 4.17: Offer Dependent Expected Profits and Feasible Contracting Constraints of a Private Debt Contract – Case IId (EM, Lender Optimization)

• Lender offers both borrower types a contract $(D^{\bullet(2)})$	
$PLD_{L,IId}^{\bullet(2)}$	$\mu + (1-\widehat{\lambda}_B)vc\sqrt{\dfrac{nc}{\zeta_{max}\mu}} - 1 - 2nc - (1-\widehat{\lambda}_B)vc - \widehat{\lambda}_B P_B^+$
$PBD_{L,IId}^{\bullet(2)}$	$\begin{cases} P_B^+\vert 0 \text{ if } D^{\bullet(2)} \text{ only feasible for N-type lender} \\ P_B^+\vert 0 \text{ if } D^{\bullet(2)} \text{ feasible for A\&N-type lender} \end{cases}$
CDr/nr	$\zeta_{max} > \dfrac{nc}{\mu}$
BPCb/nb	$\dfrac{vc^2}{4\mu nc} < \zeta_{max} < \dfrac{P_B^+ + nc}{\mu}$
LCOC(AB)s/ns	$\mu \geq 1 + P_B^+ + 2nc + \overline{P}_L$
	table continued on next page

table continued from previous page	
• Lender only offers N-type borrower a contract $(D^{\bullet(1)})$	
$PLD_{L,IId}^{\bullet(1)}$	$(1 - \widehat{\lambda}_B)(\mu + vc\sqrt{\dfrac{nc}{\zeta_{max}\mu}} - vc - 2nc - 1) + \widehat{\lambda}_B\overline{P}_L$
$PBD_{L,IId}^{\bullet(1)}$	$\begin{cases} P_B^+|0 \text{ if } D^{\bullet(1)} \text{ only feasible for N-type lender} \\ P_B^+|0 \text{ if } D^{\bullet(1)} \text{ feasible for A\&N-type lender} \end{cases}$
CDr/nr	$\zeta_{max} > \dfrac{nc}{\mu}$
BPCb/nb	$\dfrac{vc^2}{4\mu nc} < \zeta_{max} < \dfrac{P_B^+ + nc}{\mu}$
LCOC(AB)s/ns	$\mu < 1 + P_B^+ + 2nc + \overline{P}_L$
LCOC(NB)s	$\mu \geq 1 + vc + 2nc + \overline{P}_L \vee$ $\zeta_{max} \leq \dfrac{nc}{\mu(1 + vc + 2nc + \overline{P}_L - \mu)^2}$

Note: The structure of the table is equivalent to the structure of Table 4.12.

- *Case III – Risky debt:* $y_{min}\big|_{\zeta=\zeta_{max}} < h_{L,\text{A-type borrower}}, h_{L,\text{N-type borrower}}$

  Finally, in Case III risky debt is offered to both borrower types which implies that their informational rent from interim and ex-post uncertainty $(IR(D_{II}^\circ) = \dfrac{vc^2}{4\zeta_{max}\mu})$ is identical, requiring only the examination of three subcases as in Case I.

  Moreover, if the lender only prefers to offer the N-type borrower a (risky) debt contract, the A-type's BPC requirement (BPCb/nb) becomes irrelevant for the lender's expected profit which implies that the profits of Case IIa, IIb and IIIb as well as of Case IIc, IId and IIIc coincide. However, it is still important to consider the A-type borrower's BPC (BPCb/nb) to determine whether the proclaimed cases offer feasible contracting solutions.

  - *IIIa – No BPC binding:* $\dfrac{vc^2}{4\zeta_{max}\mu} \geq P_B^+ + nc$

    If both BPC are not binding the optimal repayment obligations are identical $(h_{L,IIIa,\text{A-type borrower}}^* = h_{L,IIIa,\text{N-type borrower}}^* = \mu(1+\zeta_{max}) - vc)$ which implies that the lender either offers both borrower types a risky debt contract or none. When both BPC are not binding (assumption of Case IIIa) the lender can optimize the borrowers' repayment obligations only taking his own participation constraint and the Case

IIIa requirements into account.[56] The borrower's type does not matter. Since another requirement of Case IIIa is that both borrow types receive the same, i. e. risky, debt offers their optimal repayment obligations are identical. A reduction of the A-type borrower's repayment obligation is not necessary since his informational rent $(\dfrac{vc^2}{4\zeta_{max}\mu})$ is already large enough to compensate for his positive outside option and the occurring negotiation costs.[57] On the other hand any increase of the A or the N-type borrower's repayment obligation is unprofitable for the lender since the agency cost increase is larger than the additional gain due to the increase. Since the A and the N-type borrower's repayment obligations are identical, they both receive the same expected profit in case of contract agreement. The only difference occurs in case of no contract agreement due to the A-type's positive outside option (P_B^+) (see Table 4.18).

Table 4.18: Expected Profits and Feasible Contracting Constraints of a Private Debt Contract – Case IIIa (EM, Lender Optimization)

• Lender offers both borrower types a contract $(D^{\bullet(2)})$	
$PLD_{L,IIIa}^{\bullet(2)}$	$\mu + \dfrac{vc^2}{4\zeta_{max}\mu} - 1 - vc - nc$
$PBD_{L,IIIa}^{\bullet(2)}$	$\begin{cases} (1-\widehat{\lambda}_L)(\dfrac{vc^2}{4\zeta_{max}\mu} - nc) + \widehat{\lambda}_L P_B^+\vert & \text{if } D^{\bullet(2)} \text{ only feasible} \\ (1-\widehat{\lambda}_L)(\dfrac{vc^2}{4\zeta_{max}\mu} - nc) & \text{for N-type lender} \\[4pt] \dfrac{vc^2}{4\zeta_{max}\mu} - nc\vert \dfrac{vc^2}{4\zeta_{max}\mu} - nc & \text{if } D^{\bullet(2)} \text{ feasible for} \\ & \text{A\&N-type lender} \end{cases}$
CDr/nr	$\zeta_{max} > \dfrac{vc}{2\mu}$
BPCb/nb	$\zeta_{max} \le \dfrac{vc^2}{4\mu(P_B^+ + nc)}$
LCOC(AB)s/ns	$\mu \ge 1 + nc + \overline{P}_L + vc \vee$ $\zeta_{max} \le \dfrac{vc^2}{4(1 + \overline{P}_L + nc + vc - \mu)\mu}$

Note: The structure of the table is equivalent to the structure of Table 4.11.

---

[56] Both BPC are not binding when the borrower's informational rent from debt financing is larger that the A-type borrower's positive outside option plus occurring negotiation costs $(P_B^+ + nc)$. The N-type borrower's participation constraint is then always automatically satisfied.

[57] This finding is similar to Case Ia where also both BPC are not binding and both borrower types receive the same, i. e. riskless, debt offers. Hence, their optimal repayment obligations also coincide (see page 143).

- *IIIb – A-type borrower's BPC binding, N-type's not:* $nc \leq \dfrac{vc^2}{4\zeta_{max}\mu} <$
$P_B^+ + nc$

$h^*_{L,IIIb,\text{A-type borrower}} = \mu(1 + \zeta_{max}) - 2\sqrt{(P_B^+ + nc)\zeta_{max}\mu}$

$h^*_{L,IIIb,\text{N-type borrower}} = \mu(1 + \zeta_{max}) - vc$

Table 4.19: Offer Dependent Expected Profits and Feasible Contracting Constraints of a Private Debt Contract – Case IIIb (EM, Lender Optimization)

• Lender offers both borrower types a contract $(D^{\bullet(2)})$	
$PLD_{L,IIIb}^{\bullet(2)}$	$\mu - 1 - vc - nc + \widehat{\lambda}_B\left(vc\sqrt{\dfrac{P_B^+ + nc}{\zeta_{max}\mu}} - (P_B^+ + nc)\right)$ $+(1 - \widehat{\lambda}_B)\dfrac{vc^2}{4\zeta_{max}\mu}$
$PBD_{L,IIIb}^{\bullet(2)}$	$\begin{cases} P_B^+ \mid (1 - \widehat{\lambda}_L)(\dfrac{vc^2}{4\zeta_{max}\mu} - nc) & \text{if } D^{\bullet(2)} \text{ only feasible for N-type lender} \\ P_B^+ \mid \dfrac{vc^2}{4\zeta_{max}\mu} - nc & \text{if } D^{\bullet(2)} \text{ feasible for A\&N-type lender} \end{cases}$
CDr/nr	$\zeta_{max} > \max\{\dfrac{vc}{2\mu}; \dfrac{vc^2}{4(P_B^+ + nc)\mu}\}$
BPCb/nb	$\dfrac{(P_B^+ + nc)}{\mu} < \zeta_{max} \leq \dfrac{vc^2}{4\mu nc}$
LCOC(AB)s/ns	$\mu + vc\sqrt{\dfrac{P_B^+ + nc}{\zeta_{max}\mu}} - 1 - vc - 2nc - P_B^+ \geq \overline{P}_L$
• Lender only offers N-type borrower a contract $(D^{\bullet(1)})$	
$PLD_{L,IIIb}^{\bullet(1)}$	$(1 - \widehat{\lambda}_B)(\mu + \dfrac{vc^2}{4\zeta_{max}\mu} - vc - nc - 1) + \widehat{\lambda}_B\overline{P}_L$
$PBD_{L,IIIb}^{\bullet(1)}$	$\begin{cases} P_B^+ \mid (1 - \widehat{\lambda}_L)(\dfrac{vc^2}{4\zeta_{max}\mu} - nc) & \text{if } D^{\bullet(1)} \text{ only feasible for N-type lender} \\ P_B^+ \mid \dfrac{vc^2}{4\zeta_{max}\mu} - nc & \text{if } D^{\bullet(1)} \text{ feasible for A\&N-type lender} \end{cases}$
CDr/nr	$\zeta_{max} > \max\{\dfrac{vc}{2\mu}; \dfrac{vc^2}{4(P_B^+ + nc)\mu}\}$
BPCb/nb	$\dfrac{(P_B^+ + nc)}{\mu} < \zeta_{max} \leq \dfrac{vc^2}{4\mu nc}$
LCOC(AB)s/ns	$\mu + vc\sqrt{\dfrac{P_B^+ + nc}{\zeta_{max}\mu}} - 1 - vc - 2nc - P_B^+ < \overline{P}_L$
LCOC(NB)s	$\mu \geq 1 + vc + nc + \overline{P}_L \vee$ $\zeta_{max} \leq \dfrac{vc^2}{4\mu(1 + vc + nc + \overline{P}_L - \mu)}$

Note: The structure of the table is equivalent to the structure of Table 4.12.

- IIIc – A&N-type borrowers' BPC binding: $\dfrac{vc^2}{4\zeta_{max}\mu} < nc$

$$h^*_{L,IIIc,\text{A-type borrower}} = \mu(1 + \zeta_{max}) - 2\sqrt{(P^+_B + nc)\zeta_{max}\mu}$$

$$h^*_{L,IIIc,\text{N-type borrower}} = \mu(1 + \zeta_{max}) - 2\sqrt{nc\zeta_{max}\mu}$$

Table 4.20: Offer Dependent Expected Profits and Feasible Contracting Constraints of a Private Debt Contract – Case IIIc (EM, Lender Optimization)

• Lender offers both borrower types a contract ($D^{\bullet(2)}$)	
$PLD^{\bullet(2)}_{L,IIIc}$	$\mu - 1 - vc - nc + \widehat{\lambda}_B\left(vc\sqrt{\dfrac{P^+_B + nc}{\zeta_{max}\mu}} - (P^+_B + nc)\right)$ $+(1 - \widehat{\lambda}_B)\left(vc\sqrt{\dfrac{nc}{\zeta_{max}\mu}} - nc\right)$
$PBD^{\bullet(2)}_{L,IIIc}$	$\begin{cases} P^+_B\lvert 0 \text{ if } D^{\bullet(2)} \text{ only feasible for N-type lender} \\ P^+_B\lvert 0 \text{ if } D^{\bullet(2)} \text{ feasible for A\&N-type lender} \end{cases}$
CDr/nr	$\zeta_{max} > \dfrac{(P^+_B + nc)}{\mu}$
BPCb/nb	$\zeta_{max} > \dfrac{vc^2}{4\mu nc}$
LCOC(AB)s/ns	$\mu + vc\sqrt{\dfrac{P^+_B + nc}{\zeta_{max}\mu}} - 1 - vc - 2nc - P^+_B \geq \overline{P}_L$
• Lender only offers N-type borrower a contract ($D^{\bullet(1)}$)	
$PLD^{\bullet(1)}_{L,IIIc}$	$(1 - \widehat{\lambda}_B)(\mu + vc\sqrt{\dfrac{nc}{\zeta_{max}\mu}} - vc - 2nc - 1) + \widehat{\lambda}_B\overline{P}_L$
$PBD^{\bullet(1)}_{L,IIIc}$	$\begin{cases} P^+_B\lvert 0 \text{ if } D^{\bullet(1)} \text{ only feasible for N-type lender} \\ P^+_B\lvert 0 \text{ if } D^{\bullet(1)} \text{ feasible for A\&N-type lender} \end{cases}$
CDr/nr	$\zeta_{max} > \dfrac{(P^+_B + nc)}{\mu}$
BPCb/nb	$\zeta_{max} > \dfrac{vc^2}{4\mu nc}$
LCOC(AB)s/ns	$\mu + vc\sqrt{\dfrac{P^+_B + nc}{\zeta_{max}\mu}} - 1 - vc - 2nc - P^+_B < \overline{P}_L$
LCOC(NB)s	$\mu \geq 1 + vc + 2nc + \overline{P}_L \vee$ $\zeta_{max} \leq \dfrac{nc}{\mu(1 + vc + 2nc + \overline{P}_L - \mu)^2}$

Note: The structure of the table is equivalent to the structure of Table 4.12.

The comparison of the different private debt cases to define the lender's preferred case if more than one is feasible is postponed to Section 4.3, where the comparison is treated jointly with the determination of the preferred ct/pm-choice. Both questions can not be answered without an explicit derivation of the lender's expectations about the borrower's type. However, similar to private debt financing in the basic model, see page 72, we already know the following preferences of the lender:[58]

- $D_{L,IIIa}^{\bullet(2)} \succ D_{L,IIa}^{\bullet(2)} \succ D_{L,Ia}^{\bullet(2)}$ since the lender always prefers risky to riskless debt financing when no BPC is binding because if risky debt would not be preferable risk debt financing would be unfeasible, i. e. it is optimal to reduce h such that $h \leq y_{min}\big|_{\zeta=\zeta_{max}}$ (riskless debt);

- $D_{L,IIIa}^{\bullet(2)}$ dominates $\{D_{L,Ib}^{\bullet(2)}, D_{L,Ic}^{\bullet(2)}, D_{L,IIb}^{\bullet(2)}, D_{L,IIc}^{\bullet(2)}, D_{L,IId}^{\bullet(2)}\}$; and

- $D_{L,IIa}^{\bullet(2)}$ dominates $\{D_{L,Ib}^{\bullet(2)}, D_{L,Ic}^{\bullet(2)}\}$ since risky debt financing is preferable to riskless when no BPC is binding.[59]

- The same reasoning which applies for $D_L^{\bullet(2)}$ is also valid for $D_L^{\bullet(1)}$:

  - $D_{L,IIIb}^{\bullet(1)} \sim D_{L,IIb}^{\bullet(1)} \sim D_{L,IIa}^{\bullet(1)}$ since in all cases the N-type borrower is financed with risky debt and non binding BPC;

  - $D_{L,IIIc}^{\bullet(1)} \sim D_{L,IIc}^{\bullet(1)} \sim D_{L,IId}^{\bullet(1)}$ since in all cases the N-type borrower is financed with risky debt and binding BPC;

  - $\{D_{L,IIIb}^{\bullet(1)}, D_{L,IIb}^{\bullet(1)}, D_{L,IIa}^{\bullet(1)}\} \succ D_{L,Ib}^{\bullet(1)}$;

  - $\{D_{L,IIIb}^{\bullet(1)}, D_{L,IIb}^{\bullet(1)}, D_{L,IIa}^{\bullet(1)}\} \succ \{D_{L,IIIc}^{\bullet(1)}, D_{L,IId}^{\bullet(1)}, D_{L,IIc}^{\bullet(1)}\}$;

  - $\{D_{L,IIIb}^{\bullet(1)}, D_{L,IIIc}^{\bullet(1)}, D_{L,IId}^{\bullet(1)}, D_{L,IIc}^{\bullet(1)}, D_{L,IIb}^{\bullet(1)}, D_{L,IIa}^{\bullet(1)}\} \succ D_{L,Ic}^{\bullet(1)}$.

These preferences allow us to restrict the number of possible private debt cases which have to be considered in the inter case comparison as well as in the comparison of the alternative ct/pm-constellations.

4.2.4.2 Borrower Optimization

Finally, the borrower as the condition optimizer can set private debt's repayment obligation lender type-dependent ($h_{B,\text{A-type lender}}, h_{B,\text{N-type lender}}$). All potential ex-ante informational asymmetries are resolved, while negotiation

---

[58] Remember, by, e. g., $D_{L,IIIa}^{\bullet(2)}$ we refer to the case in which the lender offers both potential borrower types a $D_{L,IIIa}^{\bullet}$ contract, while for $D_{L,IIIa}^{\bullet(1)}$ he is only interested in contracting with the N-type.

[59] Only if the BPC is binding, which implies that the lender can not set h unrestricted, it is possible that the lender's expected profit of riskless debt financing exceeds the profit of risky debt. This is possible since agency costs for risky debt financing might not be optimally balanced against the gains from risky debt when the BPC is binding.

costs nc occur for both parties. The optimal lender type-dependent repayment obligations will satisfy $h^*_{B,\text{A-type lender}} \geq h^*_{B,\text{N-type lender}}$ due to the A-type lender's positive outside option.[60]

The optimization problem for private debt is

$$\max_{\substack{h_{B,\text{A-type lender}} \\ h_{B,\text{N-type lender}}}} \quad PBD^\bullet(h) = \widehat{\lambda}_L PBD^\bullet(h_{B,\text{A-type lender}})$$
$$+(1 - \widehat{\lambda}_L)PBD^\bullet(h_{B,\text{N-type lender}})$$

$$s.t.\ PLD^\bullet(h_{B,\text{A-type lender}}) \geq P_L^+,\ \ PLD^\bullet(h_{B,\text{N-type lender}}) \geq 0$$
$$\wedge\ PBD^\bullet(h_{B,\text{A-type lender}}) \geq \overline{P}_B,\ \ PBD^\bullet(h_{B,\text{N-type lender}}) \geq \overline{P}_B\ .$$

Since the borrower has the choice between riskless and risky debt financing, three sub cases have to be considered: Both lender types receive a riskless debt offer (Case I); only the N-type lender receives a riskless debt offer while the A-type risky debt is offered (Case II); finally risky debt contracts are offered to both lender types (Case III). No subcases have to be distinguished since the lender never receives an informational rent due to interim and ex-post uncertainty. Hence, the LPC is always binding.

- Case I – Riskless debt: $h_{B,\text{A-type lender}}, h_{B,\text{N-type lender}} \leq y_{min}\big|_{\zeta=\zeta_{max}}$

  When both lender types receive a riskless debt offer the optimal lender type-dependent repayment obligations are

  $$h^*_{B,I,\text{A-type lender}} = 1 + P_L^+ + nc$$
  $$h^*_{B,I,\text{N-type lender}} = 1 + nc\ .$$

  In private financing all ex-ante informational asymmetries are resolved which implies that the borrower can always reduce the lender to his participation constraint. Whether the borrower is actually willing to finance the project by both lender types $(D^\bullet(2))$ or only by the N-type lender $(D^{\bullet(1)})$ depends on the surplus extractable when negotiating with the A-type lender (BCOC(AL): $PBD^\bullet(h^*_{B,I,\text{A-type lender}}) = \mu - 1 - P_L^+ - 2nc$). The borrower can always extract a higher surplus when contracting with the N-type instead of the A-type lender due to the latter's positive profit alternative.

---

[60] This condition is contrary to the condition in the lender optimization ($h^*_{L,\text{A-type borrower}} \leq h^*_{L,\text{N-type borrower}}$) since in the lender optimization the lender bargains with the debtor (the borrower) with a possible positive outside option while in the borrower optimization the borrower bargains with the creditor (the lender) with a possible positive outside option.

Table 4.21: Offer Dependent Expected Profits and Feasible Contracting Constraints of a Private Debt Contract – Case I (EM, Borrower Optimization)

• Borrower offers both lender types a contract ($D^{\bullet(2)}$)	
$PBD_{B,I}^{\bullet(2)}$	$\mu - 1 - \widehat{\lambda}_L P_L^+ - 2nc$
$PLD_{B,I}^{\bullet(2)}$	$\begin{cases} P_L^+\|0 \text{ if } D^{\bullet(2)} \text{ only feasible for N-type borrower} \\ P_L^+\|0 \text{ if } D^{\bullet(2)} \text{ feasible for A\&N-type borrower} \end{cases}$
CDr/nr	$\mu(1 - \zeta_{max}) \geq 1 + P_L^+ + nc$
BCOC(AL)s/ns	$\mu \geq 1 + P_L^+ + 2nc + \overline{P}_B$
• Borrower only offers N-type lender a contract ($D^{\bullet(1)}$)	
$PBD_{B,I}^{\bullet(1)}$	$(1 - \widehat{\lambda}_L)(\mu - 1 - 2nc) + \widehat{\lambda}_L \overline{P}_B$
$PLD_{B,I}^{\bullet(1)}$	$\begin{cases} P_L^+\|0 \text{ if } D^{\bullet(1)} \text{ only feasible for N-type borrower} \\ P_L^+\|0 \text{ if } D^{\bullet(1)} \text{ feasible for A\&N-type borrower} \end{cases}$
CDr/nr	$\mu(1 - \zeta_{max}) \geq 1 + P_L^+ + nc$
BCOC(AL)s/ns	$\mu < 1 + P_L^+ + 2nc + \overline{P}_B$
BCOC(NL)s	$\mu \geq 1 + 2nc + \overline{P}_B$

Note: The borrower's as well as the lender's expected profits are stated given private debt financing. The borrower has the power to determine the contract's repayment obligation (h). The borrower's expected profit $PBD_{B,I}^{\bullet}$ and the respective conditions depend on whether the borrower just offers the N-type lender a contract or both lender types. The lender's expected profit $PLD_{B,I}^{\bullet}$ is additionally conditioned on the borrower type for which $D_{B,I}^{\bullet}$ is feasible. Since the LPC is lender type-dependent, $PLD_{B,I}^{\bullet}$ is stated lender type-dependent ($PLD_{B,I}^{\bullet}$ for A-type lender| $PLD_{B,I}^{\bullet}$ for N-type lender).

- *Case II – Riskless and risky debt:* $h_{B,\text{N-type lender}} \leq y_{min}\big|_{\zeta=\zeta_{max}} < h_{B,\text{A-type lender}}$

$$h_{B,II,\text{A-type lender}}^* =$$
$$\mu(1 + \zeta_{max}) - vc - \sqrt{4\zeta_{max}\mu(\mu - 1 - P_L^+ - nc - vc) + vc^2}$$
$$h_{B,II,\text{N-type lender}}^* = 1 + nc$$

The borrower's expected profit when he prefers only to finance the N-type lender (BCOC(AL) not satisfied) coincides with the borrower's profit $D_{B,I}^{\bullet(1)}$ since the lender receives in both cases a riskless debt offer while the A-type lender is irrelevant.

Table 4.22: Offer Dependent Expected Profits and Feasible Contracting Constraints of a Private Debt Contract – Case II (EM, Borrower Optimization)

• Borrower offers both lender types a contract $(D^{\bullet(2)})$	
$PBD_{B,II}^{\bullet(2)}$	$\widehat{\lambda}_L \dfrac{\left[vc + \sqrt{4\zeta_{max}\mu(\mu - 1 - P_L^+ - nc - vc) + vc^2} \right]^2}{4\zeta_{max}\mu}$ $+ (1 - \widehat{\lambda}_L)(\mu - 1 - nc) - nc$
$PLD_{B,II}^{\bullet(2)}$	$\begin{cases} P_L^+ \vert 0 \text{ if } D^{\bullet(2)} \text{ only feasible for N-type borrower} \\ P_L^+ \vert 0 \text{ if } D^{\bullet(2)} \text{ feasible for A\&N-type borrower} \end{cases}$
NoN	$\mu \geq 1 + P_L^+ + nc + vc \vee$ $\zeta_{max} \leq \dfrac{vc^2}{4(1 + P_L^+ + nc + vc - \mu)\mu}$
CDr/nr	$1 + nc \leq \mu(1 - \zeta_{max}) < 1 + P_L^+ + nc \wedge \zeta_{max} \geq \dfrac{vc}{2\mu}$
BCOC(AL)s/ns	$\dfrac{\left[vc + \sqrt{4\zeta_{max}\mu(\mu - 1 - P_L^+ - nc - vc) + vc^2} \right]^2}{4\zeta_{max}\mu}$ $-nc \geq \overline{P}_B$
• Borrower only offers N-type lender a contract $(D^{\bullet(1)})$	
$PBD_{B,II}^{\bullet(1)}$	$(1 - \widehat{\lambda}_L)(\mu - 1 - 2nc) + \widehat{\lambda}_L \overline{P}_B$
$PLD_{B,II}^{\bullet(1)}$	$\begin{cases} P_L^+ \vert 0 \text{ if } D^{\bullet(1)} \text{ only feasible for N-type borrower} \\ P_L^+ \vert 0 \text{ if } D^{\bullet(1)} \text{ feasible for A\&N-type borrower} \end{cases}$
CDr/nr	$1 + nc \leq \mu(1 - \zeta_{max}) < 1 + P_L^+ + nc \wedge \zeta_{max} \geq \dfrac{vc}{2\mu}$
BCOC(AL)s/ns	$\dfrac{\left[vc + \sqrt{4\zeta_{max}\mu(\mu - 1 - P_L^+ - nc - vc) + vc^2} \right]^2}{4\zeta_{max}\mu}$ $-nc < \overline{P}_B$
BCOC(NL)s	$\mu \geq 1 + 2nc + \overline{P}_B$

Note: The structure of the table is equivalent to the structure of Table 4.21.

- *Case III – Risky debt:* $y_{min}\big|_{\zeta=\zeta_{max}} < h_{B,\text{A-type lender}}, h_{B,\text{N-type lender}}$

$$h^{*}_{B,III,\text{A-type lender}} =$$
$$\mu(1+\zeta_{max}) - vc - \sqrt{4\zeta_{max}\mu(\mu-1-P_L^+ - nc - vc) + vc^2}$$
$$h^{*}_{B,III,\text{N-type lender}} =$$
$$\mu(1+\zeta_{max}) - vc - \sqrt{4\zeta_{max}\mu(\mu-1 - nc - vc) + vc^2}$$

Table 4.23: Offer Dependent Expected Profits and Feasible Contracting Constraints of a Private Debt Contract – Case III (EM, Borrower Optimization)

• Borrower offers both lender types a contract $(D^{\bullet(2)})$	
$PBD^{\bullet(2)}_{B,III}$	$\dfrac{1}{4\zeta_{max}\mu}\left[\widehat{\lambda}_L\left[vc + \sqrt{4\zeta_{max}\mu(\mu-1-P_L^+ - nc - vc) + vc^2}\right]^2\right.$
	$\left. +(1-\widehat{\lambda}_L)\left[vc + \sqrt{4\zeta_{max}\mu(\mu-1-nc-vc)+vc^2}\right]^2\right] - nc$
$PLD^{\bullet(2)}_{B,III}$	$\begin{cases} P_L^+\|0 \text{ if } D^{\bullet(2)} \text{ only feasible for N-type borrower} \\[4pt] P_L^+\|0 \text{ if } D^{\bullet(2)} \text{ feasible for A\&N-type borrower} \end{cases}$
NoN	$\mu \geq 1 + P_L^+ + nc + vc \vee \zeta_{max} \leq \dfrac{vc^2}{4\mu(1+P_L^+ + nc + vc - \mu)}$
CDr/nr	$\zeta_{max} \geq \dfrac{vc}{2\mu} \wedge \mu(1-\zeta_{max}) < 1 + nc$
BCOC(AL)s/ns	$\dfrac{\left[vc + \sqrt{4\zeta_{max}\mu(\mu-1-P_L^+ - nc - vc) + vc^2}\right]^2}{4\zeta_{max}\mu} - nc \geq \overline{P}_B$
	table continued on next page

table continued on next page

table continued from previous page

• Borrower only offers N-type lender a contract ($D^{\bullet(1)}$)	
$PBD^{\bullet(1)}$	$(1 - \widehat{\lambda}_L)\left(\dfrac{\left[vc + \sqrt{4\zeta_{max}\mu(\mu - 1 - nc - vc) + vc^2} \right]^2}{4\zeta_{max}\mu} - nc \right)$ $+\widehat{\lambda}_L \overline{P}_B$
$PLD^{\bullet(1)}$	$\begin{cases} P_L^+ \mid 0 \text{ if } D^{\bullet(1)} \text{ only feasible for N-type borrower} \\ P_L^+ \mid 0 \text{ if } D^{\bullet(1)} \text{ feasible for A\&N-type borrower} \end{cases}$
NoN	$\mu \geq 1 + nc + vc \lor \zeta_{max} \leq \dfrac{vc^2}{4(1 + nc + vc - \mu)\mu}$
CDr/nr	$\zeta_{max} \geq \dfrac{vc}{2\mu} \land \mu(1 - \zeta_{max}) < 1 + nc$
BCOC(AL)s/ns	$\dfrac{\left[vc + \sqrt{4\zeta_{max}\mu(\mu - 1 - P_L^+ - nc - vc) + vc^2} \right]^2}{4\zeta_{max}\mu} - nc < \overline{P}_B$
BCOC(NL)s	$\dfrac{\left[vc + \sqrt{4\zeta_{max}\mu(\mu - 1 - nc - vc) + vc^2} \right]^2}{4\zeta_{max}\mu} - nc \geq \overline{P}_B$

Note: The structure of the table is equivalent to the structure of Table 4.21.

No comparison between the different private debt financing cases is required, since they do not overlap. For each possible project (μ, ζ_{max}-specification) the borrower's preferred private debt case is uniquely defined.

4.2.5 Comparison of Feasible Contracting Solutions Given the Lender's and the Borrower's Private Information

The aim of this section is to examine whether certain power effects have to occur since the contracting parties possess private information. Power effects might not only result from power variations among contracting parties with different objectives (e. g. to maximize or to minimize q/h with respect to the contracting party's participation constraint), they might also emerge due to asymmetrically distributed, private information. If contracting parties possess private information their feasibility judgement for different contracts can vary and hence they might achieve different optimal contracting solutions. For example, if a particular ct/pm-constellation is feasible given the lender's private information but not the borrower's, all bargaining power effects related to this

ct/pm-constellation are significantly affected by a redistribution of the bargaining power to determine contract conditions since the ct/pm-constellation becomes feasible/unfeasible.

Therefore, it is important to realize that the feasible contract choices and hence the observable bargaining power effects depend on contracting parties' objectives as well as on their private information. The distribution of incentives as well as of information affects the outcome of the contracting game.

Contrary to the basic model without ex-ante informational asymmetry about contractual partner's type, in this extended context a ct/pm's feasibility from the condition optimizer's perspective, i. e. given his private information, does not imply that the project can definitively be financed by this ct/pm-constellation. For example, a public S2 offering can be feasible from the lender's perspective being unaware of the A-type borrower.

Many of the effects described below become already obvious by a comparison of equity's feasible contracting constraints (FCCs: LPC, BPC, LCOC, BCOC) in the lender and the borrower optimization.

Table 4.24: Equity Financing's Feasible Contracting Constraints (EM)

	Condition optimizer	
	Lender	Borrower
$E^\circ \vert S2$	$\mu \geq 1 + mc + \overline{P}_L$	$\mu \geq 1 + mc + \overline{P}_B$
$E^\circ \vert S1$	$\mu \geq 1 + P_B^+ + mc + \overline{P}_L$	$\mu \geq 1 + P_L^+ + mc + \overline{P}_B$
$E^{\bullet(2)}$	$\mu \geq 1 + P_B^+ + mc + 2nc + \overline{P}_L$	$\mu \geq 1 + P_L^+ + mc + 2nc + \overline{P}_B$
$E^{\bullet(1)}$	$1 + mc + 2nc + \overline{P}_L \leq \mu <$ $1 + P_B^+ + mc + 2nc + \overline{P}_L$	$1 + mc + 2nc + \overline{P}_B \leq \mu <$ $1 + P_L^+ + mc + 2nc + \overline{P}_B$

Ignoring ex-ante informational asymmetries for a moment, i. e. the borrower's as well as the lender's type is commonly known ($\lambda_L, \lambda_B \in \{0,1\}$) and both act accordingly,[61] the examination of the feasible contracting constraints (FCCs) for the different ct/pm-choices ($E^\circ, D^\circ, E^\bullet, D^\bullet$) reveals that, like in the basic model (see Result 1 (BM)), the range of feasible contracting solutions is *identical* in both optimizations. They are *independent* of the bargaining power distribution. They are identical since the relevant FCCs are characterized such that the lender and the borrower just receive their opportunity 0 or $P_j^+ \; \forall j \in \{L, B\}$, while for debt financing the borrower might keep an uncontractible informational rent. At the "FCC border" no surplus remains

---

[61] In public offerings no strategy choice emerges since the contractual partner's opportunity, i. e. their type, is known.

to be distributed. The power distribution becomes irrelevant for projects at the "FCC border".

Taking ex-ante informational asymmetries into account, we observe that the feasible contracting solutions *depend* on the bargaining power distribution, i. e. differ in both optimizations, due to two components (see, for example, Table 4.24):

1. the specification of the lender's and the borrower's outside option (P_L^+ and P_B^+ do not have to coincide) as well as
2. the type of the condition optimizer.

The specification of the ex-ante informational asymmetry is crucial for the feasible contracting solutions since each ct/pm-constellation offers a predefined way to cope with the uncertainty. Except for *symmetrical* outside options ($P_L^+ = P_B^+$) the uncertainty about contractual partner's opportunity varies from the lender's to the borrower's perspective. Hence, the FCCs of the individual ct/pm-constellations to cope with this uncertainty vary. For example, the N-type lender's ($\overline{P}_L = 0$) and the N-type borrower's ($\overline{P}_B = 0$) FCC requirements for $E°|S1$ differ for $P_L^+ \neq P_B^+$, see Table 4.24. Hence, the range of feasible contracting solutions depends on the specification of the lender's and the borrower's outside option. Additionally to the outside option specification, the condition optimizer's type affects the feasible contracting solutions. E. g., even for the same ct/pm and power distribution the range of feasible contracting solutions varies with the condition optimizer's type due to their differing opportunities (outside option 0 or P_j^+). Therefore, the range of feasible contracting solutions generally varies with bargaining power redistributions.

However, constellations exist where the feasible contracting solutions are independent of the bargaining power distribution. For a symmetrical outside option ($P_L^+ = P_B^+$) and a constant condition optimizer's type, for example, the feasible contracting solutions of *public and private equity* financing are *independent* of the power distribution since the agency costs associated with public and private equity financing (mc or $mc + 2nc$) are independent of q. However, in debt financing the agency costs vary with the agreed repayment (h). Therefore, basically only contracting solutions are independent of the bargaining power distribution when both parties receive the same expected profit at the "FCC border"[62], such as for $D°|S1$ from the A-type's perspective and for $D°|S2$ and $D^{•(1)}$ from the N-type's perspective. In these three circumstances both parties receive the same expected profit at the "FCC border", causing identical total agency costs. This is not possible otherwise. The only exception is $D^{•(2)}$ financing from the A-type's perspective since $D^{•(2)}$'s fea-

---

[62] The FCCs of a particular ct/pm-constellation given a specific perspective define the feasible contracting area of this particular constellation from this perspective. To the border between the feasible contracting area and the none feasible contracting area we refer to as the "FCC border".

sibility is restricted by the lender's, respectively the borrower's, COC for the A-type lender or borrower, and not by an (ex-ante) participation constraint.[63] For any other constellation the feasible contracting solutions depend on the bargaining power distribution. For example, the FCC of $D^{\bullet(1)}$ from the A-type lender's perspective is determined in a way that the lender is just compensated for his profit alternative while the borrower keeps an informational rent. The lender only offers the N-type borrower ($\overline{P}_B = 0$) a contract. The FCC of $D^{\bullet(1)}$ from the A-type borrower's perspective differs. Of course, the FCC of $D^{\bullet(1)}$ is defined analogously to the lender's perspective, but since a smaller expected profit has to be transferred from the borrower to the lender lower total agency costs arise, changing the FCC of private debt.

Result 1
Extended Model: Feasible Contracting Solutions
Since the feasible contracting solutions of $E^\circ|S1, E^\circ|S2, E^\bullet$ for the i-type ($i \in \{A; N\}$), of $D^\circ|S1$ and $D^{\bullet(2)}$ for the A-type, and of $D^\circ|S2$ and $D^{\bullet(1)}$ for the N-type condition optimizer are *independent* of the bargaining power distribution and the condition optimizer's private information when (i) the outside options are *symmetric* ($P_L^+ = P_B^+$) and (ii) the condition optimizer's type stays constant. Then changes in these particular *ct/pm*-choices do *not need* to occur due to a variation in the bargaining power distribution. □

It becomes obvious that in this extended framework bargaining power effects are not only caused by agency cost variations like in Chapter 3. They also significantly depend on the outside option specification and potential condition optimizer's type variations.

4.3 Contract Type / Placement Mode-Choices in the Bargaining Power Scenarios

After analyzing which contracts are feasible for particular project specifications in Section 4.2, the *optimal* contract choices for the four bargaining power scenarios are determined in this section, see Sections 4.3.1 to 4.3.4. To cope with the increasing number of interdependencies between the economic effects underlying our analysis, the following contract choice determination is restricted to a fixed parameter constellation:

---

[63] For $D^{\bullet(2)}$ financing the A-type lender's and the A-type borrower's expected profits can differ at the "FCC border" since the expected monetary transfer between the lender and the borrower differs and hence the occuring agency costs. However, the feasibility of $D^{\bullet(2)}$ requires that contracting with the A-type partner is profitable for the condition optimizer which implies that the feasible contracting solutions of $D^{\bullet(2)}$ from the A-type's perspective are independent of the bargaining power distribution.

PC1:
$\zeta_{max} = 0.25$, $vc = 0.4$, $mc = 0.15$, $nc = 0.05$,
$P_L^+ = P_B^+ = 0.3$, $\lambda_L = \lambda_B = 0.5$, and $\mu \in \mathbf{R}_+$.

The parameter constellation is chosen in accordance with the constellations in the basic model of Chapter 3 to keep the results comparable. The ex-ante informational asymmetry is specified symmetrically to keep the degree of uncertainty independent of the condition optimizer's perspective.[64]

Due to the restriction to PC1, the expected profits and the feasibility constraints (BPCb/nb, CDr/nr, LCOC, BCOC and NoN) of the different ct/pm-constellations, determined in the previous section, simplify to the functions stated in Tables 4.25 and 4.26.[65] Since the lender's and the borrower's outside option coincide ($P_L^+ = P_B^+$), only the condition optimizer's type matters for the feasible contracting solutions. Therefore, the lender's, respectively the borrower's, expected profits are stated, like in the previous section, type and expectation dependent. When the lender has the power to determine contracts' conditions (lender optimization) he knows his own type ($\overline{P}_L \in \{0, P_L^+\}$) but is unaware of the borrower's type. Hence, his expected profits depend on his expectations about the borrower's type ($\hat{\lambda}_B$). On the other hand, the borrower's expected profits depend on the borrower's expectation about the lender's type ($\hat{\lambda}_L$) since the lender's outside option is type-dependent. The same is true for the borrower optimization.

---

[64] The robustness of the derived optimal contract choices and, in particular, of the resulting bargaining power effects due to changes in the underlying fixed parameter constellation is examined in Section 4.4.5.

[65] We deliberately did not include the LPC respectively the BPC since they are examined at the end when the lender's and the borrower's expected profits from financial contracting are determined.

Table 4.25: Expected Profits – Lender Optimization (EM, PC1)

ct/pm	$PL(\mu, \widehat{\lambda}_B)$	$PB(\mu, \widehat{\lambda}_L)$	BPCb/nb; CDr/nr; LCOC(AB)
$E_L^\circ\|S1$	$\mu - 1.45$	$\left\{\begin{array}{l} 0.3\|(1-\widehat{\lambda}_L)0.3 \\ 0.3\|0.3 \end{array}\right.$	—
$E_L^\circ\|S2$	$(1-\widehat{\lambda}_B)(\mu - 1.15) + \widehat{\lambda}_B \overline{P}_L$	$0.3\|0$	—
$D_{L,Ia}^\circ$	$0.75\mu - 1$	$\left\{\begin{array}{l} (1-\widehat{\lambda}_L)0.25\mu + \widehat{\lambda}_L0.3\|(1-\widehat{\lambda}_L)0.25\mu \\ 0.25\mu\|0.25\mu \end{array}\right.$	$\mu \geq 1.2$
$D_{L,Ib}^\circ\|S1$	$\mu - 1.3$	$\left\{\begin{array}{l} 0.3\|(1-\widehat{\lambda}_L)0.3 \\ 0.3\|0.3 \end{array}\right.$	$\mu < 1.2$
$D_{L,Ib}^\circ\|S2$	$(1-\widehat{\lambda}_B)(0.75\mu - 1) + \widehat{\lambda}_B \overline{P}_L$	$\left\{\begin{array}{l} 0.3\|(1-\widehat{\lambda}_L)0.25\mu \\ 0.3\|0.25\mu \end{array}\right.$	$\mu < 1.2$
$D_{L,IIa}^\circ{}^\dagger$	$\mu + \frac{0.16}{\mu} - 1.4$	$\left\{\begin{array}{l} (1-\widehat{\lambda}_L)\frac{0.16}{\mu}\|\frac{0.16}{\mu} + \widehat{\lambda}_L0.3\|(1-\widehat{\lambda}_L)\frac{0.16}{\mu} \\ \frac{0.16}{\mu}\|\frac{0.16}{\mu} \end{array}\right.$	$0.8 < \mu \leq 0.5\overline{3}$
$D_{L,IIb}^\circ\|S1$	$\mu + \frac{0.438}{\sqrt{\mu}} - 1.7$	$\left\{\begin{array}{l} 0.3\|(1-\widehat{\lambda}_L)0.3 \\ 0.3\|0.3 \end{array}\right.$	$\mu > 1.2$
$D_{L,IIb}^\circ\|S2$	$(1-\widehat{\lambda}_B)(\mu + \frac{0.16}{\mu} - 1.4) + \widehat{\lambda}_B \overline{P}_L$	$\left\{\begin{array}{l} 0.3\|(1-\widehat{\lambda}_L)\frac{0.16}{\mu} \\ 0.3\|\frac{0.16}{\mu} \end{array}\right.$	$\mu > 0.8$
$E_L^{\bullet(2)}$	$\mu - \widehat{\lambda}_B0.3 - 1.25$	$0.3\|0$	$\mu \geq 1.55 + \overline{P}_L$

table continued on next page

table continued from previous page

$E_L^{\bullet(1)}$	$(1-\widehat{\lambda}_B)(\mu - 1.25) + \widehat{\lambda}_B \overline{P}_L$	$0.3\vert 0$	$\mu < 1.55 + \overline{P}_L$
$D_{L,Ia}^{\bullet(2)}$	$0.75\mu - 1.05$	$\left\{\begin{array}{l}(1-\widehat{\lambda}_L)(0.25\mu - 0.05) + \widehat{\lambda}_L 0.3\vert \\ (1-\widehat{\lambda}_L)(0.25\mu - 0.05)\end{array}\right.$ $0.25\mu - 0.05\vert 0.25\mu - 0.05$	$\mu \geq 1.4 + 1.3\overline{P}_L$
$D_{L,Ib}^{\bullet(2)\dagger}$	$\widehat{\lambda}_B(\mu - 0.35) + (1-\widehat{\lambda}_B)0.75\mu - 1.05$	$\left\{\begin{array}{l}0.3\vert(1-\widehat{\lambda}_L)(0.25\mu - 0.05) \\ 0.3\vert 0.25\mu - 0.05\end{array}\right.$	$1.4 + \overline{P}_L \leq \mu < 1.4$
$D_{L,Ib}^{\bullet(1)}$	$(1-\widehat{\lambda}_B)(0.75\mu - 1.05) + \widehat{\lambda}_B \overline{P}_L$	$\left\{\begin{array}{l}0.3\vert(1-\widehat{\lambda}_L)(0.25\mu - 0.05) \\ 0.3\vert 0.25\mu - 0.05\end{array}\right.$	$0.2 \leq \mu < 1.4$
$D_{L,Ic}^{\bullet(2)\dagger}$	$\mu - \widehat{\lambda}_B 0.3 - 1.1$	$0.3\vert 0$	$1.4 + \overline{P}_L \leq \mu < 0.2$
$D_{L,Ic}^{\bullet(1)\dagger}$	$(1-\widehat{\lambda}_B)(\mu - 1.1) + \widehat{\lambda}_B \overline{P}_L$	$0.3\vert 0$	$1.1 + \overline{P}_L \leq \mu < 0.2$
$D_{L,IIa}^{\bullet(2)}$	$\widehat{\lambda}_B 0.75\mu + (1-\widehat{\lambda}_B)(\mu + \frac{0.16}{\mu} - 0.4) - 1.05$	$\left\{\begin{array}{l}(1-\widehat{\lambda}_L)(0.25\mu - 0.05) + \widehat{\lambda}_L 0.3\vert \\ (1-\widehat{\lambda}_L)(\frac{0.16}{\mu} - 0.05)\end{array}\right.$ $0.25\mu - 0.05\vert\frac{0.16}{\mu} - 0.05$	$1.4 + 1.3\overline{P}_L \leq \mu \leq 3.2$
$D_{L,IIa}^{\bullet(1)}$	$(1-\widehat{\lambda}_B)(\mu + \frac{0.16}{\mu} - 1.45) + \widehat{\lambda}_B \overline{P}_L$	$\left\{\begin{array}{l}0.3\vert(1-\widehat{\lambda}_L)(\frac{0.16}{\mu} - 0.05) \\ 0.3\vert\frac{0.16}{\mu} - 0.05\end{array}\right.$	$1.4 \leq \mu \leq 1.4 + 1.3\overline{P}_L$
$D_{L,IIb}^{\bullet(2)\dagger}$	$\mu + (1-\widehat{\lambda}_B)\frac{0.16}{\mu} + \widehat{\lambda}_B 0.05 - 1.45$	$\left\{\begin{array}{l}0.3\vert(1-\widehat{\lambda}_L)(\frac{0.16}{\mu} - 0.05) \\ 0.3\vert\frac{0.16}{\mu} - 0.05\end{array}\right.$	$1.4 + \overline{P}_L < \mu < 1.4$
$D_{L,IIb}^{\bullet(1)}$	$(1-\widehat{\lambda}_B)(\mu + \frac{0.16}{\mu} - 1.45) + \widehat{\lambda}_B \overline{P}_L$	$\left\{\begin{array}{l}0.3\vert(1-\widehat{\lambda}_L)(\frac{0.16}{\mu} - 0.05) \\ 0.3\vert\frac{0.16}{\mu} - 0.05\end{array}\right.$	$0.8 < \mu < 1.4$

table continued on next page

table continued from previous page

	$PL(\mu,\widehat{\lambda}_B)$	$PB(\mu,\widehat{\lambda}_L)$	condition
$D_{L,IIc}^{\bullet(2)}$	$\widehat{\lambda}_B 0.75\mu + (1-\widehat{\lambda}_B)(\mu + \frac{0.179}{\sqrt{\mu}} - 0.45) - 1.05$	$\begin{cases}(1-\widehat{\lambda}_L)(0.25\mu - 0.05) + \widehat{\lambda}_L 0.3\,\vert\,0\\ 0.25\mu - 0.05\,\vert\,0\end{cases}$	$\mu > 3.2$
$D_{L,IIc}^{\bullet(1)\dagger}$	$(1-\widehat{\lambda}_B)(\mu + \frac{0.179}{\sqrt{\mu}} - 1.5) + \widehat{\lambda}_B \overline{P}_L$	$0.3\,\vert\,0$	$3.2 < \mu < 1.4 + 1.3\overline{P}_L$
$D_{L,IId}^{\bullet(2)\dagger}$	$\mu + (1-\widehat{\lambda}_B)(\frac{0.179}{\sqrt{\mu}} - 0.4) - \widehat{\lambda}_B 0.3 - 1.1$	$0.3\,\vert\,0$	$3.2 < \mu < 1.4$
$D_{L,IId}^{\bullet(1)\dagger}$	$(1-\widehat{\lambda}_B)(\mu + \frac{0.179}{\sqrt{\mu}} - 1.5) + \widehat{\lambda}_B \overline{P}_L$	$0.3\,\vert\,0$	$3.2 < \mu < 1.4$
$D_{L,IIIa}^{\bullet(2)\dagger}$	$\mu + \frac{0.16}{\mu} - 1.45$	$\begin{cases}(1-\widehat{\lambda}_L)(\frac{0.16}{\mu} - 0.05) + \widehat{\lambda}_L 0.3\,\vert\,(1-\widehat{\lambda}_L)(\frac{0.16}{\mu} - 0.05)\\ \frac{0.16}{\mu} - 0.05\,\vert\,\frac{0.16}{\mu} - 0.05\end{cases}$	$0.8 < \mu \le 0.457$
$D_{L,IIIb}^{\bullet(2)}$	$\mu + (1-\widehat{\lambda}_B)\frac{0.16}{\mu} + \widehat{\lambda}_B\frac{0.473}{\sqrt{\mu}} - \widehat{\lambda}_B 0.35 - 1.45$	$\begin{cases}0.3(1-\widehat{\lambda}_L)(\frac{0.16}{\mu} - 0.05)\\ 0.3\frac{0.16}{\mu} - 0.05\end{cases}$	$1.4 + 1.138\overline{P}_L \le \mu \le 3.2$
$D_{L,IIIb}^{\bullet(1)}$	$(1-\widehat{\lambda}_B)(\mu + \frac{0.16}{\mu} - 1.45) + \widehat{\lambda}_B \overline{P}_L$	$\begin{cases}0.3(1-\widehat{\lambda}_L)(\frac{0.16}{\mu} - 0.05)\\ 0.3\frac{0.16}{\mu} - 0.05\end{cases}$	$1.4 < \mu < 1.4 + 1.138\overline{P}_L$
$D_{L,IIIc}^{\bullet(2)}$	$\mu + \frac{(1-\widehat{\lambda}_B)0.179+\widehat{\lambda}_B 0.473}{\sqrt{\mu}} - \widehat{\lambda}_B 0.3 - 1.5$	$0.3\,\vert\,0$	$\mu > 3.2$
$D_{L,IIIc}^{\bullet(1)\dagger}$	$(1-\widehat{\lambda}_B)(\mu + \frac{0.179}{\sqrt{\mu}} - 1.5) + \widehat{\lambda}_B \overline{P}_L$	$0.3\,\vert\,0$	$3.2 < \mu < 1.4 + 1.138\overline{P}_L$

Note: The lender's and the borrower's expected profits are stated $(PL(\mu,\widehat{\lambda}_B), PB(\mu,\widehat{\lambda}_L))$ with the respective feasible contracting constraints. The power to determine contracts' conditions is given to the lender. The lender's profits' type-dependency is expressed by $\overline{P}_L (\overline{P}_L \in \{0, P_L^+\})$, while the borrower's profits' type-dependency is illustrated by the vertical line (PB for A-type borrower|PB for N-type borrower). The borrower's expected profits are additionally (if required) conditioned on the lender type for which the respective ct/pm-choice is feasible $\left(\begin{cases} PB \text{ if the } ct/pm\text{-choice is only feasible for the N-type lender}\\ PB \text{ if the } ct/pm\text{-choice is feasible for the A\&N-type lenders}\end{cases}\right)$.

† = not feasible due to the BPCb/nb, the CDr/nr and the LCOC(AB) given PC1.

Table 4.26: Expected Profits – Borrower Optimization (EM, PC1)

ct/pm	$PB(\mu, \widehat{\lambda}_L)$	$PL(\mu, \widehat{\lambda}_B)$	NoN; BCOC(AL)	CDr/nr;
$E^{\circ}_B\lvert S1$	$\mu - 1.45$	$\begin{cases} 0.3\lvert(1-\widehat{\lambda}_B)0.3 \\ 0.3\lvert0.3 \end{cases}$	—	
$E^{\circ}_B\lvert S2$	$(1-\widehat{\lambda}_L)(\mu - 1.15) + \widehat{\lambda}_L \overline{P}_B$	$0.3\lvert0$	—	
$D^{\circ}_{B,I}\lvert S1$	$\mu - 1.3$	$\begin{cases} 0.3\lvert(1-\widehat{\lambda}_B)0.3 \\ 0.3\lvert0.3 \end{cases}$	$\mu \geq 1.7\overline{3}$	
$D^{\circ}_{B,I}\lvert S2$	$(1-\widehat{\lambda}_L)(\mu - 1) + \widehat{\lambda}_L \overline{P}_B$	$0.3\lvert0$	$\mu \geq 1.\overline{3}$	
$D^{\circ}_{B,II}\lvert S1$	$\mu + \frac{0.32}{\mu} + \frac{0.8}{\mu}\sqrt{(\mu-1.6)(\mu-0.1)} - 1.7$	$\begin{cases} 0.3\lvert(1-\widehat{\lambda}_B)0.3 \\ 0.3\lvert0.3 \end{cases}$	$1.6 \leq \mu < 1.7\overline{3}$	
$D^{\circ}_{B,II}\lvert S2$	$(1-\widehat{\lambda}_L)\left[\mu + \frac{0.32}{\mu} + \frac{0.8}{\mu}\sqrt{(\mu-1.274)(\mu-0.126)} - 1.4\right] + \widehat{\lambda}_L \overline{P}_B$	$0.3\lvert0$	$1.274 \leq \mu < 1.\overline{3}$	
$E^{\bullet(2)}_B$	$\mu - \widehat{\lambda}_L 0.3 - 1.25$	$0.3\lvert0$	$\mu \geq 1.55 + \overline{P}_B$	
$E^{\bullet(1)}_B$	$(1-\widehat{\lambda}_L)(\mu - 1.25) + \widehat{\lambda}_L \overline{P}_B$	$0.3\lvert0$	$\mu < 1.55 + \overline{P}_B$	
$D^{\bullet(2)}_{B,I}$	$\mu - \widehat{\lambda}_L 0.3 - 1.1$	$0.3\lvert0$	$\mu \geq 1.8$	
$D^{\bullet(1)}_{B,I}{}^{\dagger}$	$(1-\widehat{\lambda}_L)(\mu - 1.1) + \widehat{\lambda}_L \overline{P}_B$	$0.3\lvert0$	$1.8 \leq \mu < 1.4 + \overline{P}_B$	

table continued on next page

table continued from previous page

$D^{\bullet(2)}_{B,JI}$	$\mu + \hat{\lambda}_L \dfrac{0.32}{\mu} + \hat{\lambda}_L \dfrac{0.8}{\mu}\sqrt{(\mu - 1.653)(\mu - 0.097)} - \hat{\lambda}_L 0.7 - 1.1$	0.3|0	$1.653 + 0.294\overline{P}_B \le \mu < 1.8$
$D^{\bullet(1)}_{B,JI}$	$(1 - \hat{\lambda}_L)(\mu - 1.1) + \hat{\lambda}_L \overline{P}_B$	0.3|0	$1.4 \le \mu < 1.653 + 0.294\overline{P}_B$
$D^{\bullet(2)\dagger}_{B,JII}$	$\mu + \dfrac{0.32}{\mu} + \dfrac{0.8}{\mu}\left[\hat{\lambda}_L\sqrt{(\mu - 1.653)(\mu - 0.097)} + (1 - \hat{\lambda}_L)\sqrt{(\mu - 1.330)(\mu - 0.120)}\right] - \hat{\lambda}_L 0.3 - 1.5$	0.3|0	$1.653 + 0.294\overline{P}_B \le \mu < 1.4$
$D^{\bullet(1)}_{B,JII}$	$(1 - \hat{\lambda}_L)(\mu + \dfrac{0.32}{\mu} + \dfrac{0.8}{\mu}\sqrt{(\mu - 0.120)(\mu - 1.330)} - 1.5) + \hat{\lambda}_L \overline{P}_B$	0.3|0	$1.330 \le \mu < 1.4$

Note: The lender's and the borrower's expected profits $(PL(\mu, \hat{\lambda}_B), PB(\mu, \hat{\lambda}_L))$ with the respective feasible contracting constraints are stated for the case that the borrower has the power to determine contracts' conditions. The borrower's profits' type-dependency is expressed by $\overline{P}_B(\overline{P}_B \in \{0, P_B^+\})$, while the lender's profits' type-dependency is illustrated by the vertical line (PL for A-type lender|PL for N-type lender). The lender's expected profits are additionally (if required) conditioned on the borrower type for the respective ct/pm-choice is feasible $\left(\begin{cases} PL \text{ if the } ct/pm\text{-choice is only feasible for the N-type borrower} \\ PL \text{ if the } ct/pm\text{-choice is feasible for the A\&N-type borrowers} \end{cases}\right)$.

$\dagger$ = not feasible due to the NoN, the CDr/nr and the BCOC(AL) given PC1.

Based on these results the *optimal ct/pm*-choices in the different power scenarios, which are summarized as a preview in Table 4.27, are determined in the remainder of this section.

Table 4.27: Summary of Contract Type / Placement Mode-Choices (EM, PC1)

Interval	μ	LS^{a}		LS^{r}		BS^{r}		BS^{a}	
		A-type lender	N-type lender	A-type lender	N-type lender	A-type borrower	N-type borrower	A-type borrower	N-type borrower
I	[1-1.15)		−	−		−		−	
II	[1.15-1.25)				E°	E°			E°
III	[1.25-1.274)				$E^{\circ}, E^{\bullet}$		−	−	
IV	[1.274-1.318)				E°, D° $E^{\bullet}$				
V	[1.318-1.3)	−	E°	−					
VI	[1.3-1.368)				D°				
VII	[1.368-1.4)					D°	D°		D°
VIII	[1.4-1.45)				$D^{\circ}, D^{\bullet}$				
IX	[1.45-1.545)				E°, D°	$D^{\bullet}$			
X	[1.545-1.55)				$D^{\bullet}$				
XI	[1.55-1.6)			$E^{\bullet}$				D°	
XII	[1.6-1.653)				E°, D°	$E^{\bullet}, D^{\bullet}$			
XIII	[1.653-1.659)	E°	D°	$E^{\bullet}, D^{\bullet}$	$E^{\bullet}, D^{\bullet}$				
XIV	[1.659-1.741)								
XV	[1.741-1.75)					$D^{\bullet}$	D°		
XVI	[1.75-1.8)			$D^{\bullet}$	E°, D°	$D^{\circ}, D^{\bullet}$			
XVII	[1.8-1.85)				$D^{\bullet}$				$D^{\bullet}$
XVIII	[1.85-1.917)					$D^{\circ}, E^{\bullet}$			
XIX	[1.917-2.05)	D°		$E^{\bullet}, D^{\bullet}$	E°, D°	$D^{\bullet}$		$D^{\bullet}$	
XX	[2.05-2.132)				$E^{\bullet}, D^{\bullet}$	E°, D°	E°, D°		
XXI	[2.132-∞)	$E^{\bullet}$	$E^{\bullet}$			$E^{\bullet}, D^{\bullet}$			

Note: This table contains the preferred *ct/pm*-choices conditioned on the *ct/pm* optimizer's (the first mover's) type for the four alternative bargaining power scenarios. If there is more than one *ct/pm*-constellation stated for a particular contracting interval, the *ct/pm*-optimizer (first mover) is indifferent between them.

Among the preferred *ct/pm*-choices in the alternative power scenarios, Table 4.27 reveals that the feasible contracting areas, determined in Section 4.2, can be summarized to contracting intervals since ζ_{max} is fixed to 0.25 with μ being kept variable. Also the first mover's (*ct/pm* optimizer's) type becomes relevant for the preferred choice since the *ct/pm* optimizer has the power to determine *ct/pm*.[66]

4.3.1 Lender Scenario with Absolute Power (LS^a)

The *lender scenario with absolute power* (LS^a) is characterized by a lender who can propose a take-it-or-leave-it offer concerning *ct/pm* and contract conditions to the borrower, which the latter can accept or reject. If the borrower accepts, he receives the necessary funds to finance the project and proceeds. Otherwise the game ends and the lender as well as the borrower are left with their outside option.[67]

To determine the contract $(E^\circ, D^\circ, E^\bullet, D^\bullet)$ the lender prefers in this scenario, his expected profits $PL(\mu, \widehat{\lambda}_B)$ for the different contract choices must be compared. To make the lender's profits, stated in Table 4.25, comparable, the lender's expectation about the borrower's type $(\widehat{\lambda}_B)$ has to be defined at first. Since in this scenario the borrower cannot respond to the lender's *ct/pm*-choice as the lender determines contracts' conditions as well,[68] the lender expects to be confronted with the A-type borrower with probability $\widehat{\lambda}_B = \lambda_B = 0.5$ for all feasible *ct/pm*-choices.[69] The lender's profits are now comparable. However, before we can compare the different contract choices to determine the optimal one, we have to derive the optimal contracts given a particular *ct/pm*-constellation since so far the condition optimizer's optimal behavior, e. g. his strategy choice, for a given *ct/pm* has not been determined. Therefore, we now examine the lender's optimization behavior as the condition optimizer who expects $\widehat{\lambda}_B = \lambda_B$ (Stage 2 of the contracting game), before analyzing the lender's behavior as the *ct/pm* optimizer (Stage 1 of the contracting game).

Each profit comparison is a two step procedure. Firstly, the lender's type-dependent feasible contracting intervals are determined. Secondly, the optimal

---

[66] For public offerings no strategy choice is indicated since the strategy choice belongs to the second mover (the condition optimizer) who not necessarily coincides with the first mover (the *ct/pm* optimizer).

[67] In the absolute power scenarios no distinction between the *ct/pm* and the condition optimizer is necessary since they coincide. Hence, any condition depending on the condition optimizer's type depends in the same way on the *ct/pm* optimizer's type and vice versa.

[68] In the LS^a scenario the lender is both the first and the second mover. The borrower is "only" the third mover who can accept or reject the proposed contract offer.

[69] According to Assumption 6 λ_B is common knowledge.

contract choice is derived by a pair-wise (binary) comparison of the lender's expected profits of the feasible contract choices. Since all lender's (continuously differentiable) profit functions satisfy the conditions $0 \leq \dfrac{\partial PL(\mu)}{\partial \mu} \leq 1$, $\dfrac{\partial^2 PL(\mu)}{\partial \mu^2} \geq 0$, and $\dfrac{\partial^3 PL(\mu)}{\partial \mu^3} \leq 0$ a pair-wise comparison of the lender's "start" and "end" profits as well as the lender's "start" and "end" profit slopes, i. e. $\dfrac{\partial PL(\mu)}{\partial \mu}$, is able to reveal the lender's preference.[70] "Start" refers to the lowest value of μ where both compared contract choices are feasible while "end" refers to the highest value of such μ.[71] For all conceivable constellations of "start" and "end" profits and profit slopes the compared profit functions, here illustrated by $PL1(\mu)$ and $PL2(\mu)$, maximally intersect once, twice or not at all (see Table 4.28).[72]

---

[70] All lender's (continuously differentiable) expected profit functions satisfy these conditions since an increase in μ increases the extractable surplus ($\mu - 1$) as well as the spread of the return distribution ($2\zeta_{max}\mu$). However, PL cannot increase overproportionally ($\dfrac{\partial PL(\mu)}{\partial \mu} \not> 1$) because maximally the whole surplus increase can be distributed to the lender since the agency costs related to the respective ct/pm-choice never decline. Additionally, the agency costs associated with debt financing can increase ($\dfrac{\partial^2 PL(\mu)}{\partial \mu^2} > 0$ with $\dfrac{\partial^3 PL(\mu)}{\partial \mu^3} < 0$) while the agency costs of equity financing stay constant ($\dfrac{\partial^2 PL(\mu)}{\partial \mu^2} = 0$ with $\dfrac{\partial^3 PL(\mu)}{\partial \mu^3} = 0$). The agency costs of riskless debt stay constant, like the costs of equity, while the costs of risky debt increase with μ (e. g. $AC(D^\circ_{L,II}) = vc \cdot DP = vc \cdot \dfrac{h - y_{min}\big|_{\zeta=\zeta_{max}}}{2\zeta_{max}\mu} = vc\left(1 - \dfrac{vc}{2\zeta_{max}\mu}\right)$; $\dfrac{\partial DP}{\partial \mu} = \dfrac{vc}{2\zeta_{max}\mu^2} \geq 0$) but their severity decreases $\dfrac{\partial^2 DP}{\partial \mu^2} \leq 0$. The agency costs of risky debt are bounded at vc for public debt and at $vc + 2nc$ for private debt. Finally, since $\dfrac{\partial^3 DP}{\partial \mu^3} \geq 0$ $\dfrac{\partial^3 PL(\mu)}{\partial \mu^3} \leq 0$.

[71] Since μ is unbounded, it is possible that no real "end" exists when μ where both compared contract choices are feasible goes to infinity.

[72] Since we are (only) interested in an ordinal comparison (larger/equal or smaller), $2^4 = 16$ constellations of profit and profit slope comparisons are theoretically possible. But two of them turn out to be impossible and, therefore, 14 constellations have to be examined further. (The impossible constellations are: $PL1(Start) \geq PL2(Start) \wedge PL1(End) < PL2(End) \wedge \partial PL1(Start)/\partial \mu \geq \partial PL2(Start)/\partial \mu \wedge \partial PL1(End)/\partial \mu \geq \partial PL2(End)/\partial \mu$ as well as $PL1(Start) < PL2(Start) \wedge PL1(End) \geq PL2(End) \wedge \partial PL1(Start)/\partial \mu < \partial PL2(Start)/\partial \mu \wedge \partial PL1(End)/\partial \mu < \partial PL2(End)/\partial \mu$.)

Table 4.28: Conceivable Constellations of Lender's Profit Functions' Characteristics and Their Implications for the Profit Comparison (EM)

Number of intersections	$PL1(\text{Start})$ vs. $PL2(\text{Start})$	$PL1(\text{End})$ vs. $PL2(\text{End})$	$\partial PL1(\text{Start})/\partial\mu$ vs. $\partial PL2(\text{Start})/\partial\mu$	$\partial PL1(\text{End})/\partial\mu$ vs. $\partial PL2(\text{End})/\partial\mu$
none	$\geq$	$\geq$	$\geq$	$\geq$
	$<$	$<$	$\geq$	$\geq$
	$<$	$<$	$<$	$\geq$
	$\geq$	$\geq$	$\geq$	$<$
	$\geq$	$\geq$	$<$	$<$
	$<$	$<$	$<$	$<$
one	$<$	$\geq$	$\geq$	$\geq$
	$<$	$\geq$	$<$	$\geq$
	$\geq$	$<$	$<$	$\geq$
	$<$	$\geq$	$\geq$	$<$
	$\geq$	$<$	$\geq$	$<$
	$\geq$	$<$	$<$	$<$
two possible[73]	$\geq$	$\geq$	$<$	$\geq$
	$<$	$<$	$\geq$	$<$

The implications of these constellations for our profit comparison are straight forward. Firstly, when $PL1$ and $PL2$ can not intersect in the feasible contracting interval where both compared contract choices are feasible, a simple comparison of the "start" or the "end" profit reveals the lender's preference among these two contract choices for the examined interval. Secondly, when $PL1$ and $PL2$ intersect once the lender's preference among the compared contract choices changes from the "start" to the "end" and hence one contract choice is preferable in the "pre-intersection" interval and the other one in the "post-intersection" interval. In this case a comparison of the "start" profits reveals which contract choice is preferable for the "pre" and also for the "post-intersection" interval. Of course, the exact intersection point has to be determined to define the "pre" and the "post-intersection" interval explicitly. Finally, in case two intersections are *possible* the comparison is a bit more

---

[73] Whether the profit functions actually intersect twice in the examined interval can not be derived from the stated conditions. The conditions only indicate that two intersections are possible. Given the conditions are satisfied, it is also possible that the profit functions do not intersect at all or have one osculation point.

complex. At first it has to be examined whether the profit functions actually intersect since the conditions in Table 4.28 only imply that the functions can intersect twice, but they do not have to. If they do not intersect in the examined interval, we are back in the first case. But if they do the examined interval has to be divided into three subintervals. The first subinterval reaches up to the first intersection point; the second lies between both intersection points and the last one reaches from the second intersection point to the "end". According to two intersections the lender's preference among the compared contract choices also changes twice in the examined interval and a comparison of the "start" profit reveals the lender's order / preference.

In each profit comparison we distinguish between the *dominance* and the *preference* criteria. The former refers to a particular case of the latter. A contract choice dominates another choice if the former is preferable for the *whole* interval where both choices are feasible and this interval coincides with the latter's feasible interval. If a dominance relationship is found, the dominated contract choice can be excluded from the (remaining) comparison. A contract choice is (only) preferable to another choice if either the favorability of the former choice depends on the project specification (interval of μ) or the feasible interval of the latter choice is not a subset of the former.

THE LENDER'S BEHAVIOR AS THE CONDITION OPTIMIZER $(\widehat{\lambda}_B = \lambda_B)$

Public equity financing

1. Feasible contracting intervals
   - A-type lender's perspective $(PL \geq P_L^+)$
     $E_L^\circ|S1 : [1.75, \infty);\ E_L^\circ|S2 : [1.45, \infty)$
   - N-type lender's perspective $(PL \geq 0)$
     $E_L^\circ|S1 : [1.45, \infty);\ E_L^\circ|S2 : [1.15, \infty)$
2. Profit comparison
   - A-type lender's perspective $(\overline{P}_L = P_L^+)$
     $E_L^\circ|S1$ is preferable to $E_L^\circ|S2$ for $\mu \geq 2.05$ and for $\mu < 2.05$ the other way round.
   - N-type lender's perspective $(\overline{P}_L = 0)$
     $E_L^\circ|S1$ is preferable to $E_L^\circ|S2$ for $\mu \geq 1.75$ and for $\mu < 1.75$ the other way round.
3. Preferred contract choice given public equity financing
   - A-type lender
     For $\mu < 1.45$: no choice; for $1.45 \leq \mu < 2.05$: $E_L^\circ|S2$ and for $\mu \geq 2.05$: $E_L^\circ|S1$.
   - N-type lender
     For $\mu < 1.15$: no choice; for $1.15 \leq \mu < 1.75$: $E_L^\circ|S2$ and for $\mu \geq 1.75$: $E_L^\circ|S1$.

Public debt financing

1. Feasible contracting intervals
   - A-type lender's perspective $(PL \geq P_L^+)$
     $D_{L,Ia}^\circ : [1.7\bar{3}, \infty); D_{L,IIb}^\circ | S1 : [1.660, \infty); D_{L,IIb}^\circ | S2 : [1.6, \infty)$
   - N-type lender's perspective $(PL \geq 0)$
     $D_{L,Ia}^\circ : [1.\bar{3}, \infty); D_{L,IIb}^\circ | S1 : [1.319, \infty); D_{L,IIb}^\circ | S2 : [1.274, \infty)$
2. Profit comparison
   - A-type lender's perspective $(\overline{P}_L = P_L^+)$
     $D_{L,IIb}^\circ | S1$ dominates $D_{L,Ia}^\circ$; $D_{L,IIb}^\circ | S1$ is preferable to $D_{L,IIb}^\circ | S2$ for $\mu \geq 1.726$ and for $\mu < 1.726$ the other way round.
   - N-type lender's perspective $(\overline{P}_L = 0)$
     $D_{L,IIb}^\circ | S1$ dominates $D_{L,Ia}^\circ$; $D_{L,IIb}^\circ | S1$ is preferable to $D_{L,IIb}^\circ | S2$ for $\mu \geq 1.368$ and for $\mu < 1.368$ the other way round.
3. Preferred contract choice given public debt financing
   - A-type lender
     For $\mu < 1.6$: no choice; for $1.6 \leq \mu < 1.726$: $D_{L,IIb}^\circ | S2$ and for $\mu \geq 1.726$: $D_{L,IIb}^\circ | S1$.
   - N-type lender
     For 1.274: no choice; for $1.274 \leq \mu < 1.368$: $D_{L,IIb}^\circ | S2$ and for $\mu \geq 1.368$: $D_{L,IIb}^\circ | S1$.

Private equity financing

1. Feasible contracting intervals
   - A-type lender's perspective $(PL \geq P_L^+)$
     $E_L^{\bullet(2)} : [1.85, \infty); E_L^{\bullet(1)} : [1.55, 1.85)$
   - N-type lender's perspective $(PL \geq 0)$
     $E_L^{\bullet(2)} : [1.55, \infty); E_L^{\bullet(1)} : [1.25, 1.55)$
2. No profit comparison required
3. Preferred contract choice given private equity financing
   - A-type lender
     For $\mu < 1.55$: no choice; for $1.55 \leq \mu < 1.85$: $E_L^{\bullet(1)}$ and for $\mu \geq 1.85$: $E_L^{\bullet(2)}$.
   - N-type lender
     For $\mu < 1.25$: no choice; for $1.25 \leq \mu < 1.55$: $E_L^{\bullet(1)}$ and for $\mu \geq 1.55$: $E_L^{\bullet(2)}$.

Private debt financing

1. Feasible contracting intervals
   - A-type lender's perspective $(PL \geq P_L^+)$
     $D_{L,Ia}^{\bullet(2)} : [1.8, \infty); D_{L,IIa}^{\bullet(2)} : [1.8, 3.2); D_{L,IIa}^{\bullet(1)} : [1.653, 1.8); D_{L,IIc}^{\bullet(2)} : [3.2, \infty); D_{L,IIIb}^{\bullet(2)} : [1.741, 3.2]; D_{L,IIIb}^{\bullet(1)} : [1.653, 1.741); D_{L,IIIc}^{\bullet(2)} : (3.2, \infty)$

- N-type lender's perspective $(PL \geq 0)$
  $D_{L,Ia}^{\bullet(2)}$: $[1.4, \infty)$; $D_{L,IIa}^{\bullet(2)}$: $[1.4, 3.2)$; $D_{L,IIb}^{\bullet(1)}$: $(1.329, 1.4)$; $D_{L,IIc}^{\bullet(2)}$: $[3.2, \infty)$; $D_{L,IIIb}^{\bullet(2)}$: $[1.4, 3.2]$; $D_{L,IIIc}^{\bullet(2)}$: $(3.2, \infty)$

2. Profit comparison[74]
   - A-type lender's perspective $(\overline{P}_L = P_L^+)$
     $D_{L,IIIc}^{\bullet(2)}$ dominates $D_{L,IIc}^{\bullet(2)}$; $D_{L,IIIb}^{\bullet(2)}$ and $D_{L,IIIc}^{\bullet(2)}$ dominate $D_{L,Ia}^{\bullet(2)}$; $D_{L,IIIb}^{\bullet(2)}$ dominates $D_{L,IIa}^{\bullet(2)}$; $D_{L,IIIb}^{\bullet(2)}$ is preferable to $D_{L,IIa}^{\bullet(1)}$; $D_{L,IIa}^{\bullet(1)}$ and $D_{L,IIIb}^{\bullet(1)}$ generate the same expected profit.
   - N-type lender's perspective $(\overline{P}_L = 0)$
     $D_{L,IIIc}^{\bullet(2)}$ dominates $D_{L,IIc}^{\bullet(2)}$; $D_{L,IIIb}^{\bullet(2)}$ and $D_{L,IIIc}^{\bullet(2)}$ dominate $D_{L,Ia}^{\bullet(2)}$; $D_{L,IIIb}^{\bullet(2)}$ dominates $D_{L,IIa}^{\bullet(2)}$.

3. Preferred contract choice given private debt financing
   - A-type lender
     For $\mu < 1.653$: no choice; for $1.653 \leq \mu < 1.741$: $D_{L,IIa}^{\bullet(1)}$ or $D_{L,IIIb}^{\bullet(1)}$; for $1.741 \leq \mu \leq 3.2$: $D_{L,IIIb}^{\bullet(2)}$ and for $\mu > 3.2$: $D_{L,IIIc}^{\bullet(2)}$.
   - N-type lender
     For $\mu < 1.329$: no choice; for $1.329 \geq \mu < 1.4$: $D_{L,IIb}^{\bullet(1)}$; for $1.4 \leq \mu \leq 3.2$: $D_{L,IIIb}^{\bullet(2)}$ and for $\mu > 3.2$: $D_{L,IIIc}^{\bullet(2)}$.

THE LENDER'S BEHAVIOR AS THE ct/pm OPTIMIZER ($\widehat{\lambda}_B = \lambda_B$)

1. Feasible contracting intervals
   - A-type lender's perspective $(PL \geq P_L^+)$
     $E_L^\circ|S2$: $[1.45, 2.05)$; $E_L^\circ|S1$: $[2.05, \infty)$; $D_{L,IIb}^\circ|S2$: $[1.6, 1.726)$; $D_{L,IIb}^\circ|S1$: $[1.726, \infty)$; $E_L^{\bullet(1)}$: $[1.55, 1.85)$; $E_L^{\bullet(2)}$: $[1.85, \infty)$; $D_{L,IIa}^{\bullet(1)}$ and $D_{L,IIIb}^{\bullet(1)}$: $[1.653, 1.741)$; $D_{L,IIIb}^{\bullet(2)}$: $[1.741, 3.2]$; $D_{L,IIIc}^{\bullet(2)}$: $(3.2, \infty)$
   - N-type lender's perspective $(PL \geq 0)$
     $E_L^\circ|S2$: $[1.15, 1.75)$; $E_L^\circ|S1$: $[1.75, \infty)$; $D_{L,IIb}^\circ|S2$: $[1.274, 1.368)$; $D_{L,IIb}^\circ|S1$: $[1.368, \infty)$; $E_L^{\bullet(1)}$: $[1.25, 1.55)$; $E_L^{\bullet(2)}$: $[1.55, \infty)$; $D_{L,IIb}^{\bullet(1)}$: $[1.329, 1.4)$; $D_{L,IIIb}^{\bullet(2)}$: $[1.4, 3.2]$; $D_{L,IIIc}^{\bullet(2)}$: $(3.2, \infty)$

2. Profit comparison
   - A-type lender's perspective $(\overline{P}_L = P_L^+)$
     $E_L^\circ|S2$ dominates: $E_L^{\bullet(1)}$, $D_{L,IIb}^\circ|S2$, $D_{L,IIa}^{\bullet(1)}$ and $D_{L,IIIb}^{\bullet(1)}$; $E_L^{\bullet(2)}$ dominates: $E_L^\circ|S1$ and $D_{L,IIIc}^{\bullet(2)}$; $E_L^{\bullet(2)}$ is preferable to $D_{L,IIb}^\circ|S1$ for $\mu \geq 2.132$; $E_L^{\bullet(2)}$ is preferable to $D_{L,IIIb}^{\bullet(2)}$ for $\mu \geq 1.85$; $D_{L,IIb}^\circ|S2$ is preferable to $E_L^\circ|S2$ for $\mu \geq 1.917$; $D_{L,IIb}^\circ|S1$ is preferable to $E_L^{\bullet(1)}$ for

---

[74] The results of this profit comparison coincide with the preferences stated for private debt financing on page 157.

$\mu \geq 1.796$; $D^{\circ}_{L,IIb}|S1$ is preferable to $D^{\bullet(1)}_{L,IIa}$ for $\mu \geq 1.726$; $E^{\circ}_{L}|S2$ is preferable to $D^{\bullet(2)}_{L,IIIb}$ for $\mu < 1.983$.

- N-type lender's perspective ($\overline{P}_L = 0$)
  $E^{\circ}_{L}|S2$ dominates: $D^{\circ}_{L,IIb}|S2$, $E^{\bullet(1)}_{L}$ and $D^{\bullet(1)}_{L,IIb}$; $E^{\bullet(2)}_{L}$ dominates: $E^{\circ}_{L}|S1$ and $D^{\bullet(2)}_{L,IIIc}$; $D^{\circ}_{L,IIb}|S1$ is preferable to $E^{\circ}_{L}|S2$ for $\mu \geq 1.545$; $D^{\bullet(1)}_{L,IIIb}$ is preferable to $E^{\circ}_{L}|S2$ for $\mu \geq 1.631$; $E^{\bullet(2)}_{L}$ is preferable to $D^{\circ}_{L,IIb}|S1$ for $\mu \geq 2.132$; $E^{\circ}_{L}|S1$ is preferable to $D^{\circ}_{L,IIb}|S1$ for $\mu \geq 3.072$.

3. Preferred *ct/pm*-choices

Table 4.29: Contract Type / Placement Mode-Choices and Agreement Probabilities in the Lender Scenario with Absolute Power (EM, PC1)

Interval	μ	*Ct/pm*-choice and agreement probabilities			
		A-type lender		N-type lender	
I	$[1; 1.15)$	–	(0)	–	(0)
$II - VIII$	$[1.15; 1.45)$	–	(0)	$E^{\circ}_{L}|S2$	$(\frac{1}{2})$
IX	$[1.45; 1.545)$	$E^{\circ}_{L}|S2$	$(\frac{1}{2})$	$E^{\circ}_{L}|S2$	$(\frac{1}{2})$
$X - XVIII$	$[1.545; 1.917)$	$E^{\circ}_{L}|S2$	$(\frac{1}{2})$	$D^{\circ}_{L,IIb}|S1$	(1)
$XIX - XX$	$[1.917; 2.132)$	$D^{\circ}_{L,IIb}|S1$	(1)	$D^{\circ}_{L,IIb}|S1$	(1)
XXI	$[2.132; \infty)$	$E^{\bullet(2)}_{L}$	(1)	$E^{\bullet(2)}_{L}$	(1)

Note: This table contains the lender's *ct/pm*-choice together with the related contract agreement probabilities stated in brackets.

From the results stated in Table 4.29 it becomes obvious that the A and the N-type lender might favor the same *ct/pm*-choice for a particular contracting interval but they do not have to.[75] While for $\mu < 1.15$ no contract will be proposed or even agreed by either the A- or the N-type lender since the arising costs are higher than the extractable surplus, the N-type lender prefers to finance projects characterized by $1.15 \leq \mu < 1.545$ and PC1 with public S2 equity, i. e. incorporating placement risk, since the saved expected informational rent payment exceeds the expected loss from S2's placement risk. For $1.545 \leq \mu < 2.132$ the N-type lender favors public debt issues without the placement risk (S1), while for $\mu \geq 2.132$ he prefers private equity. On the other hand, the A-type lender still refuses to finance projects characterized by $1.15 \leq \mu < 1.45$ and PC1, i. e. the extractable surplus ($\mu - 1$) is insufficient to cover the arising costs, since the extractable profit is lower than

---

[75] The interval definitions in Table 4.29 are identical to the definitions in Table 4.27 summarizing all *ct/pm*-choices in the alternative bargaining power scenarios.

the profit of his business alternative. But if the A-type lender participates in project financing, he favors, like the N-type lender, to finance projects generating only a small (extra) profit ($1.45 \leq \mu < 1.917$) by public S2 equity before switching to public S1 debt ($1.917 \leq \mu < 2.132$) or privately placed equity ($\mu \geq 2.132$).[76]

As expected, for (feasible) projects with a small μ the lender chooses a public S2 offering since no severe loss ($PL - \overline{P}_L$) occurs from S2's placement risk.[77] However, due to an increase in the expected project return the placement risk of S2 becomes more severe. Public S1 offerings gain attractiveness. The lender might become willing to pay the N-type borrower an (ex-ante) informational rent to guarantee the A-type's participation. In our example, this strategy change is combined with a ct-change which does not have to occur but is likely to occur since the additional amount the lender has to pay to guarantee the A-type borrower's participation is lower for debt than for equity financing due to the borrower's informational rent from interim and ex-post uncertainty. For projects with even higher expected returns the lender prefers private equity since the total agency costs are lower and no placement risk is involved. The agency costs of private equity financing are lower since, firstly, the costs to cope with the ex-ante uncertainty are fixed in a private equity placement ($2nc$) while the costs of public debt increase. The costs of public debt increase because the borrower's informational rent from interim and ex-post uncertainty decreases in μ. Secondly, the agency costs to cope with the interim and ex-post uncertainty increase for the debt financing while the costs of equity financing stay constant.

The observation that equity contracts become favorable when μ increases is known from the basic model of Chapter 3 without an ex-ante uncertainty. This is due to the fact that their agency costs are independent of the agreed contract condition, here q, while for debt financing the agency costs from the default risk increase with h.

---

[76] In addition to the lender's ct/pm and strategy choice the resulting contract agreement probabilities are stated in Table 4.29. These probabilities depend for this scenario simply on the lender's strategy choice, but we state them since they enable us to compare the optimal contract choices in this absolute power scenario with the choices in the restricted power scenarios. In the restricted power scenarios the ct/pm and the condition optimizer do not coincide. Hence, in the restricted power scenarios the ct/pm optimizer can choose among different ct/pm-constellations with implied contract agreement probabilities due to the (anticipated) condition optimizer's behavior.

[77] If the offer is rejected by the borrower, the lender and the borrower are left with their outside option ($\overline{P}_j \; \forall j \in \{B, L\}$).

4.3.2 Lender Scenario with Restricted Power (LS^r)

In the *lender scenario with restricted power* (LS^r) the lender can choose the ct/pm while the borrower determines contracts' conditions. If these conditions are accepted by the lender, the borrower's project is financed. Otherwise the game ends and the lender as well as the borrower are left with their outside option.

To determine which contract choice ($E°, D°, E^\bullet, D^\bullet$) the lender prefers, his expected profits $PL(\mu, \widehat{\lambda}_B)$ must be compared. Unfortunately, contrary to the LS^a scenario, in this scenario the lender's (ct/pm optimizer's) profits depend directly on the borrower's optimization behavior. While in the LS^a scenario the borrower's behavior only affects the lender's expected profits in a marginal way, i. e. the borrower can choose to participate or not, in the LS^r scenario the borrower's behavior directly affects the lender's profits since the borrower has the right to set contracts' conditions. In public placements, for example, the borrower can choose among two strategies to cope with the uncertainty about the lender's investment alternative, where each strategy leaves a different profit to the lender. Hence, solving the contract negotiation game in this scenario is more difficult than in the LS^a scenario since the lender's and the borrower's optimization behaviors are interrelated. The lender's (ct/pm optimizer's) optimal ct/pm-choice is affected by the borrower's (condition optimizer's) contract determination, while on the other hand the borrower's contract determination is influenced by the lender's ct/pm-choice because in order to optimize contracts' conditions the borrower tries to anticipate the lender's type due to the latter's ct/pm-choice. However, the contract negotiation game can still be solved by backward induction, only a further assumption is necessary to define how the borrower derives his expectations about the lender's type (see below).

Of course, the borrower knows λ_L (Assumption 6), but in order to incorporate all information given about the lender, i. e. about the lender's type, when setting contracts' conditions he forms *conditional* expectations concerning the lender's type based on the latter's ct/pm-choice ($\widehat{\lambda}_L | x \; \forall x \in \{E°, D°, E^\bullet, D^\bullet\}$). These conditional expectations determine the borrower's (condition optimizer's) optimal behavior at Stage 2 of the contracting game, e. g. his strategy choice. According to the backward-induction-principle[78] the borrower's conditional expectations affect the lender's (ct/pm optimizer's) preferred ct/pm-choice (Stage 1 of the contracting game) as well, which is anticipated by the borrower and vice versa.[79] In such a situation, it is not possible to say anything about one player's optimal rational behavior, based on the player's expectations, without *simultaneously* studying the other player's rational behavior.

---

[78] See, for example, Varian (1992) or Mas-Colell et al. (1995).

[79] The lender's expectation about the borrower's type is uniquely defined ($\widehat{\lambda}_B = \lambda_B$) since the lender (first mover) has no additional information about the borrower's type when determining ct/pm.

To solve the (sequential) negotiation game it is necessary to determine

- the borrower's conditional expectations about the lender's type $(\widehat{\lambda}_L|x)$, and
- the lender's expectations about the borrower's expectations about the lender's type

since they are the last unknown elements of the game. The former expectations $(\widehat{\lambda}_L|x)$ affect the borrower's optimization behavior while the latter's are relevant for the lender's optimization (anticipating the borrower's (second mover's) behavior). Two alternatives exist to determine the borrower's expectations about the lender's type and, hence, the lender's expectations about the borrower's expectations. Firstly, an adjustment process for the borrower's expectations can be specified to derive stable expectations which the lender can anticipate.[80] However, this alternative incorporates a major disadvantage. The stable expectations depend significantly on the assumed adjustment process. An alternative process probably results in other stable expectations and it is not obvious which process is appropriate in the described context. Therefore, we favor the second alternative, the axiomatic approach, i. e. the borrower's expectations have to satisfy certain criteria. A reasonable requirement for the borrower's expectations in this context is that the expectations have to be consistent according to the following definition.[81]

Definition
Consistent Expectations
Expectations are consistent if they are (ex-post) confirmed by the optimization behavior based on these expectations.

The borrower's expectations $(\widehat{\lambda}_L|x)$ are consistent when they have reached a *fixpoint* in the sense that the lender's optimization behavior based on these expectations (the lender's ct/pm-choice) confirms the borrower's expectations about the lender's behavior. No explicit adjustment process is considered. The lender anticipates the borrower's expectations. The borrower's consistent expectations are determined in Section 4.3.2.1 while Section 4.3.2.2 defines the lender's preferred contract choice given these expectations.

---

[80] Expectations can be called stable if the adjustment process does not require a (further) variation of the borrower's expectations.

[81] We refer to these expectations as consistent expectations and not as rational expectations since the definition of rational expectations is broader. Basically, expectations are called rational when all available information is used to make the best forecast (cf., e. g., Romer (2000)). However, in our case also all information available is used to make the forecast but the forecast is primarily driven by an additional (new) constraint, i. e. the consistency requirement.

4.3.2.1 Derivation of Consistent Expectations

The idea behind the consistency requirement is illustrated in Figure 4.4.

Fig. 4.4: Idea of Consistency Requirement for the Borrower's Expectations in the Lender Scenario with Restricted Power (EM)

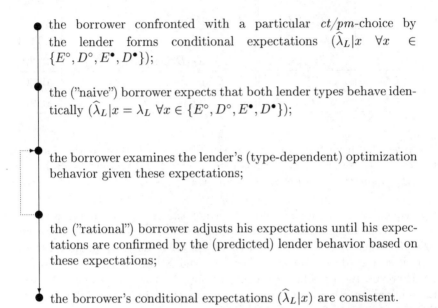

- the borrower confronted with a particular *ct/pm*-choice by the lender forms conditional expectations ($\widehat{\lambda}_L | x \ \forall x \in \{E^\circ, D^\circ, E^\bullet, D^\bullet\}$);

- the ("naive") borrower expects that both lender types behave identically ($\widehat{\lambda}_L | x = \lambda_L \ \forall x \in \{E^\circ, D^\circ, E^\bullet, D^\bullet\}$);

- the borrower examines the lender's (type-dependent) optimization behavior given these expectations;

- the ("rational") borrower adjusts his expectations until his expectations are confirmed by the (predicted) lender behavior based on these expectations;

- the borrower's conditional expectations ($\widehat{\lambda}_L | x$) are consistent.

This idea can be illustrated by the following example assuming $\mu = 2$. Given $\mu = 2$ the lender can choose between all four available *ct/pm*-constellations while for public offerings the borrower can affect the lender's as well as his own profit by his strategy choice. All possible profit allocations are illustrated below.[82]

---

[82] The expected profits are from Table 4.26. The expected profits not explicitly stated in Table 4.30 are not feasible for $\mu = 2$. Since the lender can not affect the borrower's type nor has he any additional information about the borrower's type when choosing *ct/pm*, he expects $\widehat{\lambda}_B = \lambda_B = 0.5$. Since the lender's profit is still lender type dependent, this dependency is again illustrated by the vertical line (*PL* for A-type lender | *PL* for N-type lender).

Table 4.30: Feasible Contract Choices and Expected Profits Given $\mu = 2$ (EM, Borrower Optimization)

Lender's ct/pm-choice	Available choices for the borrower	PB	PL
E°	$E_B^\circ\|S1$	0.55	0.3|0.3
	$E_B^\circ\|S2$	$(1 - \widehat{\lambda}_L\|E^\circ) \cdot 0.85 + \widehat{\lambda}_L\|E^\circ \cdot \overline{P}_B$	0.3|0
D°	$D_{B,I}^\circ\|S1$	0.7	0.3|0.3
	$D_{B,I}^\circ\|S2$	$(1 - \widehat{\lambda}_L\|D^\circ) + \widehat{\lambda}_L\|D^\circ \cdot \overline{P}_B$	0.3|0
$E^\bullet$	$E_B^{\bullet(2)}$	$0.75 - \widehat{\lambda}_L\|E^\bullet \cdot 0.3$	0.3|0
$D^\bullet$	$D_{B,I}^{\bullet(2)}$	$0.9 - \widehat{\lambda}_L\|D^\bullet \cdot 0.3$	0.3|0

Given the profit functions summarized in Table 4.30 the derivation of the borrower's consistent expectations can be nicely illustrated. Assuming as a starting point that the borrower being confronted with a particular ct/pm-choice is "naïve" and hence expects $\widehat{\lambda}_L\|x = \lambda_L = 0.5 \ \forall x \in \{E^\circ, D^\circ, E^\bullet, D^\bullet\}$. Consequently, for a public equity contract the A-type borrower prefers S2 to S1 due to his positive outside option in case of no contract agreement. S2's placement risk is larger for the N-type borrower since he does not have a positive outside option and, therefore, the N-type borrower favors S1 to S2. However, for a public debt contract both borrower types favor S1 to S2. Examining now the lender's (type-dependent) optimization behavior, given these expectations, reveals the following:

- The A-type lender (expecting $\widehat{\lambda}_B = \lambda_B = 0.5$) favors $D^\circ, E^\bullet$ and $D^\bullet$ to E° despite the fact that for all four contract choices he is left with the same expected profit ($PL = 0.3$) since he values contract agreement over no contract agreement (same assumption as in Chapter 3, see footnote 32). For $D^\circ, E^\bullet$ and $D^\bullet$ a contract agreement is sure while for E° the A-type lender might be confronted with an A-type borrower and hence no agreement will be reached.
- The N-type lender (expecting $\widehat{\lambda}_B = \lambda_B = 0.5$) definitively prefers D° to $E^\circ, E^\bullet$ and $D^\bullet$.[83]

When the borrower is not "naive", as we assumed, he will anticipate the lender's behavior and adjust his expectations accordingly: $\widehat{\lambda}_L\|E^\circ = -$, i. e. the borrower expects no public equity choice (neither from the A-type nor from the N-type lender); $\widehat{\lambda}_L\|D^\circ = \lambda_L = 0.5$; $\widehat{\lambda}_L\|E^\bullet = 1$ and $\widehat{\lambda}_L\|D^\bullet = 1$. However,

---

[83] Assuming $\widehat{\lambda}_L\|x = \lambda_L = 0.5 \ \forall x \in \{E^\circ, D^\circ, E^\bullet, D^\bullet\}$ and the borrower's resulting optimization behavior, the N-type lender's expected profit for E° is 0.15, for D° 0.3 and for both private placements 0.

given these adjusted expectations the borrower's optimization behavior, i. e. his strategy choice for public placements, might change. Consequently, also the lender's preferred ct/pm-choice might change which the borrower again anticipates and so on. No more changes will occur when the borrower's expectations are consistent in the sense that they have reached a fixpoint, i. e. the lender's optimization behavior based on these expectations confirms the borrower's expectations about the lender's behavior. Finally, two sets of consistent expectations are possible outcomes of the borrower's expectation formation:[84]

Set 1: $\widehat{\lambda}_L|E^\circ = -\widehat{\lambda}_L|D^\circ = \lambda_L \ \widehat{\lambda}_L|E^\bullet = 1 \ \widehat{\lambda}_L|D^\bullet = 1$
Set 2: $\widehat{\lambda}_L|E^\circ = 0 \ \widehat{\lambda}_L|D^\circ = 0 \ \ \widehat{\lambda}_L|E^\bullet = \lambda_L \ \widehat{\lambda}_L|D^\bullet = \lambda_L$

In order to derive the borrower's final / unique expectations, we assume that the borrower anticipates (again) the lender's behavior being confronted with these two possible sets of consistent expectations and adjusts his expectations accordingly.

An A-type lender being confronted with these two possible sets of consistent expectations definitively prefers private placements to public offerings since in case of $\widehat{\lambda}_L|E^\circ = 0$ or $\widehat{\lambda}_L|D^\circ = 0$ the borrower favors S2 to S1. On the other side the N-type lender favors public debt in the hope for an informational rent due to $\widehat{\lambda}_L|D^\circ = \lambda_L$. However, this lender behavior is also anticipated and, therefore, the borrower's final consistent expectations are $\widehat{\lambda}_L|E^\circ = 0$; $\widehat{\lambda}_L|D^\circ = 0$; $\widehat{\lambda}_L|E^\bullet = \lambda_L$, and $\widehat{\lambda}_L|D^\bullet = \lambda_L$ for $\mu = 2$.[85]

To determine the borrower's consistent conditional expectations, i. e. a set of consistent conditional expectations (SCCE), we proceed as follows:[86]

- STEP 1: Firstly, based on intuitive arguments all plausible SCCE candidates are selected from all possible constellations of conditional expectations.
- STEP 2: Secondly, the contract negotiation game is solved for the SCCE candidates to examine whether the conditional expectations are actually consistent, i. e. confirmed by the lender's and the borrower's optimization behavior, or not.

---

[84] How these conceivable candidates are found, is described in detail below. In principal all feasible candidates are defined and then it is examined whether the borrower's and hence the lender's optimization behavior are consistent with the expectations or not.

[85] Since the N-type borrower does not (anymore) obtains an informational rent for public debt financing, he becomes indifferent between all four ct/pm-constellations.

[86] A set of consistent conditional expectations (SCCE) has to be defined since the consistency of the borrower's expectation for a particular ct/pm-constellation hinges on the borrower's expectations for the other feasible constellations.

- STEP 3: Finally, if more than one SCCE candidate is consistent, the lender's optimization behavior is examined when being confronted with more than one possible SCCE. Thereby, the SCCE candidate the borrower prefers is determined since this candidate represents the lender's predicted behavior.

STEP 1 OF EXPECTATION DERIVATION

Being confronted with a particular ct/pm-choice x ($x \in \{E^\circ, D^\circ, E^\bullet, D^\bullet\}$) (by the lender) the borrower expects either that: Only the N-type lender chooses the particular ct/pm ($\widehat{\lambda}_L|x = 0$); only the A-type lender ($\widehat{\lambda}_L|x = 1$); or both lender types ($\widehat{\lambda}_L|x = \lambda_L$). Therefore, $3^4 = 81$ possible constellations of conditional expectations emerge.[87]

Three additional observations help to restrict the number of possible constellations of consistent conditional expectations (SCCE).[88]

- $\widehat{\lambda}_L|E^\circ = 1$ and $\widehat{\lambda}_L|D^\circ = 1$ can *not* be consistent expectations since for $\widehat{\lambda}_L|E^\circ = \widehat{\lambda}_L|D^\circ = 1$ the borrower's preferred strategy is S1 which attracts the A- and the N-type lender and not only the A-type. The N-type lender even achieves an ex-ante informational rent, the highest profit he can extract for each possible ct/pm-choice. Moreover, the A-type lender's feasible set of E°, respectively D°, is a subset of the N-type's which implies that if E° or D° is a feasible choice for the A-type (implicit requirement for $\widehat{\lambda}_L|x = 1 \ \forall x \in \{E^\circ, D^\circ\}$), the N-type's FCCs are satisfied as well.
- $\widehat{\lambda}_L|E^\circ = \lambda_L$ can *not* be consistently combined with $\widehat{\lambda}_L|D^\circ = 0, \widehat{\lambda}_L|E^\bullet \in \{0, \lambda_L\}$ and $\widehat{\lambda}_L|D^\bullet \in \{0, \lambda_L\}$; and $\widehat{\lambda}_L|D^\circ = \lambda_L$ can *not* be consistently combined with $\widehat{\lambda}_L|E^\circ = 0, \widehat{\lambda}_L|E^\bullet \in \{0, \lambda_L\}$ and $\widehat{\lambda}_L|D^\bullet \in \{0, \lambda_L\}$. These constellations can not be consistent since for each *public* offering (E°, D°) when the borrower expects λ_L ($\widehat{\lambda}_L|x = \lambda_L \ \forall x \in \{E^\circ, D^\circ\}$) S1 is the borrower's preferred strategy, otherwise the A-type lender is unwilling to participate. However, if S1 is optimal for E°, respectively D°, the N-type lender will *never* choose a private placement or an alternative public S2 offering. The N-type lender will prefer the A-type lender's public ct-choice to achieve an ex-ante informational rent. Therefore, the stated expectations can not be consistently combined.
- $\widehat{\lambda}_L|E^\bullet = 0$ and $\widehat{\lambda}_L|D^\bullet = 0$ can only be consistent if the borrower's contract offer condition for the A-type lender (BCOC(AL)) for $E^\bullet$, respectively $D^\bullet$,

---

[87] For each of the four possible ct/pm-choices ($E^\circ, D^\circ, E^\bullet, D^\bullet$) the borrower expects either $\widehat{\lambda}_L|x = 0, \widehat{\lambda}_L|x = \lambda_L$, or $\widehat{\lambda}_L|x = 1$.

[88] The observations are due to the fact that the lender has an interim and ex-post informational disadvantage. Therefore, the lender can only achieve a positive informational rent from his ex-ante informational advantage. Hence, the N-type lender favors public offerings to private placements to keep the chance of an ex-ante rent.

is not satisfied. Once private placements are feasible for the A-type lender, i. e. the borrower offers acceptable contract conditions, the A-type lender either prefers $E^\bullet$ or $D^\bullet$, or he favors a public offering which is then also preferable for the N-type. Hence, $\widehat{\lambda}_L|E^\bullet = 0$ and $\widehat{\lambda}_L|D^\bullet = 0$ can only be consistent as long as the borrower's BCOC(AL) is not satisfied. Moreover, since both borrower types have identical information about the lender, i. e. the latter's ct/pm-choice and λ_L, their expectations about the lender's type coincide and, hence, for $\widehat{\lambda}_L|E^\bullet = \widehat{\lambda}_L|D^\bullet = 0$ to be consistent the A and the N-type borrower's BCOC(AL) has to be violated.[89]

Due to these observations, only 10 reasonable sets of probably consistent conditional expectations (SCCE candidates) remain. The expectations are of course only relevant when the respective ct/pm is feasible.[90]

Table 4.31: Possible Sets of Borrower's Consistent Conditional Expectations (SCCE) in the Lender Scenario with Restricted Power (EM)

SCCE	C1	C2	C3	C4	C5	C6*	C7*	C8*	C9*	C10*
$\widehat{\lambda}_L|E^\circ$	0	0	0	0	λ_L	0	0	0	0	0
$\widehat{\lambda}_L|D^\circ$	0	0	0	0	λ_L	0	0	0	0	0
$\widehat{\lambda}_L|E^\bullet$	λ_L	1	λ_L	1	1	0	λ_L	0	1	0
$\widehat{\lambda}_L|D^\bullet$	λ_L	λ_L	1	1	1	0	0	λ_L	0	1

* = N-type borrower's BCOC(AL) for $E^\bullet$, respectively $D^\bullet$, has to be violated.

After the derivation of SCCE candidates their consistency is examined by solving the negotiation game for each candidate.

STEP 2 OF EXPECTATION DERIVATION

To solve the contract negotiation game for each SCCE candidate and PC1, we firstly determine the borrower's optimization behavior for each possible ct/pm-choice given these expectations.[91] By setting contracts' conditions the

---

[89] Furthermore, since the N-type borrower possesses contrary to the A-type no outside option the former's BCOC(AL) defines the interval where $\widehat{\lambda}_L|E^\bullet = \widehat{\lambda}_L|D^\bullet = 0$ can be consistent. The N-type's BCOC(AL) is not as strict as the A-type's.

[90] Reading example: Given SCCE candidate 1 (C1), the borrower expects for each public offering chosen by the lender to be confronted with the N-type lender ($\widehat{\lambda}_L|E^\circ = \widehat{\lambda}_L|D^\circ = 0$) while for each private placement choice the borrower expects the A-type lender with probability λ_L and the N-type with $1 - \lambda_L$. (A lender's private placement choice does not reveal additional information about the lender's type.)

[91] The question whether a particular ct/pm-choice is at all feasible from the borrower's perspective, i. e. given his private information, belongs to the

borrower can choose for public offerings among S1 and S2, whereas for private placements he can choose whether to offer both lender types acceptable contract conditions or only the N-type ($E^{\bullet(2)}, D^{\bullet(2)}$ vs. $E^{\bullet(1)}, D^{\bullet(1)}$). Moreover, the borrower can choose whether to offer risky or riskless debt contracts. Since the borrower's optimization behavior for a given ct/pm-choice depends on his private information, i. e. his type, each borrower type is considered separately in the following analysis.

Public equity financing ($\widehat{\lambda}_L | E^\circ \in \{0, \lambda_L\}$):

- for $\widehat{\lambda}_L | E^\circ = 0$
  the A-type borrower favors S2 to S1. Hence, he chooses $E^\circ_B | S2$ for $\mu \geq 1.45$;
  the N-type borrower favors S2 to S1. Hence, he chooses $E^\circ_B | S2$ for $\mu \geq 1.15$.
- for $\widehat{\lambda}_L | E^\circ = \lambda_L$
  the A-type borrower favors S1 to S2 for $\mu \geq 2.05$ and for $\mu < 2.05$ the other way round. Hence, he chooses $E^\circ_B | S2$ for $1.45 \leq \mu < 2.05$ and $E^\circ_B | S1$ for $\mu \geq 2.05$;
  the N-type borrower favors S1 to S2 for $\mu \geq 1.75$ and for $\mu < 1.75$ the other way round. Hence, he chooses $E^\circ_B | S2$ for $1.15 \leq \mu < 1.75$ and $E^\circ_B | S1$ for $\mu \geq 1.75$.

Public debt financing ($\widehat{\lambda}_L | D^\circ \in \{0, \lambda_L\}$):

- for $\widehat{\lambda}_L | D^\circ = 0$
  the A-type borrower favors S2 to S1. Hence, he chooses $D^\circ_{B,II} | S2$ for $1.318 \leq \mu < 1.\bar{3}$ and $D^\circ_{B,I} | S2$ for $\mu \geq 1.\bar{3}$;
  the N-type borrower favors S2 to S1. Hence, he chooses $D^\circ_{B,II} | S2$ for $1.274 \leq \mu < 1.\bar{3}$ and $D^\circ_{B,I} | S2$ for $\mu \geq 1.\bar{3}$.
- for $\widehat{\lambda}_L | D^\circ = \lambda_L$
  the A-type borrower favors $D^\circ_{B,I} | S1$ to $D^\circ_{B,I} | S2$ for $\mu \geq 1.9$ and $D^\circ_{B,II} | S1$ is dominated by $D^\circ_{B,I} | S2$. Hence, he chooses $D^\circ_{B,II} | S2$ for $1.318 \leq \mu < 1.\bar{3}$, $D^\circ_{B,I} | S2$ for $1.\bar{3} \leq \mu < 1.9$ and $D^\circ_{B,I} | S1$ for $\mu \geq 1.9$;
  the N-type borrower favors $D^\circ_{B,I} | S1$ to $D^\circ_{B,I} | S2$ for $\mu \geq 1.6$ and $D^\circ_{B,II} | S1$ to $D^\circ_{B,I} | S2$ for $\mu \geq 1.680$. Hence, he chooses $D^\circ_{B,II} | S2$ for $1.274 \leq \mu < 1.\bar{3}$, $D^\circ_{B,I} | S2$ for $1.\bar{3} \leq \mu < 1.680$, $D^\circ_{B,II} | S1$ for $1.680 \leq \mu < 1.7\bar{3}$ and $D^\circ_{B,I} | S1$ for $\mu \geq 1.7\bar{3}$.

---

borrower's optimization behavior. If a ct/pm-choice is not feasible from the borrower's perspective, he refuses to participate and the game ends. The lender and the borrower are left with their outside option. However, a contract's feasibility from the borrower's perspective does not imply that the project can be financed by the respective contract since the borrower's feasibility judgement depends on the borrower's expectation about the lender's type while the borrower is unaware about the actual lender type.

Private equity financing ($\widehat{\lambda}_L|E^\bullet \in \{\lambda_L, 1\}$ and $\widehat{\lambda}_L|E^\bullet = 0$ if the N-type borrower's BCOC(AL) is violated):

- for $\widehat{\lambda}_L|E^\bullet = 0$[92]
  the N-type borrower chooses $E_B^{\bullet(1)}$ for $1.25 \leq \mu < 1.55$.

- for $\widehat{\lambda}_L|E^\bullet = \lambda_L$
  the A-type borrower chooses $E_B^{\bullet(1)}$ for $1.55 \leq \mu < 1.85$ and $E_B^{\bullet(2)}$ for $\mu \geq 1.85$;
  the N-type borrower chooses $E_B^{\bullet(1)}$ for $1.25 \leq \mu < 1.55$ and $E_B^{\bullet(2)}$ for $\mu \geq 1.55$.

- for $\widehat{\lambda}_L|E^\bullet = 1$[93]
  the A-type borrower only chooses $E_B^{\bullet(2)}$ for $\mu \geq 1.85$;
  the N-type borrower only chooses $E_B^{\bullet(2)}$ for $\mu \geq 1.55$.

Private debt financing ($\widehat{\lambda}_L|D^\bullet \in \{\lambda_L, 1\}$ and $\widehat{\lambda}_L|D^\bullet = 0$ if the N-type borrower's BCOC(AL) is not satisfied):

- for $\widehat{\lambda}_L|D^\bullet = 0$
  the A-type borrower chooses $D_{B,II}^{\bullet(1)}$ for $1.4 \leq \mu < 1.653$;[94]
  the N-type borrower chooses $D_{B,III}^{\bullet(1)}$ for $1.330 \leq \mu < 1.4$ and $D_{B,II}^{\bullet(1)}$ for $1.4 \leq \mu < 1.653$

- for $\widehat{\lambda}_L|D^\bullet = \lambda_L$.
  the A-type borrower chooses $D_{B,II}^{\bullet(1)}$ for $1.4 \leq \mu < 1.741$, $D_{B,II}^{\bullet(2)}$ for $1.741 \leq \mu < 1.8$ and $D_{B,I}^{\bullet(2)}$ for $\mu \geq 1.8$;
  the N-type borrower chooses $D_{B,III}^{\bullet(1)}$ for $1.330 \leq \mu < 1.4$, $D_{B,II}^{\bullet(1)}$ for $1.4 \leq \mu < 1.653$, $D_{B,II}^{\bullet(2)}$ for $1.653 \leq \mu < 1.8$ and $D_{B,I}^{\bullet(2)}$ for $\mu \geq 1.8$.

- for $\widehat{\lambda}_L|D^\bullet = 1$
  the A-type borrower chooses $D_{B,II}^{\bullet(2)}$ for $1.741 \leq \mu < 1.8$ and $D_{B,I}^{\bullet(2)}$ for $\mu \geq 1.8$;
  the N-type borrower chooses $D_{B,II}^{\bullet(2)}$ for $1.653 \leq \mu < 1.8$ and $D_{B,I}^{\bullet(2)}$ for $\mu \geq 1.8$.

---

[92] Since the consistency of $\widehat{\lambda}_L|E^\bullet = 0$ requires that the N-type borrower's BCOC(AL) is violated, i. e. the surplus from project financing ($\mu - 1$) is not sufficient to cover the agency costs of private equity financing ($mc + 2nc$) and the A-type lender's profit requirement (P_L^+), the A-type borrower never consistently expects $\widehat{\lambda}_L|E^\bullet = 0$. The A-type borrower's BPC can not be satisfied under these circumstances ($P_L^+ = P_B^+$, PC1).

[93] For $\widehat{\lambda}_L|E^\bullet = 1$ the borrower's alternative of offering $E^{\bullet(1)}$ is worthless since rejected by the (expected) A-type lender, hence only $E^{\bullet(2)}$ is feasible from the borrower's perspective.

[94] $D_{B,II}^{\bullet(1)}$ is restricted by the N-type borrower's BCOC(AL).

After examining the borrower's optimization behavior, the lender's resulting expected profits are analyzed to define the lender's respective optimization behavior, i. e. the latter's ct/pm-choice. Contracts' feasibility from the lender's perspective, i. e. given the lender's private information about his type, is not considered explicitly at this stage since the lender keeps an option to end the game until Stage 3 of the contract negotiation game (accept or reject the offered conditions). However, the contract agreement probabilities are considered since, like in the basic model of Chapter 3 (see, e. g., footnote 32 on page 53), we assume that the lender values contract agreement above no agreement. If the lender's expected profits from contract agreement or not coincide ($PL = \overline{P}_L$), the lender favors contract participation.[95] Since the lender cares about expected profits and contract agreement probabilities, both are stated (lender type-dependent) in the following table.[96]

Examining, for example, public equity financing given $\widehat{\lambda}_L|E^\circ = \lambda_L$, the borrower acts according to the defined optimization behavior which implies for the lender's expected profit for $1.15 \leq \mu < 1.75$: $0.3|0$; for $1.75 \leq \mu < 2.05$: $0.3|(1 - \lambda_B)0.3$ and for $\mu \geq 2.05$: $0.3|0.3$. However, to evaluate public equity financing appropriately a finer differentiation is required since even if the lender's expected profits are constant the (lender type-dependent) contract agreement probabilities can vary. Therefore, we obtain for public equity financing given $\widehat{\lambda}_L|E^\circ = \lambda_L$: for $1.15 \leq \mu < 1.45$: $\dfrac{0.3|0}{(0|1 - \lambda_B)}$; for $1.45 \leq \mu < 1.75$: $\dfrac{0.3|0}{(0|1)}$; for $1.75 \leq \mu < 2.05$: $\dfrac{0.3|(1 - \lambda_B)0.3}{(1 - \lambda_B|1)}$ and for $\mu \geq 2.05$: $\dfrac{0.3|0.3}{(1|1)}$. The contract agreement probabilities are stated italic and in brackets below the respective expected profits, e. g. for $1.15 \leq \mu < 1.45$ the A-type lender's profit requirement is too high and hence he is left with his outside option, i. e. $PL = 0.3$ with contract agreement probability zero, while the N-type lender will agree a contract with the N-type borrower but not with the A-type, i. e. contract agreement probability $(1 - \lambda_B)$.[97]

---

[95] Consequently, the lender favors the ct/pm-choice with the highest contract agreement probability among choices with the same expected profits. Contract agreement probabilities are only relevant when no uniquely defined optimal ct/pm-choice exist since more than one choice offers the maximum achievable profit.

[96] The borrower's potential expectations about the lender's type are not borrower type-dependent since both borrowers have at this stage the same information about the lender, i. e. the latter's ct/pm-choice and λ_L.

[97] We do not substitute λ_L by 0.5 in accordance with PC1 in Table 4.32 since the individual profit components become more obvious when we do not substitute λ_L already.

Table 4.32: Lender's Expected Profits and Contract Agreement Probabilities in the Lender Scenario with Restricted Power Depending on the Borrower's Expectation about the Lender's Type (EM, PC1)

Lender's ct/pm-choice	Borrower's expectation	Lender's expected profits and contract agreement probabilities			
		from	from	from	from
E°	$\widehat{\lambda}_L\|E^\circ = 0$	1.15 $0.3\|0$ $(0\|1-\lambda_B)$	1.45 $0.3\|0$ $(0\|1)$		
	$\widehat{\lambda}_L\|E^\circ = \lambda_L$	1.15 $0.3\|0$ $(0\|1-\lambda_B)$	1.45 $0.3\|0$ $(0\|1)$	1.75 $0.3\|(1-\lambda_B)0.3$ $(1-\lambda_B\|1)$	2.05 $0.3\|0.3$ $(1\|1)$
D°	$\widehat{\lambda}_L\|D^\circ = 0$	1.274 $0.3\|0$ $(0\|1-\lambda_B)$	1.318 $0.3\|0$ $(0\|1)$		
	$\widehat{\lambda}_L\|D^\circ = \lambda_L$	1.274 $0.3\|0$ $(0\|1-\lambda_B)$	1.318 $0.3\|0$ $(0\|1)$	1.680 $0.3\|(1-\lambda_B)0.3$ $(1-\lambda_B\|1)$	1.9 $0.3\|0.3$ $(1\|1)$
$E^\bullet$	$\widehat{\lambda}_L\|E^\bullet = 0$	1.25 $0.3\|0$ $(0\|1-\lambda_B)$	1.55 —		
	$\widehat{\lambda}_L\|E^\bullet = \lambda_L$	1.25 $0.3\|0$ $(0\|1-\lambda_B)$	1.55 $0.3\|0$ $(1-\lambda_B\|1)$	1.85 $0.3\|0$ $(1\|1)$	
	$\widehat{\lambda}_L\|E^\bullet = 1$	1.55 $0.3\|0$ $(1-\lambda_B\|1-\lambda_B)$	1.85 $0.3\|0$ $(1\|1)$		
$D^\bullet$	$\widehat{\lambda}_L\|D^\bullet = 0$	1.330 $0.3\|0$ $(0\|1-\lambda_B)$	1.4 $0.3\|0$ $(0\|1)$	1.653 —	
	$\widehat{\lambda}_L\|D^\bullet = \lambda_L$	1.330 $0.3\|0$ $(0\|1-\lambda_B)$	1.4 $0.3\|0$ $(0\|1)$	1.653 $0.3\|0$ $(1-\lambda_B\|1)$	1.741 $0.3\|0$ $(1\|1)$
	$\widehat{\lambda}_L\|D^\bullet = 1$	1.653 $0.3\|0$ $(1-\lambda_B\|1-\lambda_B)$	1.741 $0.3\|0$ $(1\|1)$		

Given these expected profits, the consistency of the SCCE candidates stated in Table 4.31 can be evaluated. Since the expectations are conditioned on the lender's ct/pm-choice, they are only relevant if the lender actually chooses the respective ct/pm-constellation. The borrower's expectations about the lender's type can either be validated or rejected if the lender chooses the respective ct/pm. The consistency of conditional expectations can not be examined if the respective ct/pm-constellation is not chosen by the lender. However, if the respective ct/pm is never chosen the expectation conditioned of this constellation becomes irrelevant.

Analyzing, for example, SCCE C1 $(\widehat{\lambda}_L|E^\circ = 0, \widehat{\lambda}_L|D^\circ = 0, \widehat{\lambda}_L|E^\bullet = \lambda_L, \widehat{\lambda}_L|D^\bullet = \lambda_L)$ the lender is confronted with the situation stated in Table 4.33.

Table 4.33: Consistency Check of SCCE C1 (EM, PC1)

μ	E°	D°	$E^\bullet$	$D^\bullet$	Lender's ct/pm-choice A-type	N-type	Confirmed expectations
1–1.15	–	–	–	–	–	–	–
1.15–1.25	0.3|0 (0|0.5)	–	–	–	–	E°	(0,-,-,-)
1.25–1.274	0.3|0 (0|0.5)	–	0.3|0 (0|0.5)	–	–	$E^\circ, E^\bullet$	–
1.274–1.318	0.3|0 (0|0.5)	0.3|0 (0|0.5)	0.3|0 (0|0.5)	–	–	E°, D° $E^\bullet$	–
1.318–1.330	0.3|0 (0|0.5)	0.3|0 (0|1)	0.3|0 (0|0.5)	–	–	D°	(-,0,-,-)
1.330–1.4	0.3|0 (0|0.5)	0.3|0 (0|1)	0.3|0 (0|0.5)	0.3|0 (0|0.5)	–	D°	(-,0,-,-)
1.4–1.45	0.3|0 (0|0.5)	0.3|0 (0|1)	0.3|0 (0|0.5)	0.3|0 (0|1)	–	$D^\circ, D^\bullet$	–
1.45–1.55	0.3|0 (0|1)	0.3|0 (0|1)	0.3|0 (0|0.5)	0.3|0 (0|1)	–	E°, D° $D^\bullet$	–
1.55–1.653	0.3|0 (0|1)	0.3|0 (0|1)	0.3|0 (0.5|1)	0.3|0 (0|1)	$E^\bullet$	E°, D° $E^\bullet, D^\bullet$	–
1.653–1.741	0.3|0 (0|1)	0.3|0 (0|1)	0.3|0 (0.5|1)	0.3|0 (0.5|1)	$E^\bullet, D^\bullet$	E°, D° $E^\bullet, D^\bullet$	$(0,0,\lambda_L,\lambda_L)$
1.741–1.85	0.3|0 (0|1)	0.3|0 (0|1)	0.3|0 (0.5|1)	0.3|0 (1|1)	$D^\bullet$	E°, D° $E^\bullet, D^\bullet$	–
1.85–∞	0.3|0 (0|1)	0.3|0 (0|1)	0.3|0 (1|1)	0.3|0 (0.5|1)	$E^\bullet, D^\bullet$	E°, D° $E^\bullet, D^\bullet$	$(0,0,\lambda_L,\lambda_L)$

Note: The lender's (type-dependent) expected profits and contract agreement probabilities are stated for all four possible ct/pm-choices ($E^\circ,D^\circ,E^\bullet,D^\bullet$). This information is evaluated to determine the lender's preferred ct/pm-choice given SCCE C1 ($\widehat{\lambda}_L|E^\circ = 0, \widehat{\lambda}_L|D^\circ = 0, \ \widehat{\lambda}_L|E^\bullet = \lambda_L, \widehat{\lambda}_L|D^\bullet = \lambda_L$) in order to examine whether the borrower's expectations (SCCE C1) are confirmed by the lender's optimization behavior based on these expectations.

Table 4.33 illustrates the lender's situation if the borrower's expectations coincide with SCCE C1. The lender's optimization behavior, i. e. his ct/pm-choice, is defined to examine the consistency of the borrower's expectations. Since in this constellation the N-type lender never receives an informational rent, the available profits (PL) stay constant while the contract agreement probabilities (AP) vary and define the lender's ct/pm-choice. The consistent expectations for all ten SCCE candidates are summarized in Table 4.34.

Table 4.34: Borrower's Confirmed Expectations of Borrower's Possible Sets of Consistent Conditional Expectations (SCCE) in the Lender Scenario with Restricted Power (EM, PC1)

SCCE C1 $(0,0,\lambda_L,\lambda_L)$		SCCE C6 $(0,0,0,0)$	
$(0,-,-,-)$	$1.15 \leq \mu < 1.25$	$(0,-,-,-)$	$1.15 \leq \mu < 1.25$
$(-,0,-,-)$	$1.318 \leq \mu < 1.4$	$(0,-,0,-)$	$1.25 \leq \mu < 1.274$
$(0,0,\lambda_L,\lambda_L)$	$1.653 \leq \mu < 1.741$	$(0,0,0,-)$	$1.274 \leq \mu < 1.318$
	$\wedge\mu \geq 1.85$	$(-,0,-,-)$	$1.318 \leq \mu < 1.4$
SCCE C2 $(0,0,1,\lambda_L)$		**SCCE C7 $(0,0,\lambda_L,0)$**	
$(0,-,-,-)$	$1.15 \leq \mu < 1.274$	$(0,-,-,-)$	$1.15 \leq \mu < 1.25$
$(0,0,-,-)$	$1.274 \leq \mu < 1.318$	$(-,0,-,-)$	$1.318 \leq \mu < 1.4$
$(-,0,-,-)$	$1.318 \leq \mu < 1.4$	$(-,0,-,0)$	$1.4 \leq \mu < 1.45$
$(0,0,1,\lambda_L)$	$1.653 \leq \mu < 1.741$	$(0,0,-,0)$	$1.45 \leq \mu < 1.55$
$(0,0,-,\lambda_L)$	$1.741 \leq \mu < 1.85$	$(0,0,\lambda_L,0)$	$1.55 \leq \mu < 1.653$
SCCE C3 $(0,0,\lambda_L,1)$		**SCCE C8 $(0,0,0,\lambda_L)$**	
$(0,-,-,-)$	$1.15 \leq \mu < 1.25$	$(0,-,-,-)$	$1.15 \leq \mu < 1.25$
$(-,0,-,-)$	$1.318 \leq \mu < 1.45$	$(0,-,0,-)$	$1.25 \leq \mu < 1.274$
$(0,0,-,-)$	$1.45 \leq \mu < 1.55$	$(0,0,0,-)$	$1.274 \leq \mu < 1.318$
$(0,0,\lambda_L,-)$	$1.55 \leq \mu < 1.653$	$(-,0,-,-)$	$1.318 \leq \mu < 1.4$
$(0,0,\lambda_L,1)$	$1.653 \leq \mu < 1.741$		
SCCE C4 $(0,0,1,1)$		**SCCE C9 $(0,0,1,0)$**	
$(0,-,-,-)$	$1.15 \leq \mu < 1.274$	$(0,-,-,-)$	$1.15 \leq \mu < 1.274$
$(0,0,-,-)$	$1.274 \leq \mu < 1.318$	$(0,0,-,-)$	$1.274 \leq \mu < 1.318$
	$\wedge 1.45 \leq \mu < 1.55$	$(-,0,-,-)$	$1.318 \leq \mu < 1.4$
$(-,0,-,-)$	$1.318 \leq \mu < 1.45$	$(-,0,-,0)$	$1.4 \leq \mu < 1.45$
$(0,0,1,-)$	$1.55 \leq \mu < 1.653$	$(0,0,-,0)$	$1.45 \leq \mu < 1.55$
$(0,0,1,1)$	$1.653 \leq \mu < 1.741$	$(0,0,1,0)$	$1.55 \leq \mu < 1.653$
SCCE C5 $(\lambda_L,\lambda_L,1,1)$		**SCCE C10 $(0,0,0,1)$**	
$(-,\lambda_L,1,1)$	$1.680 \leq \mu < 1.741$	$(0,-,-,-)$	$1.15 \leq \mu < 1.25$
	$\wedge 1.9 \leq \mu < 2.05$	$(0,-,0,-)$	$1.25 \leq \mu < 1.274$
$(\lambda_L,\lambda_L,1,1)$	$\mu \geq 2.05$	$(0,0,0,-)$	$1.274 \leq \mu < 1.318$
		$(-,0,-,-)$	$1.318 \leq \mu < 1.45$
		$(0,0,-,-)$	$1.45 \leq \mu < 1.55$

After determining the expectations of the SCCE candidates (C1-C10) which are confirmed by the lender's optimization behavior, the consistency of these conditional expectations is examined.[98] Obviously, if all four expectations of a SCCE candidate are confirmed by the lender's behavior, a set of consistent conditional expectations is found. However, to evaluate the consistency of partly confirmed SCCE candidates, i. e. not all four expectations are confirmed, is more complicated since their confirmation hinges on the borrower's expectations for the (by the lender) neglected *ct/pm*-choices. For example, the consistency of $(0, 0, -, -)$ hinges on the borrower's expectations about the lender's type for $E^\bullet$ and $D^\bullet$ since $(0, 0, -, -)$ is confirmed by the lender's behavior for $\widehat{\lambda}_L|E^\bullet = 1; \widehat{\lambda}_L|D^\bullet = \lambda_L$ (SCCE C2) but not for $\widehat{\lambda}_L|E^\bullet = \widehat{\lambda}_L|D^\bullet = \lambda_L$ (SCCE C1). Therefore, the evaluation of partly confirmed SCCE's consistency requires an examination of the impact of the borrower's unconfirmed expectations on the lender's optimization behavior. Hence, partly confirmed SCCEs are consistent if either the borrower's unconfirmed expectations possess no impact on the lender's behavior or the latters' impact vanishes when being anticipated by the borrower. E. g., an examination of $(0, -, -, -)$ for $1.15 \leq \mu < 1.25$ reveals that $\widehat{\lambda}_L|E^\circ = 0$ is consistent for $1.15 \leq \mu < 1.25$ since $\widehat{\lambda}_L|D^\circ$, $\widehat{\lambda}_L|E^\bullet$ and $\widehat{\lambda}_L|D^\bullet$ can take all potentially consistent expectations without affecting the result $(0, -, -, -)$. On the other hand, the consistency of $(0, 0, -, \lambda_L)$ for $1.741 \leq \mu < 1.85$ requires an anticipation of the lender's optimization behavior when $\widehat{\lambda}_L|E^\bullet$ deviates from one. For the SCCE C1 $\left(\widehat{\lambda}_L|E^\bullet = \lambda_L \right)$, for example, $E^\bullet$ is only chosen by the N-type lender, hence $\widehat{\lambda}_L|E^\bullet = \lambda_L$ is not confirmed. Moreover, $\widehat{\lambda}_L|E^\bullet = 0$ can not be consistent for $1.741 \leq \mu < 1.85$ since the N-type borrower's BCOC(AL) is no longer violated.

In total, the following consistent expectations are found:

---

[98] See page 182 for the definition of consistent expectations.

Table 4.35: Sets of Borrower's Consistent Conditional Expectations (SCCE) in the Lender Scenario with Restricted Power (EM, PC1)

μ	Sets of consistent conditional expectations
$[1.15, 1.25)$	$(0, -, -, -)$
$[1.25, 1.274)$	$(0, -, 0, -)$
$[1.274, 1.318)$	$(0, 0, 0, -)$
$[1.318, 1.4)$	$(-, 0-, -, -)$
$[1.4, 1.45)$	$(-, 0, -, 0)$
$[1.45, 1.55)$	$(0, 0, -, 0)$
$[1.55, 1.653)$	$(0, 0, \lambda_L, 0), (0, 0, 1, 0)$
$[1.653, 1.680)$	$(0, 0, \lambda_L, \lambda_L), (0, 0, 1, \lambda_L), (0, 0, \lambda_L, 1), (0, 0, 1, 1)$
$[1.680, 1.741)$	$(0, 0, \lambda_L, \lambda_L), (0, 0, 1, \lambda_L), (0, 0, \lambda_L, 1), (0, 0, 1, 1), (-, \lambda_L, 1, 1)$
$[1.741, 1.85)$	$(0, 0, -, \lambda_L)$
$[1.85, 1.9)$	$(0, 0, \lambda_L, \lambda_L)$
$[1.9, 2.05)$	$(0, 0, \lambda_L, \lambda_L), (-, \lambda_L, 1, 1)$
$[2.05, \infty)$	$(0, 0, \lambda_L, \lambda_L), (\lambda_L, \lambda_L, 1, 1)$

STEP 3 OF EXPECTATION DERIVATION

Finally, in the case of multiple SCCE for a contracting interval, the SCCE chosen by the borrower must be determined. The borrower's choice is based on his anticipation of the lender's behavior being confronted with these multiple SCCEs, i. e. the lender has no uniquely defined prediction about the borrower's expectations. The lender, of course, again anticipates the borrower's SCCE selection.

- $1.55 \leq \mu < 1.653$: $(0, 0, \lambda_L, 0)$ vs. $(0, 0, 1, 0)$
  In this situation the A-type lender prefers a private equity placement to secure contract participation while the N-type lender is indifferent between all four ct/pm-choices. The N-type lender is indifferent between all four choices since no choice offers him a profit above his outside option ($\overline{P}_L = 0$). For both public offerings the borrower expects the N-type lender which implies that S2 is the borrower's preferred strategy, leaving no ex-ante rent to the N-type lender. Both private placements leave independent of the borrower's expectations no rent to the N-type lender.[99] Consequently, the N-type lender is indifferent between $E^\bullet$ and $D^\bullet$. Hence, the borrower expects $(0, 0, \lambda_L, 0)$ for $1.55 \leq \mu < 1.653$.

---

[99] Due to the lender's interim and ex-post informational disadvantage, he can only receive a rent because of ex-ante uncertainty.

- $1.653 \leq \mu < 1.680$: $(0,0,\lambda_L,\lambda_L)$ vs. $(0,0,1,\lambda_L)$ vs. $(0,0,\lambda_L,1)$ vs. $(0,0,1,1)$

  In this situation the A-type lender favors private placements to secure participation. The N-type lender never receives an informational rent for any ct/pm-choice and is, therefore, indifferent between them. For $1.653 \leq \mu < 1.680$ the borrower expects $(0,0,\lambda_L,\lambda_L)$.

- $1.680 \leq \mu < 1.741$: $(0,0,\lambda_L,\lambda_L)$ vs. $(0,0,1,\lambda_L)$ vs. $(0,0,\lambda_L,1)$ vs. $(0,0,1,1)$ vs. $(-,\lambda_L,1,1)$

  Like in the previous situation the A-type lender favors private financing to secure participation. However, contrary to the previous situation, the N-type lender is not indifferent between all ct/pm-choices. The N-type lender favors public debt in the hope of an ex-ante informational rent $(\widehat{\lambda}_L|D^\circ = \lambda_L)$. But unfortunately for the N-type lender, the borrower anticipates the N-type lender's behavior and adjusts $\widehat{\lambda}_L|D^\circ = 0$. Hence, the N-type lender is again indifferent between all four ct/pm-choices which implies that the borrower expects $(0,0,\lambda_L,\lambda_L)$ for $1.680 \leq \mu < 1.741$.

- $1.9 \leq \mu < 2.05$: $0,0,\lambda_L,\lambda_L$ vs. $(-,\lambda_L,1,1)$ and $\mu \geq 2.05$: $(0,0,\lambda_L,\lambda_L)$ vs. $(\lambda_L,\lambda_L,1,1)$

  By the same reasoning it can be shown that the borrower expects $(0,0,\lambda_L,\lambda_L)$ for $\mu \geq 1.9$.

4.3.2.2 Contract Type / Placement Mode-Choice Predictions

After determining the borrower's consistent expectations about the lender's type, the borrower's, i. e. the condition optimizer's, optimal behavior is uniquely defined. Anticipating this behavior, the lender as the ct/pm optimizer prefers the ct/pm-choices stated in Table 4.36. Table 4.36 also states the contract agreement probabilities implied by the lender's ct/pm-choice indicating the borrower's (condition optimizer's) behavior due to the lender's ct/pm-choice. The borrower has to decide whether to participate at all and, if he participates, how he determines contracts conditions. For public offerings he can choose among S1 and S2 while for private placements he might refuse to offer both potential lender types acceptable conditions. As we have observed the borrower's optimal behavior is also borrower type-dependent. We, therefore, refrain from explicitly stating the borrower's optimization behavior, e. g. his strategy choice, in Table 4.36 since each lender type might contract for the same interval with different borrower types preferring different strategies.[100]

Overall we can say that due to the consistency requirement of the borrower's expectations, the lender's expected ct/pm-choice is also the lender's

---

[100] For example, for $1.55 \leq \mu < 1.653$ the borrower expects $(0,0,\lambda_L,0)$, hence the A-type borrower favors S2 for E°, respectively D° and $E^{\bullet(1)}$, $D^{\bullet(1)}$ for $E^\bullet, D^\bullet$. The N-type borrower's contract determination differs with respect to $E^\bullet$. The N-type borrower chooses $E^{\bullet(2)}$ instead of $E^{\bullet(1)}$.

optimal ct/pm-choice. Falsely pretending to be an A-type, respectively a N-type, lender by acting accordingly is unprofitable and, therefore, undesirable for both lender types.

Table 4.36: Contract Type / Placement Mode-Choices and Agreement Probabilities in the Lender Scenario with Restricted Power (EM, PC1)

Interval	μ	Ct/pm-choice and agreement probabilities	
		A-type lender	N-type lender
I	$[1,1.15)$	$-$ (0)	$-$ (0)
II	$[1.15,1.25)$	$-$ (0)	E° $(\frac{1}{2})$
III	$[1.25,1.274)$	$-$ (0)	$E^\circ, E^\bullet$ $(\frac{1}{2})$
IV	$[1.274,1.318)$	$-$ (0)	$E^\circ, D^\circ, E^\bullet$ $(\frac{1}{2})$
$V - VII$	$[1.318,1.4)$	$-$ (0)	D° (1)
VIII	$[1.4,1.45)$	$-$ (0)	$D^\circ, D^\bullet$ (1)
$IX - X$	$[1.45,1.55)$	$-$ (0)	$E^\circ, D^\circ, D^\bullet$ (1)
$XI - XII$	$[1.55,1.653)$	$E^\bullet$ $(\frac{1}{2})$	$E^\circ, D^\circ, E^\bullet, D^\bullet$ (1)
$XIII - XIV$	$[1.653,1.741)$	$E^\bullet, D^\bullet$ $(\frac{1}{2})$	$E^\circ, D^\circ, E^\bullet, D^\bullet$ (1)
$XV - XVII$	$[1.741,1.85)$	$D^\bullet$ (1)	$E^\circ, D^\circ, D^\bullet$ (1)
$XVIII - XXI$	$[1.85,\infty)$	$E^\bullet, D^\bullet$ (1)	$E^\circ, D^\circ, E^\bullet, D^\bullet$ (1)

Note: This table contains the lender's ct/pm-choice together with the related contract agreement probabilities stated in brackets.

From the results derived so far, it is obvious that the A-type lender only participates in bilateral financing negotiations when the expected return μ is sufficiently high ($\mu \geq 1.55$). The N-type lender is willing to offer contracts for projects with a lower μ ($\mu \geq 1.15$) as his profit alternative is lower.

Due to the lender's interim and ex-post informational disadvantage, he never receives an informational rent from these uncertainties. Hence, the A-type lender never receives a rent at all while the N-type lender has still a chance to receive a rent from the ex-ante informational asymmetry. Since the A-type lender never receives an informational rent, he only cares about contract agreement, i. e. he chooses the ct/pm-constellation which maximizes his chance of contract participation. Anticipating this behavior the borrower can try to separate both lender types by his strategy choice for public placements and, hence, prevent any informational rent payment. If the borrower favors S2 for all public offerings, the A-type lender is forced to choose private placements while the N-type borrower becomes indifferent between all feasible *ct/pm*-choices since he also does not receive an informational rent. Both lender types only care about contract agreement.

However, there are costs for the borrower to "separate out" the A-type lender and hence prevent any informational rent payment. Private placements cause higher agency costs than public offerings and hence reduce the surplus extractable by the borrower. However, these costs can not be prevented in the described context since the lender can not credible promise not to play S2 for public offerings when the A-type lender has the choice between both *pms*. The borrower can only credible deviate from S2 if from the A-type lender's perspective only public S1 offerings are feasible and the borrower's gain from contract agreement with the A-type lender exceeds his expected informational rent payment to the N-type lender. In this situation the borrower expects that both lender types choose the respective public offering, hence he plays S1 and leaves thereby an (ex-ante) informational rent to the N-type lender.

For PC1 we observe that the A-type lender prefers private financing whenever feasible to guarantee contract participation. The A-type's preference for private financing is independent of the *ct* since the A-type lender never receives an informational rent anyway. He is squeezed to his participation constraint. Due to the A-type lender's private placement focus, the N-type lender has no chance to obtain an ex-ante informational rent. For all public offerings the borrower can stick to the risky placement strategy (S2) without incorporating any placement risk. The A-type lender sticks to private financing anyway.

To sum it up, despite the lender's power to determine *ct/pm*, he is for PC1 always squeezed to his type-dependent profit alternative, since as soon as the A-type lender participates the A-type favors private placements whenever feasible. Only contract agreement matters for the lender. The N-type even becomes indifferent between public and private debt or equity placements as long as they offer the same contract agreement probability. The *pm*-choice plays a crucial role in the limited power scenarios due to its strategic importance, i. e. to secure contract participation or to gamble for an informational rent.

4.3.3 Borrower Scenario with Restricted Power (BS^r)

In the *borrower scenario with restricted power* (BS^r) the borrower can choose the *ct/pm*, while the lender determines contracts' conditions. The offered conditions can either be accepted or rejected by the borrower. If the borrower accepts the conditions, he receives the funds required to finance the project. Otherwise, the game ends and the lender as well as the borrower are left with their outside option.

To determine which contract choice ($E^\circ, D^\circ, E^\bullet, D^\bullet$) the borrower prefers, his expected profits $PB(\mu, \widehat{\lambda}_L)$ for the different choices have to be compared. Unfortunately, like in the LS^r scenario, in this scenario the borrower's (*ct/pm* optimizer's) profits cannot be compared before the lender's (condition optimizer's) optimization behavior is determined, while the lender's behavior again depends on the borrower's (ct/pm optimizer's) optimization. Due to this interdependency between both optimizations, like in the LS^r scenario, the lender's and the borrower's optimization behavior have to be treated *jointly*. Since in the BS^r scenario the lender as the condition optimizer tries to optimally respond to the borrower's (ct/pm optimizer's) ct/pm-choice, the lender conditions his expectations about the borrower's type on the latter's ct/pm-choice ($\widehat{\lambda}_B|x \ \forall x \in \{E^\circ, D^\circ, E^\bullet, D^\bullet\}$). These conditional expectations determine the lender's (condition optimizer's) optimal behavior at Stage 2 of the contracting game. According to the backward-induction-principle the lender's conditional expectations also affect the borrower's (ct/pm optimizer's) preferred ct/pm-choice (Stage 1 of the contracting game). This interdependency is anticipated by the lender and the borrower.

Therefore, to solve the (sequential) contract negotiation game it is necessary to determine

- the lender's conditional expectations about the borrower's type ($\widehat{\lambda}_B|x$), and
- the borrower's expectations about the lender's expectations about the borrower's type

since they are the last unknown elements of the game. The former expectations ($\widehat{\lambda}_B|x$) affect the lender's optimization behavior, while the latters are relevant for the borrower's optimization, trying to anticipate the lender's (second mover's) behavior. To determine these expectations we follow the procedure described for the derivation of conditional expectations in the LS^r scenario.

In Section 4.3.3.1 the lender's consistent expectations are determined while Section B.3.2 defines the borrower's preferred contract choices given these expectations.

4.3.3.1 Derivation of Consistent Expectations

The lender's expectations ($\widehat{\lambda}_B|x$) are consistent when they have reached a *fixpoint* in the sense that the borrower's optimization behavior given these expectations confirm the lender's expectations about the borrower's behavior. The borrower anticipates the lender's expectations. The lender's consistent conditional expectations, i. e. a set of consistent conditional expectations (SCCE), are determined as follows:

- STEP 1: Firstly, based on intuitive arguments all plausible SCCE candidates are selected from all possible constellations of conditional expectations.
- STEP 2: Secondly, the contract negotiation game is solved for the SCCE candidates to examined whether the conditional expectations are actually consistent, i. e. confirmed by the lender's and the borrower's optimization behavior, or not.

- STEP 3: Finally, if more than one SCCE candidate is consistent, the borrower's optimization behavior is examined when being confronted with more than one possible SCCE. Thereby, the SCCE candidate preferred from the borrower's perspective is determined because the candidate represents the borrower's predicted behavior.

STEP 1 OF EXPECTATION DERIVATION

To reduce the number of possible SCCE candidates from $3^4 = 81$ to 9,[101] we compare the informational rents the borrower might achieve for specific ct/pm-choices x ($x \in \{E^\circ, D^\circ, E^\bullet, D^\bullet\}$). For public debt financing, for example, the borrower always achieves a positive informational rent at least due to his interim and ex-post advantage. Ex-ante uncertainty is another reason for a possible informational rent. Whether the borrower keeps an informational rent for private debt financing depends on the relationship of the informational rent from interim and ex-post uncertainty, private placement's negotiation costs, and the borrower's financing alternative. Equity financing does not offer an informational rent to the borrower due to his interim and ex-post informational advantage. However, the borrower might achieve an informational rent due to his ex-ante advantage.

The following observations help to restrict the number of possible SCCE candidates.

- $\widehat{\lambda}_B|E^\circ = 1$ and $\widehat{\lambda}_B|D^\circ = 1$ can *not* be consistent expectations since for $\widehat{\lambda}_B|E^\circ = \widehat{\lambda}_B|D^\circ = 1$ the lender's preferred strategy is S1 which attracts the A- and the N-type borrower and not only the A-type. The N-type borrower even achieves an ex-ante informational rent. Moreover, the A-type borrower's feasible set of E°, respectively D°, is a subset of the N-type's which implies that if E° or D° is a feasible choice for the A-type (implicit requirement for $\widehat{\lambda}_B|x = 1 \ \forall x \in \{E^\circ, D^\circ\}$), the N-type's FCCs are satisfied as well.
- $\widehat{\lambda}_B|E^\circ = \lambda_B$ can *not* be consistently combined with $\widehat{\lambda}_B|E^\bullet \in \{0, \lambda_B\}$; and $\widehat{\lambda}_B|D^\circ = \lambda_B$ can *not* be combined with $\widehat{\lambda}_B|E^\circ = 0, \widehat{\lambda}_B|E^\bullet \in \{0, \lambda_B\}$ *and* $\widehat{\lambda}_B|D^\bullet \in \{0, \lambda_B\}$.[102] For $\widehat{\lambda}_B|E^\circ = \lambda_L$ S1 is the lender's optimal strategy since otherwise the A-type borrower is unwilling to participate. But if S1 is optimal for E°, the N-type borrower will never favor $E^\bullet$ ($\widehat{\lambda}_B|E^\bullet \notin \{0, \lambda_B\}$). Equivalently, for $\widehat{\lambda}_B|D^\circ = \lambda_L$ S1 is the lender's optimal strategy for D°, but, similar to equity financing, in such a situation the N-type

---

[101] Possible conditional expectations for each ct/pm-constellation ($x \in \{E^\circ, D^\circ, E^\bullet, D^\bullet\}$) are $\widehat{\lambda}_B|x \in \{0, \lambda_B, 1\}$.

[102] This condition is not as strict as in the LS^r scenario since for debt financing the borrower keeps an informational rent from interim and ex-post uncertainty which might overcompensate the ex-ante informational rent discussed here. This makes it *im*possible to rule out combinations like $\widehat{\lambda}_B|E^\circ = \lambda_B$ and $\widehat{\lambda}_B|D^\circ = 0$.

borrower will never prefer a private or public S2 equity placement with no informational rent $(\widehat{\lambda}_B|E° \neq 0, \widehat{\lambda}_B|D° \notin \{0, \lambda_B\})$. Moreover, $\widehat{\lambda}_B|D° = \lambda_B$ cannot be consistently combined with $\widehat{\lambda}_B|D^{\bullet} \in \{0, \lambda_B\}$ since in private placements the borrower's debt financing informational rent is diminished in comparison to public offerings.

- $\widehat{\lambda}_B|D° = 0$ can *not* be consistently combined with $\widehat{\lambda}_B|E° = 0$ since one ct/pm-constellation offers the N-type borrower an informational rent due to his interim and ex-post advantage even for S2 while the other one does not.

- $\widehat{\lambda}_B|E^{\bullet} = \widehat{\lambda}_B|D^{\bullet} = 0$ are only possible as long as the lender's LCOC(AB) for $E^{\bullet}$, respectively $D^{\bullet}$, is not satisfied. Once a contract is offered to the A-type borrower, the A-type either prefers $E^{\bullet}/D^{\bullet}$ or even favors another public contract choice which in this case is also preferable for the N-type. Therefore, $\widehat{\lambda}_B|E^{\bullet} = \widehat{\lambda}_B|D^{\bullet} = 0$ become impossible.

Due to these observations, we are down to 9 possible SCCE candidates, which are stated in Table 4.37.

Table 4.37: Possible Sets of Lender's Consistent Conditional Expectations (SCCE) in the Borrower Scenario with Restricted Power (EM)

SCCE	C1	C2	C3	C4	C5	C6*	C7†	C8†	C9†
$\widehat{\lambda}_B|E°$	λ_B	λ_B	λ_B	λ_B	λ_B	λ_B	0	0	0
$\widehat{\lambda}_B|D°$	0	0	0	0	λ_B	0	–	–	–
$\widehat{\lambda}_B|E^{\bullet}$	λ_B	1	λ_B	1	1	1	0	λ_B	1
$\widehat{\lambda}_B|D^{\bullet}$	λ_B	λ_B	1	1	1	0	–	–	–

* = the N-type lender's LCOC(AB) for $D^{\bullet}$ is not satisfied.
† = the N-type's LPC for $D°$ and $\widehat{\lambda}_B|D° \in \{0, \lambda_B\}$ is not satisfied.[103]

After the derivation of SCCE candidates their consistency is examined by solving the negotiation game for each candidate.

STEP 2 OF EXPECTATION DERIVATION

Firstly, the lender's optimization behavior for each possible ct/pm-choice given the expectations stated in Table 4.37 is determined.

---

[103] This requirement guarantees that $D°$ and $D^{\bullet}$ are never chosen by the borrower independent of the borrower's expectations about the lender's expectations. Hence, the conditional expectations for $D°$ and $D^{\bullet}$ are irrelevant.

Public equity financing ($\widehat{\lambda}_B|E^\circ = \lambda_B$ and $\widehat{\lambda}_B|E^\circ = 0$ if the N-type's LPC for D° with $\widehat{\lambda}_B|D^\circ \in \{0, \lambda_B\}$ is not satisfied):

- for $\widehat{\lambda}_B|E^\circ = 0$[104]
  the N-type lender favors S2 to S1. Hence, he chooses $E_L^\circ|S2$ for $1.15 \leq \mu < 1.274$.
- for $\widehat{\lambda}_B|E^\circ = \lambda_B$
  the A-type lender favors S1 to S2 for $\mu \geq 2.05$ and for $\mu < 2.05$ the other way round. Hence, he chooses $E_L^\circ|S2$ for $1.45 \leq \mu < 2.05$ and $E_L^\circ|S1$ for $\mu \geq 2.05$;
  the N-type lender favors S1 to S2 for $\mu \geq 1.75$ and for $\mu < 1.75$ the other way round. Hence, he chooses $E_L^\circ|S2$ for $1.15 \leq \mu < 1.75$ and $E_L^\circ|S1$ for $\mu \geq 1.75$.

Public debt financing ($\widehat{\lambda}_B|D^\circ \in \{0, \lambda_B\}$):

- for $\widehat{\lambda}_B|D^\circ = 0$
  the A-type lender favors S2 to S1. Hence, he chooses $D_{L,IIb}^\circ|S2$ for $\mu \geq 1.6$;
  the N-type lender favors S2 to S1. Hence, he chooses $D_{L,IIb}^\circ|S2$ for $\mu \geq 1.274$.
- for $\widehat{\lambda}_B|D^\circ = \lambda_B$
  the A-type lender favors $D_{L,IIb}^\circ|S1$ to $D_{L,IIb}^\circ|S2$ for $\mu \geq 1.726$ and $D_{L,IIb}^\circ|S1$ dominates $D_{L,Ia}^\circ$. Hence, he chooses $D_{L,IIb}^\circ|S2$ for $1.6 \leq \mu < 1.726$ and $D_{L,IIb}^\circ|S1$ for $\mu \geq 1.726$;
  the N-type lender favors $D_{L,IIb}^\circ|S1$ to $D_{L,IIb}^\circ|S2$ for $\mu \geq 1.368$ and $D_{L,IIb}^\circ|S1$ dominates $D_{L,Ia}^\circ$. Hence, he chooses $D_{L,IIb}^\circ|S2$ for $1.274 \leq \mu < 1.368$ and $D_{L,IIb}^\circ|S1$ for $\mu \geq 1.368$.

Private equity financing ($\widehat{\lambda}_B|E^\bullet \in \{\lambda_B, 1\}$ and $\widehat{\lambda}_B|E^\bullet = 0$ if the N-type lender's participation constraint for D° and $\widehat{\lambda}_B|D^\circ \in \{0, \lambda_B\}$ is not satisfied):

- for $\widehat{\lambda}_B|E^\bullet = 0$
  not possible under the imposed constraint.
- for $\widehat{\lambda}_B|E^\bullet = \lambda_B$
  the A-type lender chooses $E_L^{\bullet(1)}$ for $1.55 \leq \mu < 1.85$ and $E_L^{\bullet(2)}$ for $\mu \geq 1.85$;
  the N-type lender chooses $E_L^{\bullet(1)}$ for $1.25 \leq \mu < 1.55$ and $E_L^{\bullet(2)}$ for $\mu \geq 1.55$.
- for $\widehat{\lambda}_B|E^\bullet = 1$
  the A-type lender only chooses $E_L^{\bullet(2)}$ for $\mu \geq 1.85$;
  the N-type lender only chooses $E_L^{\bullet(2)}$ for $\mu \geq 1.55$.

---

[104] The A-type lender never consistently expects $\widehat{\lambda}_B|E^\circ = 0$ since once $E_B^\circ|S2$ becomes feasible ($\mu \geq 1.45$) the N-type's LPC for D° with $\widehat{\lambda}_B|D^\circ \in \{0, \lambda_B\}$ is satisfied. D° with $\widehat{\lambda}_B|D^\circ \in \{0, \lambda_B\}$ is feasible if $\mu \geq 1.274$.

Private debt financing ($\widehat{\lambda}_B | D^\bullet \in \{\lambda_B, 1\}$ and $\widehat{\lambda}_B | D^\bullet = 0$ if the N-type lender's LCOC(AB) for $D^\bullet$ is violated):

- for $\widehat{\lambda}_B | D^\bullet = 0$
  the A-type lender favors to finance the expected N-type borrower with risky instead of riskless debt, $D_{L,IIa}^{\bullet(1)}$ is dominated. Hence, he chooses $D_{L,IIa}^{\bullet(1)}$ or $D_{L,IIIb}^{\bullet(1)}$ for $1.653 \leq \mu < 1.741$;
  the N-type lender chooses $D_{L,IIb}^{\bullet(1)}$ for $1.329 \leq \mu < 1.4$.

- for $\widehat{\lambda}_B | D^\bullet = \lambda_B$
  the lender can either offer both borrower types a contract, only the N-type borrower, or no contract at all. Due to the lender's preference analysis for private debt financing (see page 157) we realize that
  the A-type lender chooses $D_{L,IIIb}^{\bullet(1)}$ or $D_{L,IIa}^{\bullet(1)}$ for $1.653 \leq \mu < 1.741$, $D_{L,IIIb}^{\bullet(2)}$ for $1.741 \leq \mu < 3.2$ and $D_{L,IIIc}^{\bullet(2)}$ for $\mu \geq 3.2$;
  the N-type lender chooses $D_{L,IIb}^{\bullet(1)}$ for $1.329 \leq \mu < 1.4$, $D_{L,IIIb}^{\bullet(1)}$ for $1.4 \leq \mu < 3.2$ and $D_{L,IIIc}^{\bullet(2)}$ for $\mu \geq 3.2$.

- for $\widehat{\lambda}_B | D^\bullet = 1$
  the A-type lender either finances the expected A-type borrower with riskless or risky debt, $D_{L,Ia}^{\bullet(2)}, D_{L,IIa}^{\bullet(2)}, D_{L,IIc}^{\bullet(2)}$ dominated by $D_{L,IIIb}^{\bullet(2)}, D_{L,IIIc}^{\bullet(2)}$. Hence, he chooses $D_{L,IIIb}^{\bullet(2)}$ for $1.741 \leq \mu < 3.2$ and $D_{L,IIIc}^{\bullet(2)}$ for $\mu \geq 3.2$;
  the N-type lender chooses $D_{L,IIIb}^{\bullet(2)}$ for $1.4 \leq \mu < 3.2$ and $D_{L,IIIc}^{\bullet(2)}$ for $\mu \geq 3.2$.

After examining the lender's optimization behavior the borrower's resulting expected profits are analyzed to define the borrower's respective optimization behavior, i. e. the latter's ct/pm-choice. The borrower's expected profits and contract agreement probabilities are stated in the following table.[105, 106]

---

[105] The lender's potential expectations about the borrower's type are not lender type-dependent since both lenders have at this stage of the contracting game the same information about the borrower, i. e. the latter's ct/pm-choice and λ_B.

[106] λ_B is not substituted by 0.5 in accordance with PC1 since the individual profit components become more obvious when λ_B is not substituted.

Table 4.38: Borrower's Expected Profits and Contract Agreement Probabilities in the Borrower Scenario with Restricted Power Depending on the Lender's Expectation About the Borrower's Type (EM, PC1)

Borrower's ct/pm-choice	Lender's expectation	Borrower's expected profits and contract agreement probabilities			
		from	from	from	from
E°	$\widehat{\lambda}_B\|E^\circ = 0$	$1.15\quad \dfrac{0.3\|0}{(0\|1-\lambda_L)}$		$1.274\quad —$	
	$\widehat{\lambda}_B\|E^\circ = \lambda_B$	$1.15\quad \dfrac{0.3\|0}{(0\|1-\lambda_L)}$ $\;2.05\quad \dfrac{0.3\|0.3}{(1\|1)}$		$1.45\quad \dfrac{0.3\|0}{(0\|1)}$	$1.75\quad \dfrac{0.3\|(1-\lambda_L)0.3}{(1-\lambda_L\|1)}$
D°	$\widehat{\lambda}_B\|D^\circ = 0$	$1.274\quad \dfrac{0.3\|(1-\lambda_L)\frac{0.16}{\mu}}{(0\|1-\lambda_L)}$		$1.6\quad \dfrac{0.3\|\frac{0.16}{\mu}}{(0\|1^{\mu})}$	
	$\widehat{\lambda}_B\|D^\circ = \lambda_B$	$1.274\quad \dfrac{0.3\|(1-\lambda_L)\frac{0.16}{\mu}}{(0\|1-\lambda_L)}$ $\;1.726\quad \dfrac{0.3\|0.3}{(1\|1)}$		$1.368\quad \dfrac{0.3\|(1-\lambda_L)0.3}{(1-\lambda_L\|1-\lambda_L)}$	$1.6\quad \dfrac{0.3\|(1-\lambda_L)0.3+\lambda_L\frac{0.16}{\mu}}{(1-\lambda_L\|1)}$
$E^\bullet$	$\widehat{\lambda}_B\|E^\bullet = \lambda_B$	$1.25\quad \dfrac{0.3\|0}{(0\|1-\lambda_L)}$		$1.55\quad \dfrac{0.3\|0}{(0\|1-\lambda_L)}$	$1.85\quad \dfrac{0.3\|0}{(1-\lambda_L\|1)}$
	$\widehat{\lambda}_B\|E^\bullet = 1$	$1.55\quad \dfrac{0.3\|0}{(1-\lambda_L\|1-\lambda_L)}$		$1.85\quad \dfrac{0.3\|0}{(1\|1)}$	$1.85\quad \dfrac{0.3\|0}{(1\|1)}$
$D^\bullet$	$\widehat{\lambda}_B\|D^\bullet = 0$	$1.329\quad \dfrac{0.3\|(1-\lambda_L)(\frac{0.16}{\mu}-0.05)}{(0\|1-\lambda_L)}$		$1.4\quad —$	
	$\widehat{\lambda}_B\|D^\bullet = \lambda_B$	$1.329\quad \dfrac{0.3\|(1-\lambda_L)(\frac{0.16}{\mu}-0.05)}{(0\|1-\lambda_L)}$		$1.4\quad \dfrac{0.3\|(1-\lambda_L)(\frac{0.16}{\mu}-0.05)}{(1-\lambda_L\|1-\lambda_L)}$	$1.653\quad \dfrac{0.3\|\frac{0.16}{\mu}-0.05}{(1-\lambda_L\|1)}$
	$\widehat{\lambda}_B\|D^\bullet = 1$	$1.741\quad \dfrac{0.3\|\frac{0.16}{\mu}-0.05}{(1\|1^{\mu})}$	$3.2\quad \dfrac{0.3\|0}{(1\|1)}$	$1.741\quad \dfrac{0.3\|\frac{0.16}{\mu}-0.05}{(1\|1^{\mu})}$	$3.2\quad \dfrac{0.3\|\frac{0.16}{\mu}-0.05}{(1\|1)}$

Given these expected profits, the consistency of the SCCE candidates stated in Table 4.37 can be evaluated. In total, the following consistent expectations are found:

Table 4.39: Sets of Lender's Consistent Conditional Expectations (SCCE) in the Borrower Scenario with Restricted Power (EM, PC1)

μ	Sets of consistent conditional expectations
$[1.15, 1.274)$	$(0, -, -, -)$
$[1.274, 1.368)$	$(-, 0, -, -)$
$[1.368, 1.4)$	$(-, 0, -, -), (-, \lambda_B, -, -)$
$[1.4, 1.55)$	$(-, 0, -, 1), (-, \lambda_B, -, 1)$
$[1.55, 1.726)$	$(-, 0, 1, 1), (-, \lambda_B, 1, 1)$
$[1.726, 1.741)$	$(-, 0, 1, 1), (-, \lambda_B, -, -)$
$[1.741, 1.75)$	$(-, 0, -, 1), (-, \lambda_B, -, 1)$
$[1.75, 1.85)$	$(-, \lambda_B, -, 1)$
$[1.85, 2.05)$	$(-, \lambda_B, 1, 1)$
$[2.05, \infty)$	$(\lambda_B, \lambda_B, 1, 1)$

STEP 3 OF EXPECTATION DERIVATION

Finally, in the case of multiple SCCE for a contracting interval, the SCCE chosen by the lender has to be defined.

- $1.368 \leq \mu < 1.4$: $(-, 0, -, -)$ vs. $(-, \lambda_B, -, -)$
  In this situation the A-type borrower favors public debt since he has nothing to lose. Anticipating the borrower's behavior the lender expects $(-, \lambda_B, -, -)$ (which leaves the N-type borrower with an (additional) ex-ante informational rent).
- $1.4 \leq \mu < 1.55$: $(-, 0, -, 1)$ vs. $(-, \lambda_B, -, 1)$
  In this situation the A-type borrower prefers private debt maximizing his chance of contract participation while the N-type favors public debt in the hope of an (additional) ex-ante rent.[107] The lender anticipates the

---

[107] The profits the A-type borrower can achieve for D° with $\widehat{\lambda}_B | D^\circ \in \{0, \lambda_B\}$ and for $D^\bullet$ with $\widehat{\lambda}_B | D^\bullet = 1$ coincide. Hence, the borrower tries to maximize the contract agreement probability and, therefore, favors $D^\bullet$ to D°. The borrower only cares about contract agreement when his profit is unaffected by his ct/pm-choice. Otherwise, he always favors the ct/pm-choice maximizing his expected profit.

borrower's behavior and expects $(-,0,-,1)$. (Hence, the lender can prevent to pay the borrower and extra rent due to ex-ante uncertainty.)

- $1.55 \leq \mu < 1.726$: $(-,0,1,1)$ vs. $(-,\lambda_B,1,1)$ and $1.726 \leq \mu < 1.741$: $(-,0,1,1)$ vs. $(-,\lambda_B,-,-)$
  Due to equivalent arguments stated above the lender expects $(-,0,1,1)$.
- $1.741 \leq \mu < 1.75$: $(-,0,-,1)$ vs. $(-,\lambda_B,-,1)$
  In this situation the lender expects $(-,0,-,1)$.

4.3.3.2 Contract Type / Placement Mode-Choice Predictions

Equivalently to the LS^r scenario, the pm-choice plays a crucial role in the BS^r scenarios. The borrower's ct/pm-choices and the implied contract agreement probabilities are summarized in Table 4.40.

Table 4.40: Contract Type / Placement Mode-Choices and Agreement Probabilities in the Borrower Scenario with Restricted Power (EM, PC1)

Interval	μ	Ct/pm-choice and agreement probabilities			
		A-type borrower		N-type borrower	
I	$[1,1.15)$	$-$	(0)	$-$	(0)
$II - III$	$[1.15,1.274)$	$-$	(0)	E°	$(\frac{1}{2})$
$IV - VI$	$[1.274,1.368)$	$-$	(0)	D°	$(\frac{1}{2})$
VII	$[1.368,1.4)$	D°	$(\frac{1}{2})$	D°	$(\frac{1}{2})$
$VIII - X$	$[1.4,1.55)$	$D^\bullet$	$(\frac{1}{2})$	D°	$(\frac{1}{2})$
XI	$[1.55,1.6)$	$E^\bullet, D^\bullet$	$(\frac{1}{2})$	D°	$(\frac{1}{2})$
$XII - XIV$	$[1.6,1.741)$	$E^\bullet, D^\bullet$	$(\frac{1}{2})$	D°	(1)
XV	$[1.741,1.75)$	$D^\bullet$	(1)	D°	(1)
$XVI - XVII$	$[1.75,1.85)$	$D^\circ, D^\bullet$	(1)	D°	(1)
$XVIII - XIX$	$[1.85,2.05)$	$D^\circ, E^\bullet, D^\bullet$	(1)	D°	(1)
$XX - XXI$	$[2.05,\infty)$	$E^\circ, D^\circ, E^\bullet, D^\bullet$	(1)	E°, D°	(1)

Note: This table contains the borrower's ct/pm-choice together with the related contract agreement probabilities stated in brackets.

We observe, as expected, that the N-type borrower is willing to finance projects with lower expected returns ($\mu \geq 1.15$) than the A-type does ($\mu \geq 1.368$). This is due to their differing financing alternatives. Furthermore, the N-type borrower prefers public to private financing, while no tendency is observable for the A-type.

Due to the borrower's interim and ex-post informational advantage debt financing offers him an informational rent. This informational rent attracts the borrower to public debt financing while the rent of private debt financing is lower.[108] Now, as long as the A-type borrower is not participating, the lender chooses S2 to avoid leaving an additional rent to the borrower due to the latter's ex-ante informational advantage. However, as soon as the A-type borrower participates, the lender faces a tradeoff. He can either separate both borrower types by his strategy choice in public placements, which cause the total agency costs to increase reducing the extractable profit, or leave an additional ex-ante informational rent to the N-type borrower. Normally, the lender would choose to "separate out" the A-type borrower's when it is profitable for him to do so. However, this borrower type separation, and thereby the avoidance of an ex-ante informational rent payment, is only possible as long as (i) private financing is feasible for the A-type borrower ($\mu \geq 1.4$ for $D^\bullet$ and PC1) while (ii) public equity financing is unfeasible for $\widehat{\lambda}_B|E^\circ = \lambda_B$ from the A-type borrower's perspective ($\mu < 1.75$). As soon as one condition is violated, the lender cannot separate the borrower types anymore to prevent an ex-ante informational rent payment. The first restriction stems from the fact that if only public financing is feasible for the A-type borrower, no opportunity exists to separate both borrower types. The A-type borrower has noting to lose by choosing public financing and will, therefore, do so. If the lender still chooses S2 the borrower is left with his financing alternative (outside option). Anticipating the borrower's behavior the lender prefers S1 to S2 to guarantee the A-type's participation and leaves, therefore, an ex-ante informational rent to the N-type borrower. If now private financing becomes a feasible alternative for the A-type borrower, it is possible to separate both borrower types as long as public equity financing (E°) for $\widehat{\lambda}_B|E^\circ = \lambda_B$ is unfeasible from the A-type borrower's perspective. Since as soon as E° for $\widehat{\lambda}_B|E^\circ = \lambda_B$ becomes feasible for the A-type borrower, the lender is forced to give up his separation strategy and pays the N-type borrower an ex-ante informational rent. The lender is forced to give up his separation strategy when E° becomes feasible for $\widehat{\lambda}_B|E^\circ = \lambda_B$ from the A-type borrower's perspective since both borrower types can choose a *seemingly irrational ct/pm-*

---

[108] The borrower's public debt financing informational rent from his interim and ex-post advantage is never sufficient to satisfy the A-type's participation constraint due to PC1 ($P_B^+ > vc/2$). The borrower's public debt financing informational rent is bounded below $vc/2$ since the lender prefers risky public debt financing (borrower's informational rent: $IR(D_{L,II}^\circ) = \dfrac{vc^2}{4\zeta_{max}\mu}$) whenever feasible ($\zeta_{max} > \dfrac{vc}{2\mu}$) and riskless debt (borrower's informational rent $IR(D_{L,I}^\circ) = \zeta_{max}\mu$) otherwise. The informational rent of private debt is even lower due to the arising negotiation costs. However, for parameter constellations satisfying $P_B^+ \leq vc/2$ the borrower's debt financing informational rent might exceed P_B^+ which implies that public debt financing becomes preferable for both borrower types.

constellation contradicting the lender's separating expectations. For example, the lender who expects $(-, 0, -, 1)$ while E° for $\lambda_B|E^\circ = \lambda_B$ is feasible from the A&N-type borrower's perspective, might be faced with a E° borrower choice. This choice is *not* expected in the lender's separating attempt, where for $\widehat{\lambda}_B|E^\circ = 0$ public equity is always dominated by public debt from the N-type borrower's perspective due to the latter's positive informational rent and from the A-type borrower's perspective a E° choice implies unnecessary placement risk. The lender anticipates that the rational borrower must have a *different expectation* of the lender's expectation than $\widehat{\lambda}_B|E^\circ = 0$ when choosing E°. Therefore, the lender is not sure which borrower type he is facing and expects $\widehat{\lambda}_B|E^\circ = \lambda_B$.[109, 110]

Equivalently to the LS^r scenario, in the BS^r scenario the borrower's *pm*-choice plays a crucial role. The N-type borrower only chooses public offerings while the A-type is indifferent between both available *pms*. Furthermore, the N-type borrower favors public debt as long as the lender can separate both borrower types to achieve a rent from his interim and ex-post informational advantage. However, when the borrower can force the lender to abandon his separation strategy, the N-type borrower becomes indifferent between all public placements which offer the same contract agreement probability. When the lender abandons his separation strategy he pays the N-type borrower an (ex-ante) informational rent which implies that the profit the N-type borrower can achieve with public debt is identical to the profit of public equity. The profits are identical for both public offerings since debt financing's informational rent due to the borrower's interim and ex-post advantage is deducted from the rent from the ex-ante uncertainty.[111] The A-type borrower never receives an informational rent and, therefore, only cares about contract agreement.

4.3.4 Borrower Scenario with Absolute Power (BS^a)

In the *borrower scenario with absolute power* (BS^a) the borrower has the right to propose a take-it-or-leave-it offer to the lender concerning ct/pm and contracts' conditions which the latter can either accept or reject. If the lender

---

[109] A potential threat from the lender to stick to $\widehat{\lambda}_B|E^\circ = 0$ to prevent the borrower from choosing E° lacks subgame completeness since once the borrower has chosen E° the lender adjusts his expectation to $\widehat{\lambda}_B|E^\circ = \lambda_B$.

[110] In the LS^r scenario the borrower can avoid paying the N-type lender an ex-ante informational rent when private financing is feasible for the A-type lender since the lender cannot force the borrower to abandon his separation strategy, i. e. prevent that borrower reasonably expects $(0, 0, \lambda_L, \lambda_L)$. The lender has noting to "put at risk", for all possible ct/pm-choices he just receives his profit alternative and can, therefore, not exercise a credible threat to the borrower's separating strategy.

[111] Due to PC1 the informational rent from ex-ante uncertainty always exceeds any rent from interim and ex-post uncertainty, see footnote 108 for details.

accepts the borrower receives the necessary funds to finance the project and proceeds, otherwise the game ends. The lender and the borrower are left with their outside option.

To determine the contract $(E^\circ, D^\circ, E^\bullet, D^\bullet)$ the borrower prefers his expected profits $PB(\mu, \widehat{\lambda}_L)$ for the different contract choices must be compared. Since the lender cannot respond to the borrower's ct/pm-choice as the borrower determines the contract conditions as well, the borrower expects $\widehat{\lambda}_L = \lambda_L = 0.5$ for all possible ct/pm-constellations.[112] Therefore, we can now examine the borrower's optimization behavior as the condition optimizer who expects $\widehat{\lambda}_L = \lambda_L$ (Stage 2 of the contracting game), before analyzing the borrower's behavior as the ct/pm optimizer (Stage 1 of the contracting game).

Like in the LS^a scenario, the profit comparison in the BS^a scenario is a two step procedure. Firstly, the borrower's type-dependent feasible contracting intervals are determined. Secondly, the optimal contract choices are defined by a pair-wise (binary) comparison of the borrower's expected profits of the feasible contract choices.

THE BORROWER'S BEHAVIOR AS THE CONDITION OPTIMIZER $\left(\widehat{\lambda}_L = \lambda_L\right)$

Public equity financing

1. Feasible contracting intervals
   - A-type borrower's perspective $(PB \geq P_B^+)$
     $E_B^\circ | S1 : [1.75, \infty); \ E_B^\circ | S2 : [1.45, \infty)$
   - N-type borrower's perspective $(PB \geq 0)$
     $E_B^\circ | S1 : [1.45, \infty); \ E_B^\circ | S2 : [1.15, \infty)$
2. Profit comparison
   - A-type borrower's perspective $(\overline{P}_B = P_B^+)$
     $E_B^\circ | S1$ is preferable to $E_B^\circ | S2$ for $\mu \geq 2.05$ and for $\mu < 2.05$ the other way round.
   - N-type borrower's perspective $(\overline{P}_B = 0)$
     $E_B^\circ | S1$ is preferable to $E_B^\circ | S2$ for $\mu \geq 1.75$ and for $\mu < 2.05$ the other way round.
3. Preferred contract choice given public equity financing
   - A-type borrower
     For $\mu < 1.45$: no choice; for $1.45 \leq \mu < 2.05$: $E_B^\circ | S2$ and for $\mu \geq 2.05$: $E_B^\circ | S1$.
   - N-type borrower
     For $\mu < 1.15$: no choice; for $1.15 \leq \mu < 1.75$: $E_B^\circ | S2$ and for $\mu \geq 1.75$: $E_B^\circ | S1$.

---

[112] In the BS^a scenario the borrower is both the first and the second mover. The lender is only the third mover who can accept or reject the proposed contract offer.

Public debt financing

1. Feasible contracting intervals
   - A-type borrower's perspective ($PB \geq P_B^+$)
     $D_{B,I}^\circ|S1 : [1.7\bar{3}, \infty)$; $D_{B,I}^\circ|S2 : [1.\bar{3}, \infty)$; $D_{B,II}^\circ|S1 : [1.66, 1.7\bar{3})$;
     $D_{B,II}^\circ|S2 : [1.318, 1.\bar{3})$
   - N-type borrower's perspective ($PB \geq 0$)
     $D_{B,I}^\circ|S1 : [1.7\bar{3}, \infty)$; $D_{B,I}^\circ|S2 : [1.\bar{3}, \infty)$; $D_{B,II}^\circ|S1 : [1.6, 1.7\bar{3})$; $D_{B,II}^\circ|S2 : [1.274, 1.\bar{3})$
2. Profit comparison
   - A-type borrower's perspective ($\overline{P}_B = P_B^+$)
     $D_{B,I}^\circ|S2$ dominates $D_{B,II}^\circ|S1$; $D_{B,I}^\circ|S1$ is preferable to $D_{B,I}^\circ|S2$ for
     $\mu \geq 1.9$ and for $\mu < 1.9$ the other way round.
   - N-type borrower's perspective ($\overline{P}_B = 0$)
     $D_{B,I}^\circ|S2$ is preferable to $D_{B,II}^\circ|S1$ for $\mu \leq 1.68$ and for $\mu > 1.68$ the
     other way round; $D_{B,I}^\circ|S1$ is preferable to $D_{B,I}^\circ|S2$ for $\mu \geq 1.6$ and for
     $\mu < 1.6$ the other way round.
3. Preferred contract choice given public debt financing
   - A-type borrower
     For $\mu < 1.318$: no choice; for $1.318 \leq \mu < 1.\bar{3}$: $D_{B,II}^\circ|S2$; for $1.\bar{3} \leq \mu <$
     1.9: $D_{B,I}^\circ|S2$ and for $\mu \geq 1.9$: $D_{B,I}^\circ|S1$.
   - N-type borrower
     For $\mu < 1.274$: no choice; for $1.274 \leq \mu < 1.\bar{3}$: $D_{B,II}^\circ|S2$; for $1.\bar{3} \leq$
     $\mu < 1.68$: $D_{B,I}^\circ|S2$; for $1.68 \geq \mu < 1.7\bar{3}$: $D_{B,II}^\circ|S1$ and for $\mu \geq 1.7\bar{3}$:
     $D_{B,I}^\circ|S1$.

Private equity financing

1. Feasible contracting intervals
   - A-type borrower's perspective ($PB \geq P_B^+$)
     $E_B^{\bullet(2)} : [1.85, \infty)$; $E_B^{\bullet(1)} : [1.55, 1.85)$
   - N-type borrower's perspective ($PB \geq 0$)
     $E_B^{\bullet(2)} : [1.55, \infty)$; $E_B^{\bullet(1)} : [1.25, 1.55)$
2. No profit comparison required
3. Preferred contract choice given private equity financing
   - A-type borrower
     For $\mu < 1.55$: no choice; for $1.55 \leq \mu < 1.85$: $E_B^{\bullet(1)}$ and for $\mu \geq 1.85$:
     $E_B^{\bullet(2)}$.
   - N-type borrower
     For $\mu < 1.25$: no choice; for $1.25 \leq \mu < 1.55$: $E_B^{\bullet(1)}$ and for $\mu \geq 1.55$:
     $E_B^{\bullet(2)}$.

Private debt financing

1. Feasible contracting intervals
   - A-type borrower's perspective $(PB \geq P_B^+)$
     $D_{B,I}^{\bullet(2)} : [1.8, \infty);\ D_{B,II}^{\bullet(2)} : [1.741, 1.8);\ D_{B,II}^{\bullet(1)} : [1.4, 1.741)$
   - N-type borrower's perspective $(PB \geq 0)$
     $D_{B,I}^{\bullet(2)} : [1.8, \infty);\ D_{B,II}^{\bullet(2)} : [1.653, 1.8);\ D_{B,II}^{\bullet(1)} : [1.4, 1.653);\ D_{B,III}^{\bullet(1)} :$
     $[1.330, 1.4)$
2. No profit comparison required
3. Preferred choice given private debt financing
   - A-type borrower
     For $\mu < 1.4$: no choice; for $1.4 \leq \mu < 1.741$: $D_{B,II}^{\bullet(1)}$; for $1.741 \leq \mu < 1.8$:
     $D_{B,II}^{\bullet(2)}$ and for $\mu \geq 1.8$: $D_{B,I}^{\bullet(2)}$.
   - N-type borrower
     For $\mu < 1.330$: no choice; for $1.330 \leq \mu < 1.4$: $D_{B,III}^{\bullet(1)}$; for $1.4 \leq \mu <$
     1.653: $D_{B,II}^{\bullet(1)}$; for $1.653 \leq \mu < 1.8$: $D_{B,II}^{\bullet(2)}$ and for $\mu \geq 1.8$: $D_{B,I}^{\bullet(2)}$.

THE BORROWER'S BEHAVIOR AS THE ct/pm OPTIMIZER $(\widehat{\lambda}_L = \lambda_L)$[113]

1. Feasible contracting intervals
   - A-type borrower's perspective $(PB \geq P_B^+)$
     $E_B^{\circ}|S2 : [1.45, 2.05);\ E_B^{\circ}|S1 : [2.05, \infty);\ D_{B,II}^{\circ}|S2 : [1.318, 1.\bar{3});$
     $D_{B,I}^{\circ}|S2 : [1.\bar{3}, 1.9);\ D_{B,I}^{\circ}|S1 : [1.9, \infty);\ E_B^{\bullet(1)} : [1.55, 1.85);\ E_B^{\bullet(2)} :$
     $[1.85, \infty);\ D_{B,II}^{\bullet(1)} : [1.4, 1.741);\ D_{B,II}^{\bullet(2)} : [1.741, 1.8);\ D_{B,I}^{\bullet(2)} : [1.8, \infty)$
   - N-type borrower's perspective $(PB \geq 0)$
     $E_B^{\circ}|S2 : [1.15, 1.75);\ E_B^{\circ}|S1 : [1.75, \infty);\ D_{B,II}^{\circ}|S2 : [1.274, 1.\bar{3});$
     $D_{B,I}^{\circ}|S2 : [1.\bar{3}, 1.68);\ D_{B,II}^{\circ}|S1 : [1.68, 1.75);\ D_{B,I}^{\circ}|S1 : [1.75, \infty);$
     $E_B^{\bullet(1)} : [1.25, 1.55);\ E_B^{\bullet(2)} : [1.55, \infty);$
     $D_{B,III}^{\bullet(1)} : [1.330, 1.4);\ D_{B,II}^{\bullet(1)} : [1.4, 1.653);\ D_{B,II}^{\bullet(2)} : [1.653, 1.8);\ D_{B,I}^{\bullet(2)} :$
     $[1.8, \infty)$
2. Profit comparison
   - A-type borrower's perspective $(\overline{P}_B = P_B^+)$
     $D_{B,I}^{\circ}|S2$ dominates: $E_B^{\bullet(1)},\ D_{B,II}^{\bullet(1)},\ D_{B,II}^{\bullet(2)};\ D_{B,I}^{\bullet(2)}$ dominates: $E_B^{\bullet(2)},$
     $D_{B,I}^{\circ}|S1,\ E_B^{\circ}|S1;\ D_{B,I}^{\circ}|S2$ is preferable to $E_B^{\circ}|S2;\ D_{B,I}^{\bullet(2)}$ is preferable
     to $D_{B,I}^{\circ}|S2$ for $\mu \geq 1.8;\ D_{B,I}^{\bullet(2)}$ is preferable to $E_B^{\circ}|S2$ for $\mu \geq 1.65$.
   - N-type borrower's perspective $(\overline{P}_B = 0)$
     $E_B^{\circ}|S2$ dominates $E_B^{\bullet(1)};\ D_{B,II}^{\circ}|S2$ and $D_{B,I}^{\circ}|S2$ dominate $D_{B,III}^{\bullet(1)};$
     $D_{B,I}^{\circ}|S2$ dominates $D_{B,II}^{\bullet(1)};\ D_{B,II}^{\bullet(2)}$ dominates $D_{B,II}^{\circ}|S1;\ D_{B,II}^{\circ}|S2$ is

---

[113] We have to treat each ct/pm's contract choice interval-dependent since our profit comparison hinges on the assumption that the profit functions are continuously differentiable.

preferable to $E_B^\circ|S2$ for $\mu \geq 1.274$; $D_{B,I}^\circ|S2$ is preferable to $E_B^{\bullet(2)}$ for $\mu < 1.8$; $D_{B,II}^{\bullet(2)}$ is preferable to $D_{B,I}^\circ|S2$ for $\mu \geq 1.659$; $D_{B,II}^{\bullet(2)}$ is preferable to $E_B^{\bullet(2)}$ and to $E_B^\circ|S1$; $D_{B,II}^{\bullet(2)}$ is preferable to $D_{B,I}^\circ$ for $\mu \geq 1.687$; $D_{B,I}^{\bullet(2)}$ is preferable to: $E_B^\circ|S1$, $D_{B,I}^\circ|S1$ and to $E_B^{\bullet(2)}$.

3. Preferred *ct/pm*-choices

Table 4.41: Contract Type / Placement Mode-Choices and Agreement Probabilities in the Borrower Scenario with Absolute Power (EM, PC1)

Interval	μ	*Ct/pm*-choices and agreement probabilities			
		A-type borrower		N-type borrower	
I	$[1; 1.15)$	$-$	(0)	$-$	(0)
$II - III$	$[1.15; 1.274)$	$-$	(0)	$E_B^\circ|S2$	$(\frac{1}{2})$
IV	$[1.274; 1.318)$	$-$	(0)	$D_{B,II}^\circ|S2$	$(\frac{1}{2})$
V	$[1.318; 1.\bar{3})$	$D_{B,II}^\circ|S2$	$(\frac{1}{2}))$	$D_{B,II}^\circ|S2$	$(\frac{1}{2})$
$VI - XIII$	$[1.\bar{3}; 1.659)$	$D_{B,I}^\circ|S2$	$(\frac{1}{2})$	$D_{B,I}^\circ|S2$	$(\frac{1}{2})$
$XIV - XVI$	$[1.659; 1.8)$	$D_{B,I}^\circ|S2$	$(\frac{1}{2})$	$D_{B,II}^{\bullet(2)}$	(1)
$XVII - XXI$	$[1.8; \infty)$	$D_{B,I}^{\bullet(2)}$	(1)	$D_{B,I}^{\bullet(2)}$	(1)

Note: This table contains the borrower's *ct/pm*-choice together with the related contract agreement probabilities stated in brackets.

We observe, equivalently to the LS^a scenario, that if both borrower types propose a *ct/pm* to finance their project they might prefer the same constellation but they do not have to. Moreover, for projects with a low expected return ($\mu < 1.15$), the extractable surplus ($\mu - 1$) is not sufficient to compensate the lender for the costs arising from project financing. Hence, no contract is proposed at all by the borrower. However, the N-type borrower prefers to finance projects characterized by $1.15 \leq \mu < 1.274$ and PC1 with public equity S2 incorporating placement risk. For $1.274 \leq \mu < 1.659$ the N-type borrower favors risky public debt issues (S2), for $\mu \geq 1.659$ he prefers private debt placements. On the other hand, the A-type borrower refuses to finance projects with $\mu < 1.\bar{3}$. For $1.\bar{3} \leq \mu < 1.8$ he prefers public S2 debt before switching to private debt ($\mu \geq 1.8$).

Equivalently to the lender's behavior in the LS^a scenario, it becomes obvious that the borrower shifts from risky financing such as public S2 offerings to secure private placements as μ increases. The further μ increases the more important becomes the elimination of S2's placement risk. Furthermore, the favorability of debt financing in comparison to equity financing increases with

μ because the agency costs of debt financing decline now with μ while the agency costs of equity financing are independent of q, respectively μ.[114]

4.4 Bargaining Power Effects – Scenario Comparison

Finally, in this section bargaining power effects are determined by a pair-wise (binary) comparison of the *optimal* contract choices in the alternative bargaining power scenarios. Since bargaining power cannot vanish in our context, only the distribution among the lender and the borrower can change. Hence, the power effects are determined by a comparison of the optimal contract choices when the respective power component is redistributed.

We focus, like in the basic model of Chapter 3, on changes in the preferred ct/pm-choice, since the main objective of this chapter is to examine the robustness of the propositions determined in Chapter 3 when an ex-ante informational asymmetry is introduced. However, we have to note that contrary to the basic model the lender's and the borrower's expected profits now depend on the bargaining power distribution in a none uniquely defined way. While in the basic model without an ex-ante informational asymmetry the power redistributions $LS^a \to \{LS^r, BS^r, BS^a\}$ and $\{LS^r, BS^r\} \to BS^a$ were beneficial for the borrower ($\Delta PB \geq 0$) but disadvantageous for the lender ($\Delta PL \leq 0$) and vice versa, these power redistributions can have the opposite effects in this extended context. Contrary to the basic model, in this extended context power redistributions have two effects on the contract negotiation game. Firstly, they change the rights of the lender, respectively the borrower, in contract negotiations. But secondly, they also affect the informational structure of the game since the lender, respectively the borrower, is unaware of the contractual partner's type but knows whether he has a business opportunity or not. These changes in the informational structure of the game related to power redistributions make it quite difficult to obtain predictions about the lender's, respectively the borrower's, expected profit dependency in this extended context.

A redistribution of the bargaining power to choose ct/pm from the borrower to the lender, while the former keeps the power to determine contracts' conditions ($BS^a \to LS^r$) can, for example, have quite contrary effects on the borrower's expected profit. The borrower might even gain from this power reduction. Assume, for example, that the borrower favors private debt financing in the BS^a scenario since he is unaware of the lender's type and is

---

[114] Whether the agency costs of debt financing increase or decrease with μ depends on the condition optimizer's choice of h. Since in the lender optimization the lender tries to maximize h with respect to the borrower's participation constraint, the agency costs of debt financing increase with μ. In the borrower optimization the borrower tries to minimize h with respect to the lender's participation constraint which implies that the agency costs of debt financing decrease with μ.

unwilling to leave the potential N-type lender with an ex-ante informational rent. A redistribution of the power to choose ct/pm may now enable the potential N-type lender to reveal his type to the borrower by his ct/pm-choice, given that the borrower expects $(0, 0, \lambda_L, \lambda_L)$:

- If, for example, the N-type lender chooses a public debt contract, the borrower knows the lender's type and plays S2. The borrower can avoid private placement's negotiation costs and hence extracts a higher profit than in the BS^a scenario without incorporating any placement risk.
- If, alternatively, the N-type lender chooses a private debt contract no information about his type is revealed to the borrower and the borrower is confronted with the same situation as in the BS^a scenario.

Obviously, the borrower may gain from the discussed power redistribution (reduction), but he does not have to. Whether such a power reduction is (ex-ante) desirable from the borrower's perspective, is difficult to judge without further assumptions. The borrower is unaware whether the A or the N-type lender obtains the power to choose ct/pm, and furthermore whether the N-type lender actually would choose the (desirable) public offering, since the N-type lender is indifferent between all four ct/pm-constellations. How the lender's expected profit would respond to the discussed bargaining power variation is also not clear. The lender might gain or lose due to the power variation depending, in particular, on the borrower's expectation about the lender's type. Of course, it would be interesting to further examine the lender's and the borrower's profit dependency on bargaining power variations in this extended context, but we focus on changes in the preferred ct/pm-constellation since the main objective of this chapter is to analyze the robustness of the power effects determined in Chapter 3. And at the end of the day, the aim of this dissertation is to examine how bargaining power variations affect the preferred ct and pm-choice.

Effects caused by absolute bargaining power redistributions ($LS^a \rightleftharpoons BS^a$) are considered at first (see Section 4.4.1) as only agency cost aspects are relevant in the absolute power scenarios. The separation of the bargaining power components causes strategic aspects to become additionally pertinent. Section 4.4.2 examines effects of a variation of the power to determine ct/pm ($LS^a \rightleftharpoons BS^r, BS^a \rightleftharpoons LS^r$). Effects due to shifts of the remaining power component to set contracts' conditions ($LS^a \rightleftharpoons LS^r, BS^a \rightleftharpoons BS^r$) are analyzed in Section 4.4.3. Each potential direction of the examined power variations is considered separately since the resulting power effects can differ.[115] The different bargaining power components are evaluated from the lender's and from the borrower's perspective in Section 4.4.4, while in Section 4.4.5 the consistency and robustness of the obtained results are examined since they

---

[115] Effects due to variations of the first mover's type have been analyzed already in the previous section where the preferred contract choice in the alternative bargaining power scenarios were determined.

potentially depend significantly on the assumed underlying parameter constellation PC1 ($\zeta_{max} = 0.25, vc = 0.4, mc = 0.15, P_L^+ = P_B^+ = 0.3, \lambda_L = \lambda_B = 0.5, nc = 0.05$).

Since the preferred ct/pm-choices in the alternative bargaining power scenarios, as seen from the results in the previous section, summarized in Table 4.27, depend significantly on the first mover's (ct/pm optimizer's) type, all potentially occurring first mover's type variations are considered separately when the power effects are determined. For example, a redistribution of the absolute bargaining power or, at least, of the bargaining power to determine ct/pm can result in the following bargaining initiator's type variations: A-type $\to$ A-type, A-type $\to$ N-type, N-type $\to$ A-type, N-type $\to$ N-type. On the other hand, a redistribution of the power to set contracts' conditions does not influence the ct/pm optimizer's type.[116] The bargaining initiator's potential type variations are analyzed separately, but, of course, are unknown by the agents. The agents are unaware of the contractual partner's type due to the ex-ante informational asymmetry. This separation enables us to focus on observable power effects for a particular lender/borrower-type constellation. We thereby ignore effects which can not occur. For example, a redistribution of the absolute bargaining power might increase public S2 offering's favorability in comparison to no proposal at all, but this variation becomes irrelevant when the lender and the borrower both have a positive outside option (A-type).

Additionally to the preferred ct/pm-choices in the alternative bargaining power scenarios, Table 4.42 contains the related contract agreement probabilities. These probabilities are determined by the bargaining initiator's (ct/pm optimizer's) as well as the condition optimizer's optimization behavior given the ex-ante informational asymmetry. The condition optimizer's behavior is in particular relevant in the restricted power scenarios. The contract agreement probabilities range from zero (no agreement due to no proposal or rejection by the contractual partner) to 0.5[117] (only agreement facing the N-type contractual partner e. g. due to a risky public S2 offering) and finally to one (definite contract agreement e. g. due to a riskless public S1 offering or a private placement). The consideration of contract agreement probabilities enable us to analyze whether certain power variations might de- or even increase the chance of contract agreement, i. e. firm's financing probability.

In the basic model without an ex-ante informational asymmetry we have seen that projects are financed independently of the bargaining power distribution [Result 1 (Basic Model)]. However, as obvious from the Result 1 (Extended Model) this is not necessarily the case in this extended frame-

---

[116] A redistribution of the power to set contracts' conditions only affects the condition optimizer's type. Such potential type variations are not explicitly considered since we focus on changes in the preferred ct/pm-constellation, while, of course, the condition optimizer's contract determination affects the ct/pm optimizer's contract choice.

[117] According to PC1 $\lambda_L = \lambda_B = 0.5$.

work where the feasible contracting solutions depend on the specification of the lender's and the borrower's outside options as well as the condition optimizer's type. Of course, for certain situations, e. g. identical outside options $(P_L^+ = P_B^+)$ and a constant condition optimizer's type, the feasible contracting solutions of *some* ct/pm-constellations are independent of the bargaining power distribution but this is not true for *all* constellations. Therefore, we might even observe variations in the contract agreement probabilities for identically specified outside options and a constant condition optimizer's type.

Table 4.42: Summary of Contract Type / Placement Mode-Choices and Contract Agreement Probabilities (EM, PC1)

Interval	μ	LS^a		LS^r		BS^r		BS^a	
		A-type lender	N-type lender	A-type lender	N-type lender	A-type borrower	N-type borrower	A-type borrower	N-type borrower
I	[1-1.15)		$-(0)$		$-(0)$		$-(0)$		$-(0)$
II	[1.15-1.25)				$E^{\circ}(\frac{1}{2})$				
III	[1.25-1.274)				$E^{\circ}, E^{\bullet}(\frac{1}{2})$	$-(0)$	$E^{\circ}(\frac{1}{2})$	$-(0)$	$E^{\circ}(\frac{1}{2})$
IV	[1.274-1.318)				$E^{\circ}, D^{\circ}\ E^{\bullet}(\frac{1}{2})$				
V	[1.318-1.3)	$-(0)$	$E^{\circ}(\frac{1}{2})$	$-(0)$					
VI	[1.3-1.368)				$D^{\circ}(1)$				
VII	[1.368-1.4)					$D^{\circ}(\frac{1}{2})$	$D^{\circ}(\frac{1}{2})$		$D^{\circ}(\frac{1}{2})$
VIII	[1.4-1.45)				$D^{\circ}, D^{\bullet}(1)$				
IX	[1.45-1.545)				E°, D°	$D^{\bullet}(\frac{1}{2})$			
X	[1.545-1.55)				$D^{\bullet}(1)$				
XI	[1.55-1.6)			$E^{\bullet}(\frac{1}{2})$				$D^{\circ}(\frac{1}{2})$	
XII	[1.6-1.653)				E°, D°	$E^{\circ}, D^{\bullet}$			
XIII	[1.653-1.659)	$E^{\circ}(\frac{1}{2})$	$D^{\circ}(1)$	$E^{\bullet}, D^{\bullet}$	$E^{\bullet}, D^{\bullet}$	$(\frac{1}{2})$			
XIV	[1.659-1.741)			$(\frac{1}{2})$	(1)				
XV	[1.741-1.75)					$D^{\bullet}(1)$	$D^{\circ}(1)$		
XVI	[1.75-1.8)			$D^{\bullet}(1)$	E°, D°	$D^{\circ}, D^{\bullet}$			
XVII	[1.8-1.85)				$D^{\bullet}(1)$	(1)			$D^{\bullet}(1)$
XVIII	[1.85-1.917)					$D^{\circ}, E^{\bullet}$			
XIX	[1.917-2.05)	$D^{\circ}(1)$		$E^{\bullet}, D^{\bullet}$	E°, D°	$D^{\bullet}(1)$		$D^{\bullet}(1)$	
XX	[2.05-2.132)			(1)	$E^{\bullet}, D^{\bullet}$	E°, D°	E°, D°		
XXI	[2.132-∞)	$E^{\bullet}(1)$	$E^{\bullet}(1)$		(1)	$E^{\bullet}, D^{\bullet}(1)$	(1)		

Note: This table contains the preferred ct/pm-choices and related contract agreement probabilities conditioned on the ct/pm optimizer's (the first mover's) type in the four bargaining power scenarios. When for certain intervals more than one ct/pm-constellation is stated, the ct/pm optimizer (first mover) is indifferent between these choices. The agreement probabilities are stated in brackets.

4.4.1 Absolute Power Effects ($LS^a \rightleftharpoons BS^a$)

Power effects caused by a redistribution of the absolute bargaining power from the lender to the borrower ($LS^a \rightarrow BS^a$) are analyzed at first before examining the effects of a redistribution from the borrower to the lender ($BS^a \rightarrow LS^a$).

It is obvious from our results so far (in particular in the basic model) that a redistribution of the absolute bargaining power from the lender to the borrower has a *direct* effect on the preferred *ct*-choice. This effect is due to the bargaining power's impact on the negotiated contract conditions and, therefore, on the agency costs associated with debt and equity financing. When the power distribution varies, the agency costs of debt, respectively equity, financing respond in different ways and cause changes in the preferred *ct*-choice.[118] For example, in our context (see Assumption 3) the agency costs of debt financing ($vc \cdot DP(h)$) depend on the agreed repayment obligation h due to possible variations in the borrower's default probability $DP(h)$; the agency costs associated with equity financing (mc) are independent of the negotiated proportional return participation q. A power shift from the lender (LS^a) to the borrower (BS^a) now results in a potential reduction of q and h. Therefore, the agency costs associated with debt financing diminish in comparison to the costs of equity financing, making debt contracts more preferable.

On the other hand, a complete power variation does not have a direct effect on the preferred *pm*. However, the *pm*-choice is *indirectly* affected in two different ways:

- Firstly, the costs to avoid potential placement risk, i. e. of a public S2 offering, might differ from the lender's and from the borrower's perspective. Assume, for example, that the lender has the absolute bargaining power and favors debt financing due to lower agency costs. Debt financing leaves the borrower with a positive informational rent due to the latter's interim and ex-post advantage, which reduces the premium the lender has to pay to guarantee the A-type borrower's participation in a public S1 offering or in a private placement. Assuming alternatively that the borrower has the absolute bargaining power, the lender never receives an informational rent due to interim and ex-post uncertainty. Therefore, the borrower always has to pay the full premium to eliminate placement risk. Hence, a power shift to the borrower can increase the costs to eliminate placement risk. This implies that the lender might favor for certain projects (particular value of μ) a riskless public S1 offering or a private placement, while the borrower still chooses a risky public S2 offering. Additionally, the borrower's debt financing informational rent influences the lender's preference among the riskless placement opportunities. While for a public S1 offering in comparison to a risky S2 placement the lender only has to pay the difference between the borrower's informational rent and the

---

[118] In the absolute power scenarios no strategy aspects need to be considered.

A-type's positive outside option (P_B^+) to guarantee the A-type borrower's participation, additional negotiation costs arise with a private placement. A power shift to the borrower can, therefore, favor private placements since the borrower has to bear the full costs in both cases anyway.

• The second way in which the distribution of bargaining power affects the *pm*-choice is closely related to the power's effect on the *ct*-choice. As mentioned, an absolute power shift from the lender to the borrower can, but of course does not have to, result in an agency cost reduction when the borrower favors debt financing, which increases the extractable surplus, making the elimination of placement risk more desirable. Therefore, the borrower might prefer to finance certain projects with a riskless public S1 offering or a private placement, while the lender still sticks to a risky public S2 offering.

As obvious, the power effects on the *ct* and on the *pm* choice are *interrelated*. For example, when the agency costs of equity are lower than the costs of debt, the described effects might not occur. By this reasoning we have proven Proposition 1.

Proposition 1
Extended Model: Power Effect $LS^a \to BS^a$
A redistribution of the absolute bargaining power from the lender to the borrower increases the favorability of debt financing in comparison to equity. Private placements become more often preferable to public offerings. The contract agreement probability can de- or increase (see Table 4.42). Observable power effects dependent on the first mover's type variation for PC1:[119]
A-type lender → A-type borrower
$- \Rightarrow D^\bullet$ for $XVII - XVIII$; $D^\circ \Rightarrow D^\bullet$ for $XIX - XX$; $E^\bullet \Rightarrow D^\bullet$ for XXI.
A-type lender → N-type borrower
$E^\circ \Rightarrow -$ for $IX - XIII$; $E^\circ \Rightarrow D^\bullet$ for $XIV - XVIII$; $D^\circ \Rightarrow D^\bullet$ for $XIX - XX$; $E^\bullet \Rightarrow D^\bullet$ for XXI.
N-type lender → A-type borrower
$- \Rightarrow D^\circ$ for $V - IX$; $D^\circ \Rightarrow D^\bullet$ for $XVII - XX$; $E^\bullet \Rightarrow D^\bullet$ for XXI.
N-type lender → N-type borrower
$E^\circ \Rightarrow D^\circ$ for $IV - IX$; $D^\circ \Rightarrow D^\bullet$ for $XIV - XX$; $E^\bullet \Rightarrow D^\bullet$ for XXI.
□

Equivalently to the basic model of Chapter 3, we observe that debt financing becomes more favorable due to an absolute power shift from the

---

[119] Effects which *can* occur given PC1 are indicated by →, while ⇒ stands for effects which *definitively* occur given PC1.

lender to the borrower. However, while in the basic model the project was financed independent of the power distribution, this is not true in this extended context despite the lender's and the borrower's identical outside option. Even for a constant bargaining initiator's type the contract agreement probability can vary. For example, an absolute power shift from the A-type lender to the A-type borrower *increases* the contract agreement probability for $1.8 \leq \mu < 1.917$ $(XVII - XVIII)$. While for $1.8 \leq \mu < 1.917$ the A-type lender (still) chooses $E^\circ | S2$, the A-type borrower favors $D^{\bullet(2)}$ since the surplus extractable requires the elimination of placement risk. However, for $1.45 \leq \mu < 1.659$ $(IX - XIII)$ the opposite can be observed. When the power is shifted from the A-type lender to the N-type borrower the contract agreement probability *decreases*. While for $1.45 \leq \mu < 1.659$ the A-type lender still chooses $E^\circ | S2$, the N-type borrower favors $D^\circ | S2$. These observations confirm the described opposing power effects on the bargaining initiator's *pm* choice.

For the power variation in the opposite direction $(BS^a \rightarrow LS^a)$ the same arguments apply equivalently. Therefore, Proposition 2 emerges.

Proposition 2
Extended Model: Power Effect $BS^a \rightarrow LS^a$
A redistribution of the absolute bargaining power from the borrower to the lender decreases the favorability of debt financing in comparison to equity. Public offerings become more often preferable to private placements. The contract agreement probability can in- or decrease (see Table 4.42). Observable power effects dependent on the first mover's type variation for PC1:

A-type borrower $\rightarrow$ A-type lender
    $D^\bullet \Rightarrow -$ for $XVII - XVIII$; $D^\bullet \Rightarrow D^\circ$ for $XIX - XX$; $D^\bullet \Rightarrow E^\bullet$ for XXI.

A-type borrower $\rightarrow$ N-type lender
    $D^\circ \Rightarrow -$ for $V - IX$; $D^\bullet \Rightarrow D^\circ$ for $XVII - XX$; $D^\bullet \Rightarrow E^\bullet$ for XXI.

N-type borrower $\rightarrow$ A-type lender
    $- \Rightarrow E^\circ$ for $IX - XIII$; $D^\bullet \Rightarrow E^\circ$ for $XIV - XVIII$; $D^\bullet \Rightarrow D^\circ$ for $XIX - XX$; $D^\bullet \Rightarrow E^\bullet$ for XXI.

N-type borrower $\rightarrow$ N-type lender
    $D^\circ \Rightarrow E^\circ$ for $IV - IX$; $D^\bullet \Rightarrow D^\circ$ for $XIV - XX$; $D^\bullet \Rightarrow E^\bullet$ for XXI.

$\square$

4.4.2 Effects of the Bargaining Power to Determine Contract Type / Placement Mode ($LS^a \rightleftharpoons BS^r$, $BS^a \rightleftharpoons LS^r$)

The examination of power effects caused by a redistribution of the bargaining power to determine ct/pm requires a distinction dependent on the distribution of the power to set contracts' conditions. The power to set contracts' conditions can either belong to the lender or to the borrower. The case in which the power to determine contracts' conditions belongs to the lender ($LS^a \rightleftharpoons BS^r$) is analyzed at first, then the case in which the power is distributed to the borrower ($BS^a \rightleftharpoons LS^r$) is considered.

In addition to the agency cost and informational rent considerations for the absolute power redistribution, strategic aspects become relevant due to the separation of both bargaining power components. For example, a borrower who only can determine the ct/pm-constellation (BS^r) primarily hopes to achieve an informational rent by his ct/pm-choice and is less concerned about agency costs which basically distract the lender's profit optimization.

Due to the borrower's potential rent seeking behavior in the BS^r scenario, a redistribution of the bargaining power to determine ct/pm from the lender to the borrower while the lender keeps the power to set the contracts' conditions ($LS^a \rightarrow BS^r$) increases public debt's favorability as long as the A-type borrower is unwilling to participate. As long as the A-type borrower is unwilling to participate the N-type borrower has no chance to achieve a higher informational rent than in public debt financing due to his interim as well as ex-post informational advantage. When private financing becomes feasible for the A-type borrower the lender can separate the borrower types so that each type sticks to the respective pm. While the N-type borrower prefers public debt, the A-type favors private debt or equity to secure participation. The lender can choose S2 for public offerings without incorporating any placement risk. He can "separate" both borrower types.

However, when only public financing is feasible for the A-type borrower, the lender's separating opportunity vanishes and both borrower types stick to the respective ct which results in an (ex-ante) informational rent payment for the N-type borrower. The lender's separating opportunity also vanishes when E° is feasible for $\widehat{\lambda}_B|E^\circ = \lambda_B$ from the A-type borrower's perspective. In this case, the borrower can counteract the lender's separating effort and force the lender to pay an (ex-ante) informational rent to the N-type borrower. Power shifts in such a situation cause debt and equity financing to become equally beneficial when they offer the same contract agreement probability, while the N-type borrower favors public placements. If power is shifted to the A-type borrower both placement modes are equally likely.

The contract agreement probability can either fall or increase:

- The contract agreement probability can fall due to the borrower's informational rent seeking behavior, which might cause an agency cost suboptimal ct/pm-choice.

- The contract agreement probability can increase since if the borrower has no chance to obtain an informational rent, he favors the ct/pm-choice which maximizes his chance of contract agreement, while the lender might deviate from this ct/pm-choice since the lender tries to maximize his expected profit.

Proposition 3
Extended Model: Power Effect $LS^a \rightarrow BS^r$
A redistribution of the bargaining power to determine ct/pm from the lender to the borrower while the former keeps the power to set the contracts' conditions increases the favorability of public debt financing in comparison to public equity for the N-type borrower as long as E° for $\widehat{\lambda}_B | E^\circ = \lambda_B$ is unfeasible for both borrower types. If E° for $\widehat{\lambda}_B | E^\circ = \lambda_B$ is feasible for both borrower types, the N-type becomes indifferent between all public offerings with the same contract agreement probability, while the A-type additionally considers private placements. The contract agreement probability can fall or increase (see Table 4.42). Observable power effects dependent on the first mover's type variation for PC1:
A-type lender → A-type borrower
  $-\ \Rightarrow\ D^\bullet$ for XV; $-\ \Rightarrow\ \{D^\circ, D^\bullet\}$ for $XVI - XVII$; $-\ \Rightarrow$ $\{D^\circ, E^\bullet, D^\bullet\}$ for XVIII; $D^\circ\ \rightarrow\ \{E^\bullet, D^\bullet\}$ for XIX; $D^\circ\ \rightarrow$ $\{E^\circ, E^\bullet, D^\bullet\}$ for XX; $E^\bullet \rightarrow \{E^\circ, D^\circ, D^\bullet\}$ for XXI.
A-type lender → N-type borrower
  $E^\circ \Rightarrow\ -$ for $IX - XI$; $E^\circ \Rightarrow D^\circ$ for $XII - XVIII$; $D^\circ \rightarrow E^\circ$ for XX; $E^\bullet \Rightarrow \{E^\circ, D^\circ\}$ for XXI.
N-type lender → A-type borrower
  $-\ \Rightarrow\ D^\circ$ for VII; $-\ \Rightarrow\ D^\bullet$ for $VIII - IX$; $D^\circ \Rightarrow D^\bullet$ for X and XV; $D^\circ \Rightarrow \{E^\bullet, D^\bullet\}$ for $XI - XIV$; $D^\circ \rightarrow D^\bullet$ for $XVI - XVII$; $D^\circ \rightarrow \{E^\bullet, D^\bullet\}$ for $XVIII - XIX$; $D^\circ \rightarrow \{E^\circ, E^\bullet, D^\bullet\}$ for XX; $E^\bullet \rightarrow \{E^\circ, D^\circ, D^\bullet\}$ for XXI.
N-type lender → N-type borrower
  $E^\circ \Rightarrow D^\circ$ for $IV - IX$; $D^\circ \rightarrow E^\circ$ for XX; $E^\bullet \Rightarrow \{E^\circ, D^\circ\}$ for XXI.
□

If the power to determine ct/pm is shifted from the borrower to the lender while the latter can choose the contracts' conditions ($BS^r \rightarrow LS^a$), the borrower loses his strategic position to secure an informational rent and agency aspects gain importance. Therefore, public equity's favorability rises, as stated in Proposition 4.

Proposition 4

Extended Model: Power Effect $BS^r \to LS^a$

A redistribution of the bargaining power to determine ct/pm from the borrower to the lender while the latter keeps the power to set the contracts' conditions decreases the favorability of public and private debt. The contract agreement probability can in- or decreases (see Table 4.42). Observable power effects dependent on the first mover's type variation for PC1:

A-type borrower $\to$ A-type lender

$D^\bullet \Rightarrow -$ for XV; $\{D^\circ, D^\bullet\} \Rightarrow -$ for $XVI-XVII$; $\{D^\circ, E^\bullet, D^\bullet\} \Rightarrow -$ for XVIII; $\{E^\bullet, D^\bullet\} \Rightarrow D^\circ$ for XIX; $\{E^\circ, E^\bullet, D^\bullet\} \Rightarrow D^\circ$ for XX; $\{E^\circ, D^\circ, D^\bullet\} \Rightarrow E^\bullet$ for XXI.

A-type borrower $\to$ N-type lender

$D^\circ \Rightarrow -$ for VII; $D^\bullet \Rightarrow -$ for $VIII - IX$; $D^\bullet \Rightarrow D^\circ$ for X and $XV - XVII$; $\{E^\bullet, D^\bullet\} \Rightarrow D^\circ$ for $XI - XIV$ and $XVIII - XIX$; $\{E^\circ, E^\bullet, D^\bullet\} \Rightarrow D^\circ$ for XX; $\{E^\circ, D^\circ, D^\bullet\} \Rightarrow E^\bullet$ for XXI.

N-type borrower $\to$ A-type lender

$- \Rightarrow E^\circ$ for $IX - XI$; $D^\circ \Rightarrow E^\circ$ for $XII - XVIII$; $E^\circ \Rightarrow D^\circ$ for XX; $\{E^\circ, D^\circ\} \Rightarrow E^\bullet$ for XXI.

N-type borrower $\to$ N-type lender

$D^\circ \Rightarrow E^\circ$ for $IV - IX$; $E^\circ \Rightarrow D^\circ$ for XX; $\{E^\circ, D^\circ\} \Rightarrow E^\bullet$ for XXI.

$\square$

When, in comparison to $LS^a \to BS^r$, the power to determine the ct/pm is redistributed from the borrower to the lender, while the former keeps the power to determine contracts' conditions ($BS^a \to LS^r$), the borrower remains able to squeeze the lender to the latter's reservation profit if either the A-type lender does not participate at all, or if private placements are feasible for the A-type. The borrower keeps this opportunity due to his interim and ex-post informational advantage. Anticipating this profit restriction the lender attempts to maximize his chance of contract agreement by his ct/pm-choice. When more than one ct/pm-constellation offers the same chance of contract agreement, the lender is indifferent between them. Therefore, public debt contracts lose their attractiveness and private financing becomes inevitable if the power is obtained by the A-type lender. The A-type lender favors private to public placements when both are feasible to secure contract participation. Due to the A-type lender's private placement preference, the N-type lender cannot achieve an (ex-ante) informational rent, i. e. exploit his strategical (ct/pm optimizer) position. The N-type lender becomes indifferent between public and private debt and equity as long as they offer the same chance of contract agreement.

The contract agreement probability increases due to the lender's power increase since the lender's only objective is to maximize this probability. Of course, the borrower is also concerned about contract agreement, but he con-

siders profit opportunities as well. The additional consideration of profit opportunities can distract the borrower from his contract agreement objective,
when both objectives might favor opposing ct/pm and strategy choices. Proposition 5 emerges.

Proposition 5
Extended Model: Power Effect $BS^a \to LS^r$
A redistribution of the bargaining power to determine ct/pm from the
borrower to the lender while the former keeps the power to set contracts' conditions decreases debt financing's favorability. If the power
is obtained by the A-type lender private placement's attractivity increases. The contract agreement probability is positively affected by
the power redistribution (see Table 4.42). Observable power effects
dependent on the first mover's type variation for PC1:
A-type borrower $\to$ A-type lender
  $- \Rightarrow D^\bullet$ for $XV - XVI$; $D^\bullet \to E^\bullet$ for $XVIII - XXI$.
A-type borrower $\to$ N-type lender
  $D^\circ \to D^\bullet$ for VIII; $D^\circ \to \{E^\circ, D^\bullet\}$ for $IX - X$ and XVI; $D^\circ \to$
  $\{E^\circ, E^\bullet, D^\bullet\}$ for $XI - XV$; $D^\bullet \to \{E^\circ, D^\circ\}$ for XVII; $D^\bullet \to$
  $\{E^\circ, D^\circ, E^\bullet\}$ for $XVIII - XXI$.
N-type borrower $\to$ A-type lender
  $- \Rightarrow E^\bullet$ for $XI - XII$; $- \Rightarrow \{E^\bullet, D^\bullet\}$ for XIII; $D^\bullet \to E^\bullet$ for
  XIV and $XVIII - XXI$.
N-type borrower $\to$ N-type lender
  $E^\circ \to E^\bullet$ for III; $D^\circ \to \{E^\circ, E^\bullet\}$ for IV; $D^\circ \to D^\bullet$ for VIII;
  $D^\circ \to \{E^\circ, D^\bullet\}$ for $IX - X$; $D^\circ \to \{E^\circ, E^\bullet, D^\bullet\}$ for $XI - XIII$;
  $D^\bullet \to \{E^\circ, D^\circ, E^\bullet\}$ for $XIV - XV$ and $XVIII - XXI$; $D^\bullet \to$
  $\{E^\circ, D^\circ\}$ for $XVI - XVII$.
□

Focusing on the opposite power redistribution ($LS^r \to BS^a$), i. e. the
power to determine ct/pm is obtained by the borrower from the lender while
the former keeps the power to set the contracts' conditions, the contract agreement probability decreases due to the borrower's (additional) profit maximizing objective. Private placement's favorability from the A-type lender's perspective to secure contract participation becomes unimportant, which does not
imply that private placement's favorability decreases in general since private
placements can still be an efficient way to cope with the ex-ante uncertainty
about the contractual partner's type. Proposition 6 summarizes the main effects.

Proposition 6

Extended Model: Power Effect $LS^r \to BS^a$

A redistribution of the bargaining power to determine ct/pm from the lender to the borrower, while the latter keeps the power to set contracts' conditions potentially increases debt financing's favorability. The contract agreement probability decreases (see Table 4.42). Observable power effects dependent on the first mover's type variation for PC1:

A-type lender $\to$ A-type borrower

$D^\bullet \Rightarrow -$ for $XV - XVI$; $E^\bullet \Rightarrow D^\bullet$ for $XVIII - XXI$.

A-type lender $\to$ N-type borrower

$E^\bullet \Rightarrow -$ for $XI - XII$; $\{E^\bullet, D^\bullet\} \Rightarrow -$ for XIII; $E^\bullet \Rightarrow D^\bullet$ for XIV and $XVIII - XXI$.

N-type lender $\to$ A-type borrower

$D^\bullet \Rightarrow D^\circ$ for VIII; $\{E^\circ, D^\bullet\} \Rightarrow D^\circ$ for $IX - X$ and XVI; $\{E^\circ, E^\bullet, D^\bullet\} \Rightarrow D^\circ$ for $XI - XV$; $\{E^\circ, D^\circ\} \Rightarrow D^\bullet$ for XVII; $\{E^\circ, D^\circ, E^\bullet\} \Rightarrow D^\bullet$ for $XVIII - XXI$.

N-type lender $\to$ N-type borrower

$E^\bullet \Rightarrow E^\circ$ for III; $\{E^\circ, E^\bullet\} \Rightarrow D^\circ$ for IV; $D^\bullet \Rightarrow D^\circ$ for VIII; $\{E^\circ, D^\bullet\} \Rightarrow D^\circ$ for $IX - X$; $\{E^\circ, E^\bullet, D^\bullet\} \Rightarrow D^\circ$ for $XI-XIII$; $\{E^\circ, D^\circ, E^\bullet\} \Rightarrow D^\bullet$ for $XIV-XV$ and $XVIII-XXI$; $\{E^\circ, D^\circ\} \Rightarrow D^\bullet$ for $XVI - XVII$.

$\square$

4.4.3 Effects of the Bargaining Power to Determine Contracts' Conditions ($LS^a \rightleftharpoons LS^r$, $BS^a \rightleftharpoons BS^r$)

Analyzing the effects from a redistribution of the bargaining power to determine contracts' conditions requires a distinction dependent on the distribution of the bargaining power to choose ct/pm. The bargaining power to choose ct/pm can either belong to the lender ($LS^a \rightleftharpoons LS^r$) or to the borrower ($BS^a \rightleftharpoons BS^r$).

The bargaining power effects analyzed in this section resemble the effects of the bargaining power to choose ct/pm, determined in the previous section, since in the restricted power scenarios strategic aspects dominate pure agency cost considerations.[120] The power effects of the redistribution $LS^a \rightleftharpoons LS^r$ ($BS^a \rightleftharpoons BS^r$) are similar to the effects of $BS^a \rightleftharpoons LS^r$ ($LS^a \rightleftharpoons BS^r$) while the

---

[120] Thereby, we do not want to imply that agency cost considerations are irrelevant in the restricted power scenarios. They are, like the strategic aspects, highly significant to define the contracting solution in the restricted power scenarios. Agency cost considerations help to define which contract choices are actually feasible and which contract agreement probability is implied, while the preferred contract choice is selected according to strategic considerations, i. e. maximizing the chance of contract agreement or gamble for an informational rent.

latter redistribution has, contrary to the former, no effect on the bargaining initiator's (ct/pm optimizer's) type. Hence, in this section power effects are only distinguished according to bargaining initiator's type and not due to potential type variations.

Similar to $BS^a \to LS^r$, a shift of the bargaining power to set contracts' conditions from the lender to the borrower while the former keeps the power to choose ct/pm ($LS^a \to LS^r$) results in a strategic lender behavior, i. e. ct/pm-choice. Since the lender has no interim and ex-post informational advantage, he never receives an informational rent from these uncertainties. Therefore, with one exception the borrower can squeeze the lender to his profit reservation in the LS^r scenario which implies that the lender becomes indifferent between debt and equity financing. The lender is only concerned about contract agreement. Hence, the A-type lender prefers private placements as soon as they are feasible to avoid any potential placement risk in the public market. The N-type lender is indifferent between public and private placements when they offer the same contract agreement probability. The only exception in which the N-type lender is able to achieve an ex-ante informational rent in the LS^r scenario is when public placements are feasible for the A-type lender while private placements are not. In such a situation the A-type lender has "nothing to lose" from choosing a public placement which the borrower anticipates and leaves the N-type lender with an ex-ante rent to guarantee the A-type's participation.

Therefore, the power redistribution $LS^a \to LS^r$ has lender type-dependent effects.

- The N-type lender becomes indifferent between all feasible ct/pm-choices with the same contract agreement probability as long as the A-type lender does not participate at all or private placements are feasible for him.
- The A-type lender prefers private placements as soon as they are feasible to guarantee participation. The A-type lender is indifferent between private debt and equity financing as long as they offer the same chance of contract agreement.

The only exception arises if no private placement but a public offering is feasible for the A-type lender, since in this situation the N-type lender favors the respective public offering to secure an informational rent from his his ex-ante informational advantage. However, this exception does not occur for PC1. Strategic aspects determine the preferred ct/pm-choice in the LS^r scenario while agency aspects define whether a certain ct/pm-choice is feasible or not. Therefore, Proposition 7 resembles Proposition 5.

Proposition 7
Extended Model: Power Effect $LS^a \rightarrow LS^r$
A redistribution of the bargaining power to set contracts' conditions
from the lender to the borrower while the former keeps the power to
determine ct/pm increases the favorability of private placements for
the A-type lender. The N-type lender becomes basically indifferent
between public and private debt and equity as long as they offer the
same contract agreement probability. The contract agreement probability increases due to the power redistribution (see Table 4.42).
Observable power effects dependent on the first mover's type for PC1:
A-type lender
  $- \Rightarrow D^\bullet$ for $XV - XVII$; $- \Rightarrow \{E^\bullet, D^\bullet\}$ for XVIII; $D^\circ \Rightarrow$
  $\{E^\bullet, D^\bullet\}$ for $XIX - XX$; $E^\bullet \rightarrow D^\bullet$ for XXI.
N-type lender
  $E^\circ \rightarrow E^\bullet$ for III; $E^\circ \rightarrow \{D^\circ, E^\bullet\}$ for IV; $E^\circ \Rightarrow D^\circ$ for $V - VII$;
  $E^\circ \Rightarrow \{D^\circ, D^\bullet\}$ for VIII; $E^\circ \rightarrow \{D^\circ, D^\bullet\}$ for IX; $D^\circ \rightarrow \{E^\circ, D^\bullet\}$
  for X and $XVI - XVII$; $D^\circ \rightarrow \{E^\circ, E^\bullet, D^\bullet\}$ for $XI - XV$ and
  $XVIII - XX$; $E^\bullet \rightarrow \{E^\circ, D^\circ, D^\bullet\}$ for XXI.
□

For the opposite power redistribution $(LS^r \rightarrow LS^a)$ the favorability of
private placements from the A-type lender's perspective due to their strategic
advantage to secure contract participation vanishes. The N-type lender's indifference between all ct/pm-constellations offering the same contract agreement
probability disappears. Profit considerations become important. Therefore the
contract agreement probability decreases, as summarized in Proposition 8.

Proposition 8
Extended Model: Power Effect $LS^r \rightarrow LS^a$
A redistribution of the bargaining power to set contracts' conditions
from the borrower to the lender while the latter keeps the power to
determine ct/pm potentially reduces private placement's favorability
from the A-type lender's perspective. The chance of contract agreement decreases (see Table 4.42). Observable power effects dependent
on the first mover's type for PC1:
A-type lender
  $D^\bullet \Rightarrow -$ for $XV - XVII$; $\{E^\bullet, D^\bullet\} \Rightarrow -$ for XVIII; $\{E^\bullet, D^\bullet\} \Rightarrow$
  D° for $XIX - XX$; $D^\bullet \Rightarrow E^\bullet$ for XXI.
N-type lender
  $E^\bullet \Rightarrow E^\circ$ for III; $\{D^\circ, E^\bullet\} \Rightarrow E^\circ$ for IV; $D^\circ \Rightarrow E^\circ$ for $V - VII$;
  $\{D^\circ, D^\bullet\} \Rightarrow E^\circ$ for VIII and IX; $\{E^\circ, D^\bullet\} \Rightarrow D^\circ$ for X and
  $XVI - XVII$; $\{E^\circ, E^\bullet, D^\bullet\} \Rightarrow D^\circ$ for $XI - XV$ and $XVIII - XX$;
  $\{E^\circ, D^\circ, D^\bullet\} \Rightarrow E^\bullet$ for XXI.
□

The same power redistribution can be analyzed when the bargaining power
to determine ct/pm is assigned to the borrower $(BS^a \rightleftharpoons BS^r)$.

When the power to determine contracts' conditions is shifted form the borrower to the lender while the former keeps the right to choose ct/pm ($BS^a \rightarrow BS^r$), the borrower tries to maximize his potential informational rent by his ct/pm-choice. Anticipating the borrower's behavior, the lender tries to maximize his expected profit by restricting the borrower's informational rent to his "old" informational rent from interim and ex-post uncertainty.[121] The lender tries to avoid to pay an additional informational rent due to the borrower's ex-ante advantage. The lender can do so by private financing or by risky public S2 offerings. The lender will try to separate both borrower types as long as it is profitable for him. If it is credible that the lender chooses S2 for all public offerings, the A-type borrower restrains from public offerings, the A-type prefers private placements. However, the lender's separation strategy is only feasible under certain conditions. Of course the lender can avoid leaving an ex-ante informational rent to the N-type borrower as long as the A-type borrower does not participate in project financing since no actual separation is required. But the lender can also separate both borrower types when private placements are feasible for the A-type while E° for $\widehat{\lambda}_B | E^\circ = \lambda_B$ is unfeasible from the A-type borrower's perspective. In this situation the lender can stick to risky public S2 offerings without incorporating any placement risk since the A-type borrower favors the opportunity of private financing. By this borrower type separation the lender saves a possible ex-ante informational rent payment but he might incorporate additional negotiation costs ($2nc$).

However, this separation opportunity vanishes when E° for $\widehat{\lambda}_B | E^\circ = \lambda_B$ becomes feasible from the A-type borrower's perspective, offering both borrower types a seemingly irrational ct/pm-choice by which they can force the lender to abandon his separating strategy and to leave the N-type borrower with an additional (ex-ante) informational rent.[122] The lender is also unable to separate both borrower types when only public offerings are feasible for the A-type, leaving the latter no alternative to a public placement. Hence, the lender is again faced with the tradeoff between S2's placement risk and S1's additional informational rent payment.

As discussed the power effects of $BS^a \rightarrow BS^r$ depend significantly on the lender's possibility to separate both borrower types in the BS^r scenario. Assuming that the lender has the opportunity to separate both borrower types, the power shift increases public debt financing's favorability from the N-type borrower's perspective due to his informational rent seeking behavior. From the A-type borrower's perspective private placement's become more preferable. However, if the lender cannot separate the borrower types, the N-type borrower achieves an informational rent from all three uncertainties which is

---

[121] The borrower's informational rent due to his interim and ex-post advantage is bounded below the borrower's potential informational rent due to all three informational asymmetries since another reason for a potential rent is added.

[122] See page 208 for the detailed derivation of this result.

identical for equity and debt financing.[123] The N-type borrower prefers the public offering which promises the highest chance of contract agreement. The A-type borrower even considers public and private placements.

Therefore, the contract agreement probability increases for the A-type borrower since the A-type borrower's objective is now only to maximize his chance of contract agreement. He never receives an informational rent anyway. The agreement probability of the N-type borrower remains unaffected since he always participates even if the lender sticks to his separating strategy. By this reasoning Proposition 9 is proven.

> **Proposition 9**
> **Extended Model: Power Effect** $BS^a \to BS^r$
> A redistribution of the bargaining power to set contracts' conditions from the borrower to the lender while the former keeps the power to determine ct/pm increases the favorability of public debt financing from the N-type borrower's and of private debt from the A-type's perspective, if either the A-type borrower does not participate in project financing or the A-type has a feasible private placement opportunity while E° for $\widehat{\lambda}_B | E^\circ = \lambda_B$ is unfeasible from his perspective, i. e. the lender can separate both borrower types. Otherwise, both borrower types become indifferent between debt and equity when they offer the same chance of contract agreement, while the A-type even considers private placements which are not preferred by the N-type. The power redistribution can increase the A-type borrower's contract agreement probability (see Table 4.42). Observable power effects dependent on the first mover's type for PC1:
> A-type borrower
> $- \Rightarrow D^\bullet$ for XV; $- \Rightarrow \{D^\circ, D^\bullet\}$ for XVI; $D^\bullet \to D^\circ$ for XVII; $D^\bullet \to \{D^\circ, E^\bullet\}$ for $XVIII - XIX$; $D^\bullet \to \{E^\circ, D^\circ, E^\bullet\}$ for $XX - XXI$.
> N-type borrower
> $D^\bullet \Rightarrow D^\circ$ for $XIV - XIX$; $D^\bullet \Rightarrow \{E^\circ, D^\circ\}$ for $XX - XXI$.
> □

The examination of the opposite power redistribution $(BS^r \to BS^a)$ reveals that again the power effects depend significantly on the lender's potential borrower type separation in the BS^r scenario, see Proposition 10.

---

[123] The informational rents of debt and equity financing are identical for the N-type borrower since due to PC1 $(vc/2 < P_B^+)$ the borrower's rent from interim and ex-post uncertainty is bounded below the total rent achievable from all three uncertainties which coincides with the rent only achievable from the ex-ante uncertainty (P_B^+).

Proposition 10

Extended Model: Power Effect $BS^r \to BS^a$

A redistribution of the bargaining power to set contracts' conditions from the lender to the borrower while the latter keeps the power to determine ct/pm decreases the favorability of public debt financing from the N-type borrower's and of private debt from the A-type's perspective, if either the A-type borrower does not participate in project financing or the A-type has a feasible private opportunity while E° for $\widehat{\lambda}_B | E^\circ = \lambda_B$ is unfeasible from his perspective. If no separating opportunity exists, the favorability of debt financing increases. The power redistribution decreases A-type borrower's contract agreement probability (see Table 4.42). Observable power effects dependent on the first mover's type for PC1:

A-type borrower

$D^\bullet \Rightarrow -$ for XV; $\{D^\circ, D^\bullet\} \Rightarrow -$ for XVI; $D^\circ \Rightarrow D^\bullet$ for XVII; $\{D^\circ, E^\bullet\} \Rightarrow D^\bullet$ for $XVIII - XIX$; $\{E^\circ, D^\circ, E^\bullet\} \Rightarrow D^\bullet$ for $XX - XXI$.

N-type borrower

$D^\circ \Rightarrow D^\bullet$ for $XIV - XIX$; $\{E^\circ, D^\circ\} \Rightarrow D^\bullet$ for $XX - XXI$.

$\square$

4.4.4 Relevance of the Bargaining Power Components

After analyzing bargaining power effects in financial contracting in Sections 4.4.1, 4.4.2 and 4.4.3, the relevance of both power components for the lender's as well as for the borrower's profit optimization is evaluated in this section. To examine which bargaining power component the lender, respectively the borrower, prefers the restricted power scenarios (LS^r, BS^r) are compared from the lender's as well as from the borrower's perspective. In the LS^r scenario the lender has the bargaining power to determine ct/pm and the borrower can set contracts' conditions, while in the BS^r scenario the borrower chooses ct/pm and the lender sets the conditions.

Equivalently to the basic model without an ex-ante informational asymmetry, in the LS^r scenario the lender (ct/pm optimizer) is generally squeezed by the borrower (condition optimizer) to his profit opportunity. The only exception where the lender might obtain an informational rent due to his ex-ante advantage is when only public financing is feasible for the A-type lender (not possible given PC1). In this situation the borrower is confronted with the tradeoff between S1's informational rent payment to the N-type lender and S2's placement risk. Only under these circumstances the N-type lender might achieve an ex-ante informational rent. This opportunity does not exist for the A-type lender. The A-type is always squeezed to his profit opportunity. Table

4.43 summarizes the lender's as well as the borrower's informational rents for the LS^r, respectively the BS^r, scenario given PC1.[124]

On the other side, the power to set contracts' conditions enables the lender in the BS^r scenario to extract a surplus even for a potentially suboptimal ct/pm-choice with respect to arising agency costs. The choice might be suboptimal because of the borrower's unique focus on maximizing his informational rent. Therefore, the lender favors the power to set contracts' conditions to the power to choose ct/pm when the borrower can separate both lender types in the LS^r scenario, i. e. the borrower does not even pay an ex-ante informational rent to the N-type lender. Finally, the borrower's informational rent seeking behavior in the BS^r scenario results in a lower contract agreement probability than in the LS^r scenario where only contract agreement matters for the lender (ct/pm optimizer).

Result 2
Extended Model: Relevance of Power Components – Lender's Perspective
The A-type lender prefers the bargaining power to set contracts' conditions to the power to determine ct/pm as long as the contract agreement probability is identical in the LS^r and the BS^r scenario. The N-type lender's preference is identical to the A-type's if the borrower has the opportunity to distinguish between both lender types when determining contracts' conditions in the BS^r scenario. Otherwise, the N-type lender's preference depends on the project specification. $\square$

To evaluate both power components from the borrower's perspective is more difficult. Similar to the lender's perspective, from the borrower's perspective the power to determine contracts' conditions is more important than the power to choose the ct/pm-constellation for projects with a high expected return, since the extractable profit is larger than the profit extractable by a specific ct/pm-choice and the related informational rent. See Table 4.43 for the borrower's potential informational rent in the BS^r scenario which is bounded by 0.3. The power to determine the actual ct/pm-choice is relevant for projects just generating sufficient funds to cover the initial outlay and the arising agency costs. For such projects the informational rent achievable by the borrower's ct/pm-choice (BS^r) can exceed the surplus extractable in the LS^r scenario. For the N-type borrower the power to set ct/pm is more important than for the A-type borrower since the A-type never receives an ex-ante informational rent anyway. However, as mentioned earlier, the final contract agreement probability might, in particular for projects near the FCC, be lower

---

[124] Since for the ct/pm optimizer only his own potential informational rent is relevant when determining ct/pm, Table 4.43 only states the lender's potential informational rents in the LS^r scenario and the borrower's informational rents in the BS^r scenario.

Table 4.43: Summary of Contract Type / Placement Mode-Choices, Contract Agreement Probabilities and Informational Rents in the Restricted Power Scenarios (EM, PC1)

Interval	LS^r				BS^r			
	A-type lender's		N-type lender's		A-type borrower's		N-type borrower's	
	ct/pm-choice	IR	ct/pm-choice	IR	ct/pm-choice	IR	ct/pm-choice	IR
I			$-(0)$	$-$			$-(0)$	$-$
II			$E^°(\frac{1}{2})$	$-$				
III			$E^°, E^• (\frac{1}{2})$	$-$	$-(0)$	$-$	$E^°(\frac{1}{2})$	$-$
IV			$E^°, D^° E^•(\frac{1}{2})$	$-$				
V-VI	$-(0)$	$-$	$D^°(1)$	$-$				
VII					$D^°(\frac{1}{2})$	$-$	$D^°(\frac{1}{2})$	$\dfrac{0.08}{\mu}$
VIII			$D^°, D^• (1)$	$-$	$D^•(\frac{1}{2})$	$-$		
IX-X			$E^°, D^° D\bullet(1)$	$-$				
XI	$E^•(\frac{1}{2})$	$-$			$E^•, D^•$			
XII			$E^°, D^°$	$-$	$(\frac{1}{2})$	$-$		
XIII-XIV	$E^•, D^• (\frac{1}{2})$	$-$	$E^•, D^•$					$\dfrac{0.16}{\mu}$
XV	$D^•(1)$	$-$	(1)		$D^•(1)$	$-$	$D^°(1)$	
XVI-XVII			$E^°, D^° D\bullet(1)$	$-$	$D^°, D^• (1)$	$-$		0.3
XVIII-XIX	$E^•, D^•$	$-$	$E^°, D^°$	$-$	$D^°, E^• D^•(1)$	$-$		
XX-XXI	(1)		$E^•, D^• (1)$		$E^°, D^° E^•, D^• (1)$	$-$	$E^°, D^° (1)$	0.3

Note: This table contains the preferred ct/pm-choices and related contract agreement probabilities conditioned on the ct/pm optimizer's (the first mover's) type in both restricted power scenarios. When for certain intervals more than one ct/pm-constellation is stated, the ct/pm optimizer (first mover) is indifferent between these choices. The agreement probabilities are stated in brackets. In addition to the preferred ct/pm-choices and the agreement probabilities, it is indicated whether the first mover will obtain an informational rent (IR) or not.

in the BS^r than in the LS^r scenario due to the borrower's informational rent seeking behavior in the BS^r scenario.

Result 3

**Extended Model: Relevance of Power Components –
Borrower's Perspective**

The borrower's preference for the bargaining power components depends on the project specification. For projects with a large expected returns the borrower favors the power to set contracts' conditions since the extractable surplus exceeds any potential informational rent while contract agreement is guaranteed. The power to determine the actual ct/pm-constellation becomes relevant for projects with a small expected return where the achievable informational rent can exceed the otherwise extractable surplus. For the N-type borrower this power component is more important than for the A-type borrower since the A-type never receives an informational rent due to ex-ante uncertainty anyway. However, not for all project specifications where the potential informational rent exceeds the otherwise extractable surplus the borrower prefers the power to determine ct/pm since the contract agreement probability can be higher in the LS^r scenario than in the BS^r scenario. □

4.4.5 Consistency Checks

Finally, in this section the consistency of the determined bargaining power effects is examined.

The consistency of the power effects derived for particular contracting intervals (project specifications) can be evaluated by a graphical illustration of the determined propositions. Figures 4.5 and 4.6 show the consistency of the propositions derived in Sections 4.4.1, 4.4.2 and 4.4.3 for the A-type lender / A-type borrower constellation. The consistency of the other possible lender / borrower type constellations is presented in Appendix A.

Fig. 4.5: Power Effect's Consistency Check 1 for an A-Type Lender / A-Type Borrower (EM)

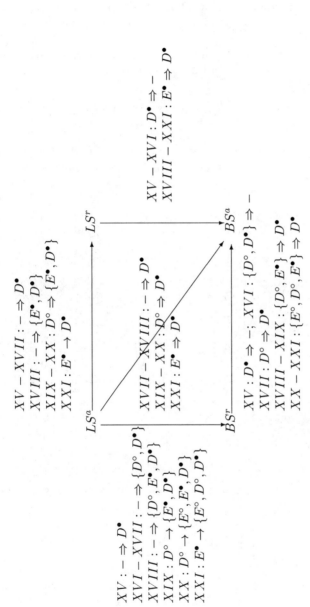

Note: This figure illustrates the consistency of the Propositions 1 ($LS^a \rightarrow BS^a$), 3 ($LS^a \rightarrow BS^r$), 6 ($LS^r \rightarrow BS^a$), 7 ($LS^r \rightarrow LS^r$) and 10 ($BS^r \rightarrow BS^a$) derived in Sections 4.4.1, 4.4.2 and 4.4.3. $\rightarrow$ refers to power effects which *can* occur given PC1, while $\Rightarrow$ indicates effects which *definitively* occur given PC1.

Fig. 4.6: Power Effect's Consistency Check 2 for an A-Type Lender / A-Type Borrower (EM)

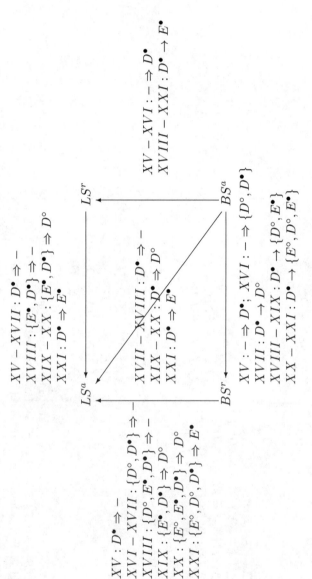

Note: This figure illustrates the consistency of the Propositions 2 ($BS^a \rightarrow LS^a$), 4 ($BS^r \rightarrow LS^a$), 5 ($BS^a \rightarrow LS^r$), 8 ($LS^r \rightarrow LS^a$) and 9 ($BS^a \rightarrow BS^r$) derived in Sections 4.4.1, 4.4.2 and 4.4.3. $\rightarrow$ refers to power effects which *can* occur given PC1, while $\Rightarrow$ indicates effects which *definitively* occur given PC1.

4.5 Robustness Checks

Since the derivation of bargaining power effects in this extended model is based on a pair-wise (binary) comparison of the optimal contract choices in the alternative power scenarios for PC1, summarized in Table 4.42, the propositions depend significantly on the underlying parameter constellation (PC1). For other parameter constellations some power effects might change or even vanish. Imagine, for example, that the monitoring costs (mc) are either 0 or ∞ which implies that either equity financing dominates debt financing for all possible project specifications or vice versa. However, since the lender's and the borrower's optimization behavior underlying our power effect derivation is independent of the parameter constellation, effects similar to those observable for PC1 are found for alternative parameter constellations (see below). Of course, the trade-offs in the lender's and the borrower's optimization behavior shift due to parameter variations, hence some observable effects (ct/pm changes) will vanish while others may arise for different project specifications. However, the economic effects underlying the derivation of the bargaining power effects do not change. For example, the lender's optimization problem being confronted with an unknown borrower type is independent of PC1. His options to cope with this uncertainty do not depend on the underlying parameter constellation. Only the significance of the trade-offs in the lender's optimization are affected by PC1 and, hence, the lender's preferred ct/pm-choice and, thereby, the related observable power effects depend on the underlying parameter constellation.

To examine the robustness of the derived bargaining power effects, we vary the underlying parameter constellation and analyze whether the changes expected due to these variation actually occur. The alternative parameter constellations we examine are stated in Table 4.44, where PC1 is stated as a benchmark.

Table 4.44: Parameter Constellations for Robustness Checks (EM)

	ζ_{max}	vc	mc	λ_B	P_B^+	λ_L	P_L^+	nc
PC1	0.25	0.4	0.15	0.5	0.3	0.5	0.3	0.05
PC2	0.25	0.4	0.3	0.5	0.3	0.5	0.3	0.05
PC3	0.75	0.4	0.15	0.5	0.3	0.5	0.3	0.05

Due to the first variation in mc (PC1: $mc = 0.15 \rightarrow$ PC2: $mc = 0.3$), we expect a decline in equity financing's favorability. The second variation in ζ_{max} (PC1: $\zeta_{max} = 0.25 \rightarrow$ PC3: $\zeta_{max} = 0.75$) probably decreases the attractiveness of debt financing for the lender but increases debt's attractiveness from the borrower's perspective since higher risk-shifting opportunities are offered. We deliberately made once (PC2) equity financing more expensive and once (PC3) debt financing more risky (instead of making both times financing more expensive) since simple cost variations may have other effects than variations in the project's return distribution. We do not consider any variation in the specification of the ex-ante informational asymmetry since, as obvious from the discussion of the feasible contracting solutions in Section 4.2.5, the implications are very difficult to predict.

The robustness checks of the power effect predictions, derived for PC1 in Sections 4.4.1, 4.4.2 and 4.4.3, are divided into two steps. Firstly, in Section 4.5.1 the implications of the discussed parameter variations (PC1 → PC2; PC1 → PC3) are analyzed for a given bargaining power scenario. Secondly, in Section 4.5.2 the actual robustness of the derived power effect predictions is examined.

4.5.1 Robustness of the Preferred Contract Type / Placement Mode-Choices

In this section we examine the implications of the discussed parameter variations for the preferred ct/pm-choices for a given bargaining power scenario. The derivation of the optimal ct/pm-choices for the alternative parameter specifications (PC2, PC3) is stated in Appendix B. Tables 4.45 to 4.48 summarize the results.

Our first prediction is that any increase in equity's monitoring costs (PC1 → PC2) reduces the attractiveness of equity financing from the lender's as well as from the borrower's perspective. This prediction is confirmed for all four bargaining power scenarios. Of course, this is not a formal prove but the results of the sample calculations match our prediction. For example, we observe the following exemplary effects due to PC1 → PC2:

- For an A-type lender in the LS^a scenario private debt contracts $(D^\bullet)$ become favorable to private equity contracts $(E^\bullet)$ for $5.026 \le \mu < 10.633$ (see Table 4.45).
- For a N-type lender in the LS^r scenario public as well as private debt contracts $(D^\circ, D^\bullet)$ gain attractiveness to public equity contracts (E°) for $1.45 \le \mu < 1.55$ (see Table 4.46).
- For an A-type borrower in the BS^r scenario private equity contracts $(E^\bullet)$ become less attractive in comparison to public as well as private debt contracts $(D^\circ, D^\bullet)$ for $1.9 \le \mu < 2.0$ (see Table 4.47).
- For a N-type borrower in the BS^a scenario the costs of public equity financing even increase so much that he prefers no contract offer and, therefore, no project financing by the lender to a public equity offer for $1.15 \le \mu < 1.274$ (see Table 4.48).

Our second prediction is that any increase in the project's return volatility (PC1 → PC3) reduces the attractiveness of debt financing from the lender's perspective since higher agency costs arise to cope with the volatility increase.

Table 4.45: Robustness of the Preferred Contract Type / Placement Mode-Choices in the Lender Scenario with Absolute Power (EM, PC1, PC2, PC3)

μ	PC1		PC2		PC3	
	A-type lender	N-type lender	A-type lender	N-type lender	A-type lender	N-type lender
[1-1.15)		– (0)		– (0)	– (0)	– (0)
[1.15-1.274)	– (0)					
[1.274-1.368)		$E^\circ(\frac{1}{2})$	– (0)	$D^\circ(\frac{1}{2})$		
[1.368-1.45)						$E^\circ(\frac{1}{2})$
[1.45-1.545)						
[1.545-1.6)						
[1.6-1.65)	$E^\circ(\frac{1}{2})$		$E^\circ(\frac{1}{2})$	$D^\circ(1)$	$E^\circ(\frac{1}{2})$	
[1.65-1.735)		$D^\circ(1)$				
[1.735-1.917)			$D^\circ(1)$			
[1.917-1.95)	$D^\circ(1)$					
[1.95-2.132)						$E^\bullet(1)$
[2.132-5.026)					$E^\bullet(1)$	
[5.026-10.633)	$E^\bullet(1)$	$E^\bullet(1)$	$D^\bullet(1)$	$D^\bullet(1)$		
[10.633-∞)			$E^\bullet(1)$	$E^\bullet(1)$		

Note: This table contains the preferred ct/pm-choices and related contract agreement probabilities in the lender scenario with absolute power given PC1, PC2 and PC3. The choices are conditioned on the lender's (the ct/pm optimizer's) type. When for certain intervals more than one ct/pm-constellation is stated, the lender is indifferent between these choices. The agreement probabilities are stated in brackets.

This agency cost increase will also reduce the favorability of debt financing from the borrower's perspective in the BS^a scenario. The only situation in which the volatility increase might improve debt financing's attractiveness for the borrower is in the BS^r scenario. Since in the BS^r scenario the borrower's ct/pm-choice is primarily driven by his attempt to maximize his informational rent, any increase in debt financing's informational rent can increase debt's attractiveness. Hence, it is important to analyze the relationship between debt financing's informational rent and the project return's volatility to obtain a prediction about the impact of PC1 $\rightarrow$ PC3 in the BS^r scenario.[125]

---

[125] The borrower's informational rent is 0.25μ for riskless and $\frac{0.16}{\mu}$ for risky debt given PC1. Given PC3 the rents are 0.75μ and $\frac{0.053}{\mu}$, respectively.

Table 4.46: Robustness of the Preferred Contract Type / Placement Mode-Choices in the Lender Scenario with Restricted Power (EM, PC1, PC2, PC3)

μ	PC1 A-type lender	PC1 N-type lender	PC2 A-type lender	PC2 N-type lender	PC3 A-type lender	PC3 N-type lender
[1-1.15)		$- (0)$				$- (0)$
[1.15-1.25)		$E^\circ(\frac{1}{2})$		$- (0)$		$E^\circ(\frac{1}{2})$
[1.25-1.274)		$E^\circ, E^\bullet$ $(\frac{1}{2})$				
[1.274-1.3)		E°, D°		$D^\circ(\frac{1}{2})$		$E^\circ, E^\bullet$
[1.3-1.318)	$- (0)$	$E^\bullet(\frac{1}{2})$		E°, D° $(\frac{1}{2})$	$- (0)$	$(\frac{1}{2})$
[1.318-1.336)		$D^\circ(1)$	$- (0)$	$D^\circ(1)$		
[1.336-1.4)						$E^\circ, E^\bullet$
[1.4-1.413)		$D^\circ, D^\bullet$				$D^\circ(\frac{1}{2})$
[1.413-1.45)		(1)				E°, D° $E^\bullet, D^\bullet$ $(\frac{1}{2})$
[1.45-1.493)		E°, D°		$D^\circ, D^\bullet$		$E^\circ(1)$
[1.493-1.55)		$D^\bullet(1)$		(1)		E°, D° (1)
[1.55-1.574)						E°, D° $E^\bullet(1)$
[1.574-1.6)	$E^\bullet(\frac{1}{2})$					
[1.6-1.653)		E°, D°		E°, D°	$E^\bullet(\frac{1}{2})$	E°, D°
[1.653-1.7)		$E^\bullet, D^\bullet$	$D^\bullet(\frac{1}{2})$	$D^\bullet(1)$		$E^\bullet, D^\bullet$
[1.7-1.721)	$E^\bullet, D^\bullet$	(1)	$E^\bullet, D^\bullet$	E°, D°		(1)
[1.721-1.741)	$(\frac{1}{2})$		$(\frac{1}{2})$	$E^\bullet, D^\bullet$ (1)	$E^\bullet, D^\bullet$	
[1.741-1.85)	$D^\bullet(1)$	E°, D° $D^\bullet(1)$	$D^\bullet(1)$	E°, D° $D^\bullet(1)$	$(\frac{1}{2})$	

table continued on next page

table continued from previous page					
[1.85-1.902)		E°, D°	$D^\bullet(1)$	E°, D° $D^\bullet(1)$	$E^\bullet(1)$	E°, D° $E^\bullet(1)$
[1.902-2.0)	$E^\bullet, D^\bullet$	$E^\bullet, D^\bullet$			$E^\bullet, D^\bullet$	E°, D°
[2.0-∞)	(1)	(1)	$E^\bullet, D^\bullet$ (1)	E°, D° $E^\bullet, D^\bullet$ (1)	(1)	$E^\bullet, D^\bullet$ (1)

Note: This table contains the preferred ct/pm-choices and related contract agreement probabilities in the lender scenario with restricted power given PC1, PC2 and PC3. The choices are conditioned on the lender's (the ct/pm optimizer's) type. When for certain intervals more than one ct/pm-constellation is stated, the lender is indifferent between these choices. The agreement probabilities are stated in brackets.

Since in our model the borrower's informational rent in the BS^r scenario only increases due to a volatility increase when the lender prefers riskless to risky debt given PC3, the probability to observe an increase in debt's favorability due to PC1 $\to$ PC3 exists. However, this probability is small since the lender's preference for riskless over risky debt decreases with any increase in the project's return volatility. The probability is even so small that we do not observe this specific effect for PC1 $\to$ PC3 in our sample calculations.

However, the overall predictions are consistent with our sample calculations. For example, we observe the following exemplary effects due to PC1 $\to$ PC3:

- For an A-type lender in the LS^a scenario private equity financing ($E^\bullet$) becomes favorable to public debt (D°) for $1.95 \leq \mu < 2.132$ (see Table 4.45).
- For a N-type lender in the LS^r scenario public as well as private equity contracts ($E^\circ, E^\bullet$) become preferable to public debt (D°) for $1.274 \leq \mu < 1.336$ (see Table 4.46).
- For an A-type borrower in the BS^r scenario the costs of public debt financing (D°) increase even so much that for $1.368 \leq \mu < 1.4$ the borrower favors no contract agreement instead of a public debt offer (see Table 4.47).
- For a N-type borrower in the BS^a scenario private debt contracts ($D^\bullet$) become less attractive than private equity ($E^\bullet$) for $1.659 \leq \mu < 1.933$ (see Table 4.48).

Table 4.47: Robustness of the Preferred Contract Type / Placement Mode-Choices in the Borrower Scenario with Restricted Power (EM, PC1, PC2, PC3)

μ	PC1		PC2		PC3	
	A-type borrower	N-type borrower	A-type borrower	N-type borrower	A-type borrower	N-type borrower
[1-1.15)		$-$ (0)		$-$ (0)		$-$ (0)
[1.15-1.274)	$-$(0)	$E^\circ(\frac{1}{2})$	$-$ (0)			$E^\circ(\frac{1}{2})$
[1.274-1.361)					$-$ (0)	
[1.361-1.368)						
[1.368-1.4)	$D^\circ(\frac{1}{2})$	$D^\circ(\frac{1}{2})$	$D^\circ(\frac{1}{2})$	$D^\circ(\frac{1}{2})$		
[1.4-1.55)	$D^\bullet(\frac{1}{2})$					$D^\circ(\frac{1}{2})$
[1.55-1.6)			$D^\bullet(\frac{1}{2})$			
[1.6-1.667)	$E^\bullet, D^\bullet$ $(\frac{1}{2})$					
[1.667-1.7)					$E^\bullet(\frac{1}{2})$	
[1.7-1.741)			$E^\bullet, D^\bullet$ $(\frac{1}{2})$			$D^\circ(1)$
[1.741-1.75)	$D^\bullet(1)$	$D^\circ(1)$				
[1.75-1.85)	$D^\circ, D^\bullet$ (1)		$D^\bullet(1)$	$D^\circ(1)$	$E^\circ, E^\bullet$ $(\frac{1}{2})$	$E^\circ(1)$
[1.85-1.9)					E°, D°	E°, D°
[1.9-1.966)	$D^\circ, E^\bullet$		$D^\circ, D^\bullet$		$E^\bullet, D^\bullet$ (1)	(1)
[1.966-2.0)	$D^\bullet(1)$		(1)		$D^\circ, E^\bullet$	$D^\circ(1)$
[2.0-2.05)			$D^\circ, E^\bullet$		$D^\bullet(1)$	
[2.05-2.2)	E°, D°	E°, D°	$D^\bullet(1)$		E°, D°	E°, D°
[2.2-∞)	$E^\bullet, D^\bullet$ (1)	(1)	E°, D° $E^\bullet, D^\bullet$ (1)	E°, D° (1)	$E^\bullet, D^\bullet$ (1)	(1)

Note: This table contains the preferred *ct/pm*-choices and related contract agreement probabilities in the borrower scenario with restricted power given PC1, PC2 and PC3. The choices are conditioned on the borrower's (the *ct/pm* optimizer's) type. When for certain intervals more than one *ct/pm*-constellation is stated, the borrower is indifferent between these choices. The agreement probabilities are stated in brackets.

Table 4.48: Robustness of the Preferred Contract Type / Placement Mode-Choices in the Borrower Scenario with Absolute Power (EM, PC1, PC2, PC3)

μ	PC1		PC2		PC3	
	A-type borrower	N-type borrower	A-type borrower	N-type borrower	A-type borrower	N-type borrower
[1-1.15)		– (0)		– (0)		– (0)
[1.15-1.274)	– (0)	$E^\circ(\frac{1}{2})$	– (0)		– (0)	
[1.274-1.318)						$E^\circ(\frac{1}{2})$
[1.318-1.45)		$D^\circ(\frac{1}{2})$		$D^\circ(\frac{1}{2})$		
[1.45-1.627)	$D^\circ(\frac{1}{2})$		$D^\circ(\frac{1}{2})$		$E^\circ(\frac{1}{2})$	
[1.627-1.659)						$D^\circ(\frac{1}{2})$
[1.659-1.8)					$D^\circ(\frac{1}{2})$	$E^\bullet(1)$
[1.8-1.933)		$D^\bullet(1)$		$D^\bullet(1)$		
[1.933-1.988)	$D^\bullet(1)$		$D^\bullet(1)$			$D^\bullet(1)$
[1.988-∞)					$D^\bullet(1)$	

Note: This table contains the preferred ct/pm-choices and related contract agreement probabilities in the borrower scenario with absolute power given PC1, PC2 and PC3. The choices are conditioned on the borrower's (the ct/pm optimizer's) type. When for certain intervals more than one ct/pm-constellation is stated, the borrower is indifferent between these choices. The agreement probabilities are stated in brackets.

4.5.2 Robustness of the Bargaining Power Effect Predictions

After analyzing the implications of the discussed parameter variations for the preferred ct/pm-choices for a given bargaining power scenario, the robustness of the bargaining power effects (derived in Sections 4.4.1, 4.4.2 and 4.4.3) is examined in this section. This examination is based on a pair-wise comparison of the lender's as well as the borrower's preferred ct/pm-choices in the alternative power scenarios for a given parameter constellation. Table 4.42 above summarizes the preferred ct/pm-choices given PC1; Table 4.49 for PC2 and Table 4.50 for PC3.

Table 4.49: Summary of Contract Type / Placement Mode-Choices and Contract Agreement Probabilities (EM, PC2)

Interval	μ	LS^a		LS^r		BS^r		BS^a	
		A-type lender	N-type lender	A-type lender	N-type lender	A-type borrower	N-type borrower	A-type borrower	N-type borrower
I	$[1\text{-}1.274)$		$-\,(0)$		$-\,(0)$		$-\,(0)$		$-\,(0)$
II	$[1.274\text{-}1.3)$				$D^\circ(\tfrac{1}{2})$			$-\,(0)$	
III	$[1.3\text{-}1.318)$		$D^\circ(\tfrac{1}{2})$		$E^\circ, D^\circ (\tfrac{1}{2})$	$-\,(0)$			
IV	$[1.318\text{-}1.\bar{3})$	$-\,(0)$		$-\,(0)$			$D^\circ(\tfrac{1}{2})$		
V	$[1.\bar{3}\text{-}1.368)$				$D^\circ(1)$				$D^\circ(\tfrac{1}{2})$
VI	$[1.368\text{-}1.4)$						$D^\circ(\tfrac{1}{2})$		
VII	$[1.4\text{-}1.6)$				$D^\circ, D^\bullet (1)$				
VIII	$[1.6\text{-}1.653)$					$D^\bullet(\tfrac{1}{2})$		$D^\circ(\tfrac{1}{2})$	
IX	$[1.653\text{-}1.659)$	$E^\circ(\tfrac{1}{2})$		$D^\bullet(\tfrac{1}{2})$	E°, D°				
X	$[1.659\text{-}1.7)$				(1)				
XI	$[1.7\text{-}1.735)$		$D^\circ(1)$	$E^\bullet, D^\bullet$	E°, D°	$E^\bullet, D^\bullet$			
XII	$[1.735\text{-}1.741)$			$(\tfrac{1}{2})$	$E^\bullet, D^\bullet (1)$	$(\tfrac{1}{2})$			
XIII	$[1.741\text{-}1.8)$					$D^\bullet(1)$	$D^\circ(1)$		
XIV	$[1.8\text{-}1.9)$	$D^\circ(1)$		$D^\bullet(1)$	E°, D°				$D^\bullet(1)$
XV	$[1.9\text{-}2.0)$				$D^\bullet(1)$	$D^\circ, D^\bullet (1)$			
XVI	$[2.0\text{-}2.2)$					$D^\circ, E^\bullet\; D^\bullet(1)$		$D^\bullet(1)$	
XVII	$[2.2\text{-}5.026)$			$E^\bullet, D^\bullet$	E°, D°	E°, D°			
XVIII	$[5.026\text{-}10.633)$	$D^\bullet(1)$	$D^\bullet(1)$	(1)	$E^\bullet, D^\bullet$	$E^\bullet, D^\bullet$	E°, D°		
XIX	$[10.633\text{-}\infty)$	$E^\bullet(1)$	$E^\bullet(1)$		(1)	(1)	(1)		

Note: This table contains the preferred ct/pm-choices and contract agreement probabilities conditioned on the ct/pm optimizer's (the first mover's) type in the four bargaining power scenarios. When for certain contracting intervals more than one ct/pm-constellation is stated, the ct/pm optimizer (first mover) is indifferent between these choices. The agreement probabilities are stated in brackets.

Table 4.50: Summary of Contract Type / Placement Mode-Choices and Contract Agreement Probabilities (EM, PC3)

Interval	μ	LS^a		LS^r		BS^r		BS^a	
		A-type lender	N-type lender	A-type lender	N-type lender	A-type borrower	N-type borrower	A-type borrower	N-type borrower
I	[1-1.15)		– (0)		– (0)		– (0)		– (0)
II	[1.15-1.25)				$E^\circ(\tfrac{1}{2})$				
III	[1.25-1.361)	– (0)			$E^\circ, E^\bullet\,(\tfrac{1}{2})$		$E^\circ(\tfrac{1}{2})$	– (0)	
IV	[1.361-1.413)			– (0)	$E^\circ, D^\circ\;E^\bullet(\tfrac{1}{2})$	– (0)			
V	[1.413-1.45)				$E^\circ, D^\circ\;E^\bullet, D^\bullet\,(\tfrac{1}{2})$				$E^\circ(\tfrac{1}{2})$
VI	[1.45-1.493)		$E^\circ(\tfrac{1}{2})$		$E^\circ(1)$				
VII	[1.493-1.55)				$E^\circ, D^\circ(1)$			$E^\circ(\tfrac{1}{2})$	
VIII	[1.55-1.574)				$E^\circ, D^\circ\;E^\bullet(1)$		$D^\circ(\tfrac{1}{2})$		
IX	[1.574-1.627)								
X	[1.627-1.637)	$E^\circ(\tfrac{1}{2})$		$E^\bullet(\tfrac{1}{2})$					
XI	[1.637-1.65)					$E^\bullet(\tfrac{1}{2})$			$D^\circ(\tfrac{1}{2})$
XII	[1.65-1.659)								
XIII	[1.659-1.667)				E°, D°				
XIV	[1.667-1.721)				$E^\bullet, D^\bullet$		$D^\circ(1)$	$D^\circ(\tfrac{1}{2})$	
XV	[1.721-1.75)	$E^\bullet(1)$		$E^\bullet, D^\bullet$	(1)				$E^\bullet(1)$
XVI	[1.75-1.85)				$(\tfrac{1}{2})$	$E^\circ, E^\bullet\,(\tfrac{1}{2})$	$E^\circ(1)$		

table continued on next page

table continued from previous page

		A	B	C	D	E	F	G	H
XVII	[1.85-1.902)			$E^\bullet(1)$	E°, D° / $E^\bullet(1)$				
XVIII	[1.902-1.933)	$E^\circ(\frac{1}{2})$				E°, D°	E°, D°	$D^\circ(\frac{1}{2})$	$E^\bullet(1)$
XIX	[1.933-1.95)		$E^\bullet(1)$	$E^\bullet, D^\bullet$	E°, D°	$E^\bullet, D^\bullet$	(1)		
XX	[1.95-1.966)			(1)	$E^\bullet, D^\bullet$	(1)			$D^\bullet(1)$
XXI	[1.966-1.988)				(1)	$D^\circ, E^\bullet$	$D^\circ(1)$		
XXII	[1.988-2.05)	$E^\bullet(1)$				$D^\bullet(1)$			
XXIII	[2.05-4.2)					E°, D°		$D^\bullet(1)$	
XXIV	[4.2-5.4)					$E^\bullet, D^\bullet$	E°, D°		
XXV	[5.4-∞)					(1)	(1)		

Note: This table contains the preferred ct/pm-choices and contract agreement probabilities conditioned on the ct/pm optimizer's (the first mover's) type in the four bargaining power scenarios. When for certain contracting intervals more than one ct/pm-constellation is stated, the ct/pm optimizer (first mover) is indifferent between these choices. The agreement probabilities are stated in brackets.

Based on these results the robustness of the bargaining power effects derived for PC1 can be examined by analyzing whether the same or similar effects are observable for all three parameter specifications. Of course, for each variation in the underlying parameter specification (PC1 → PC2, PC1 → PC3), changes in the observable bargaining power effects (ct/pm changes) will occur. However, these variations only affect the observable power effects since they influence the underlying trade-offs, but they do not affect the lender's and the borrower's optimization behavior responsible for the bargaining power's impact in financial contracting.

The question we are, therefore, analyzing is whether the general tendencies predicted by the bargaining power effect propositions are robust to the parameter variations PC1 → PC2 and PC1 → PC3. Considering Proposition 1 (LS^a → BS^a) at first, we observe that some effects observable for PC1 disappear for PC2 and / or PC3 while other effects appear. Table 4.51 summarizes the effects of an absolute bargaining power variation from the lender to the borrower (LS^a → BS^a) for all three alternative parameter specifications. Just to remind us, Proposition 1 predicts that due to this power shift debt contracts as well as private placements become more attractive while the contract agreement probability can de- or increase.

Table 4.51: Robustness of the Power Effect Predictions of Proposition 1 ($LS^a \rightarrow BS^a$, EM)

Effect	PC1	PC2	PC3
• A-type lender → A-type borrower			
$- \Rightarrow D^\bullet$	$XVII - XVIII$	$-$	$-$
$D^\circ \Rightarrow D^\bullet$	$XIX - XX$	$XIV - XVII$	$-$
$E^\bullet \Rightarrow D^\bullet$	XXI	XIX	$XXII - XXV$
$D^\circ \Rightarrow -$	$-$	$XII - XIII$	$-$
$E^\bullet \Rightarrow -$	$-$	$-$	$XX - XXI$
• A-type lender → N-type borrower			
$E^\circ \Rightarrow -$	$IX - XIII$	$VIII - IX$	$VI - XII$
$E^\circ \Rightarrow D^\bullet$	$XIV - XVIII$	$X - XI$	XIX
$D^\circ \Rightarrow D^\bullet$	$XIX - XX$	$XII - XVII$	$-$
$E^\bullet \Rightarrow D^\bullet$	XXI	XIX	$XX - XXV$
$E^\circ \Rightarrow E^\bullet$	$-$	$-$	$XIII - XVIII$
• N-type lender → A-type borrower			
$- \Rightarrow D^\circ$	$V - IX$	$IV - V$	$X - XII$
$D^\circ \Rightarrow D^\bullet$	$XVII - XX$	$XIV - XVII$	$-$
$E^\bullet \Rightarrow D^\bullet$	XXI	XIX	$XIX - XXV$
$- \Rightarrow E^\circ$	$-$	$-$	$VI - IX$
• N-type lender → N-type borrower			
$E^\circ \Rightarrow D^\circ$	$IV - IX$	$-$	$X - XII$
$D^\circ \Rightarrow D^\bullet$	$XIV - XX$	$IX - XVII$	$-$
$E^\bullet \Rightarrow D^\bullet$	XXI	XIX	$XIX - XXV$

Note: This table contains all bargaining power effects which are observable for an absolute power shift from the lender to the borrower ($LS^a \rightarrow BS^a$) given PC1, PC2 or PC3.

As predicted by Proposition 1, for all three parameter specifications we observe that debt financing as well as private placements become more attractive while the contract agreement probability can de- or increase. Additionally, in accordance with our expectations since PC1 → PC2 decreases the favorability of equity financing, bargaining power effects related to equity financing in PC1 may disappear for PC2. This expectation is confirmed. For example, for a power shift from the N-type lender to the N-type borrower $E^\circ \Rightarrow D^\circ$ is observable for PC1 ($IV - IX$) but not for PC2. Furthermore, for PC1 → PC3 we expect debt financing's favorability to decrease and, therefore, power effects related to debt financing to disappear. This expectation is also validated. For example, for a power shift from the A-type lender to the A-type borrower $D^\circ \Rightarrow D^\bullet$ is observable for PC1 $(XIX - XX)$ but not for PC3. Finally, even new effects become observable, see, for example, an absolute power redistribution from the A-type lender to the A-type borrower: $D^\circ \Rightarrow -$ for PC2 $(XII - XIII)$ and $E^\bullet \Rightarrow -$ for PC3 $(XX - XXI)$. In the same way the robustness of the general power effect tendencies as predicted by Propositions 2 to 10 can be examined (see Appendix C).

Overall we find that the power effect tendencies as predicted by Propositions 1 to 10 are quite robust to our variations in the underlying parameters. Of course, the parameter variation PC1 $\to$ PC2 has a special impact on all power effects where equity financing is involved since due to PC1 $\to$ PC2 equity financing becomes more expensive. Analogously, the parameter variation PC1 $\to$ PC3 primarily affects bargaining power effects related to debt financing since due to this variation in the project return's volatility the agency costs associated with debt financing increase while the agency costs of debt financing are unaffected.

5

Discussion of the Adopted Approach

The aim of this dissertation was to examine bargaining power effects in financial contracting (see Chapters 3 and 4). In particular, we analyzed how the (ex-ante) distribution of bargaining power between a lender and a borrower affects the chosen contract type and placement mode. Therefore, we developed a principal-agent framework in which a lender and a borrower negotiate how to finance the borrower's risky project. The lender and the borrower have to decide about the contract type (debt vs. equity), the placement mode (public offering vs. private placement) as well as about the contracts' conditions. Both, the lender and the borrower, are risk neutral and are maximizing their expected profit. To keep a clear distinction between the available contract types and placement modes, these are exogenously given (see Assumptions 3 and 4). The bargaining power to be distributed among the lender and the borrower consists of two components: The power to determine the contract type / placement mode, and the right to set contracts' conditions. This power separation enables us to assign particular bargaining power effects to certain power components.

The chosen approach is quite attractive since the derivation of the intuitive results is straight forward. Also additional aspects, such as potentially varying agency costs of equity financing à la Jensen and Meckling (1976) or the consideration of already existing claimants on the firm's, i. e. the borrower's, assets to raise capital structure issues, can easily be incorporated. However, three possible concerns might arise. Section 5.1 states the main concerns while Section 5.2 presents our methodological justifications.

5.1 Possible Concerns

Basically, three main concerns about the adopted approach to examine bargaining power effects in financial contracting are conceivable:

- Is it reasonable to exclude binding commitments between the lender and the borrower since this exclusion might cause *Pareto-inefficient* bargaining outcomes in the restricted power scenarios? Section 5.2.1 handles this concern.

- Does the assumed bargaining setting reflect real world characteristics? Section 5.2.2 copes with this question.
- Are the determined debt contracts *efficient* in the described framework? Otherwise we would have examined how bargaining power affects potentially suboptimal contract type / placement mode-choices. See Section 5.2.3 for a discussion of this issue.

All three concerns question the practical relevance of our results. But in our opinion they seem not to matter for the derived bargaining power effects, as discussed in the following section. Of course, further concerns are conceivable but in our opinion the three stated concerns are the most obvious ones and, therefore, we discuss them explicitly.

5.2 Methodological Justifications

5.2.1 Exclusion of Binding Commitments

One phenomenon which can occur in our analysis since we treat both bargaining power components separately, is that *Pareto-inefficient* bargaining outcomes might arise. This inefficiency can only occur in the *restricted power scenarios* where each agent has one power component, while in the *absolute power scenarios* both components are assigned jointly. When each agent has one power component Pareto-inefficient bargaining outcomes might occur since the agents' profits depend on the other agents' behavior (use of power) and vice versa.[1] But when one agent has the absolute power, the power of the other agent is reduced to a marginal part, i. e. the latter can only accept or reject the offer, and hence the agent's profit is not directly affected by the partner's behavior. Therefore, no Pareto-inefficient bargaining outcomes can occur in the latter situation. The potential inefficiency is due to our *non-cooperative* bargaining framework since we rule out any kind of binding commitments between the lender and the borrower to correlate their actions.[2]

This potential Pareto-inefficiency can only be resolved by either allowing binding commitments or treating both power components jointly. However, both restrictions are unrealistic for our circumstances. For example, how should a firm arrange a binding commitment in an "anonymous" capital market transaction (public placement)? Or how can a joint treatment of both

---

[1] An agent acts self-interested if he is only concerned about his expected profit without taking any (positive or negative) consequences of his behavior for the other agent into account.

[2] That Pareto-inefficient bargaining outcomes can occur in such a setting is common since the development of the famous Prisoner's Dilemma (cf. Varian (1992) and Mas-Colell et al. (1995)).

power components incorporate that often a firm still has the contract type / placement mode-choice while having to accept any potential offer? Therefore, we stick to our bargaining setting despite the possibility of Pareto-inefficient bargaining outcomes in the restricted power scenarios. Moreover, one of our results is that a separation of the bargaining power components can result in Pareto-inefficient bargaining outcomes.

Since in our opinion it is quite unrealistic to consider binding commitments in financial contracting (as just described), we refrain from any speculation about possible implications of the allowance of binding commitments for the preferred ct/pm-choices in the restricted power scenarios and the related power effects.

5.2.2 Assumed Bargaining Setting

We assumed an ultimatum bargaining game because the aim of this dissertation was to demonstrate how the ex-ante distribution of bargaining power affects a firm's financing decision. Of course, more elaborated games are imaginable:

- Firstly, for example, an alternating offer bargaining game (AOB) à la Rubinstein (1982) or one of its numerous extensions might be more appropriate to reflect the actual debt or equity raising process in particular situations.[3, 4] Such elaborated games even might allow for information transmission between bargaining parties via the bargaining process (cf. Kennan and Wilson (1993)).

- Secondly, it might be appropriate to incorporate an exogenous power disturbance in the bargaining process, such as a capital market trend, which

---

[3] In an alternating offer bargaining game à la Rubinstein (1982) in principal two players bargain how to split one dollar. At first player one has the right to propose a sharing rule. Player two can either accept or reject this proposal. When player two accepts the proposal, the game ends and the dollar is split according to the proposed rule. But if player two rejects player one's proposal, he has the right to make a counter proposal, i. e. to suggest a different sharing rule. Consequently, then player one has to decide whether to accept or to reject player two's suggestion. If player one accepts, the game ends; but if he rejects, he has the right to make a new recommendation, and so forth. The players can bargain infinite rounds but each round where no agreement is reached the present value of the dollar decreases (discounting effect).

This discounting effect encourages the players to reach a quick agreement. A potential first round agreement is that player two just receives as much as he expects to obtain when he rejects the first round offer and player one gets the rest.

[4] Recently, for example, Inderst (2002) incorporated an AOB setup in a principal-agent framework, where the principal finds a new agent when the current refuses his offer with probability p. For $p = 1$ the principal has full bargaining power while for $p = 0$ he has none. Power (p) can vary continuously.

can change the power distribution after the contract type / placement mode-choice has been made. Hence, the lender, respectively the borrower, choosing contract type / placement mode is actually not aware about the distribution of the power to set contracts' conditions.

- Finally, in a multi-period context in the spirit of Petersen and Rajan (1995), it could also be examined how firms behave strategically over time to obtain a better bargaining position in the future. For example, the attractiveness of debt financing from the borrower's perspective (due to debt financing's risk shifting opportunity) might decrease when the borrower loses his overall (long-run) power in the default case. Alternatively, the borrower might want to obtain a good (long-run) reputation and, therefore, satisfies his (short-run) obligations.

However, we stick to our setting since we see no reason why one of these suggestions to elaborate our ultimatum bargaining setup overall better reflects the real situation at a e. g. monopolistic or fully competitive capital market (LS^a or BS^a scenario).

Despite the fact that we think our bargaining setup is a good overall approximation of real capital market bargaining situations (see Section 3.1 for our justification), it is worthwhile to speculate how the discussed options to elaborate our bargaining setup might affect the derived results. We speculate about possible implications of alternative bargaining setups since even if they are not the best overall approximation of real capital market bargaining situations, they might be more realistic than our ultimatum bargaining setup in some particular circumstances / situations. Therefore, in the following we replace our ultimatum bargaining setup, where we treat the power to choose ct/pm separately from the power to define contracts' conditions, by two of the above stated alternative setups and speculate about the implications for the preferred ct/pm-choices in each bargaining power scenario and, consequently, for the bargaining power effects.[5]

Assuming at first that our ultimatum bargaining setup is replaced by an alternating offer bargaining setup à la Rubinstein, the four bargaining power scenarios can be defined as follows:

- In the "new" LS^a scenario the lender makes an offer to the borrower how to finance the borrower's risky project. This offer not only defines ct/pm but also the corresponding contract conditions. The borrower can either accept or reject the lender's offer. When the borrower accepts the offer, the contract is settled; but when the borrower rejects the offer, the lender can propose a new offer, and so forth. No time limit is given but each

---

[5] We do not analyze the implications of the third stated alternative bargaining setup since this would require to extend our current one-period principal-agent framework into a multi-period framework to incorporate potentially conflicting short- and long-run targets.

time the borrower declines the lender's offer, the expected project return decreases.[6, 7]

- In the "new" LS^r scenario at first the lender offers the borrower to finance the latter's risky project. The borrower can either accept or reject the offer. When the borrower accepts the offer, the contract is settled; but when the borrower rejects the offer, he has the right to make a counter offer. This counter offer can be accepted or can be rejected by the lender. When the lender accepts the counter offer, the contract is settled; but when the lender rejects the offer, he has the right to make a "counter" counter offer, and so forth. Each time a round of contract negotiations fails the expected project return decreases.

- Contract negotiations in the "new" BS^r scenario are equal to the negotiations in the "new" LS^r scenario despite the fact that in the "new" BS^r scenario the borrower and not the lender makes the first offer.

- In the "new" BS^a scenario the borrower makes an offer to the lender how the lender can finance the borrower's risky project. The lender can either accept or reject this offer. When the lender accepts the offer, the contract is settled; but when the lender rejects the offer, the borrower can propose a new contract, and so on. Each time a round of contract negotiations fails the expected project return decreases.

Given these power scenario definitions, we tend to expect the following implications of the discussed variations in the assumed bargaining setup for the four bargaining power scenarios:

- The lender's preferred ct/pm-choice in the "new" LS^a scenario is probably the same as in the "old" LS^a scenario since still mainly agency cost considerations matter for the lender's choice. However, the borrower's expected profit in the "new" LS^a scenario might rise above his potential informational rent in debt financing since the borrower can now threat to disturb the lender's profit maximization by rejecting the lender's first contract offer. Given the "old" ultimatum bargaining setup, the borrower does not possess such a threat option since the borrower can only once decide whether to accept or to reject the lender's offer, and when he rejects the offer, the game ends. In the "new" alternating offer bargaining setup the borrower might receive a second or even a third chance to accept an offer when rejecting the previous ones. Therefore, we expect that the lender's preferred ct/pm-choices will not vary due to this "new" bargaining setup,

---

[6] Typically it is argued that negotiations take time and that the project's expected return decreases with time, i. e. with each round of negotiations (cf. e. g. Cramton (1991)).

[7] The decrease in the expected project return when a round of contract negotiations fails is often the only reason forcing the lender and the borrower to reach a contract agreement.

but that the profit distribution between the lender and the borrower will change.

- The implications of this bargaining setup variation for the LS^r scenario are much more complicated to predict. This is since, firstly, in the lender scenario with restricted power in addition to agency cost considerations strategic aspects affect the lender's ct/pm-choice and, secondly, the alternating offer bargaining setup might allow the lender and the borrower to exchange private information via the bargaining process and not only in a private placement. Since these effects are quite complicated to predict we prefer to refrain from making explicit predictions about possible implications of the discussed bargaining setup variation for the LS^r scenario.
- Due to the same reasoning we do not make predictions about potential implications of the discussed bargaining setup variation in the BS^r scenario.
- However, we still can predict the implications for the BS^a scenario which are driven by similar effects as the implications in the LS^a scenario. The borrower's preferred ct/pm-choice in the "new" BS^a scenario is probably the same as in the "old" BS^a scenario while the lender will benefit from this bargaining setup variation since the lender obtains the opportunity to threat to disturb the borrower's profit maximization which the latter has to take into account when offering the lender a particular ct/pm with related conditions.

Therefore, we expect that the absolute power effects are mainly unaffected by the discussed bargaining setup variation while the other power effects will change.

After speculating about the potential implications when our ultimatum bargaining setup is replaced by an alternating offer bargaining setup à la Rubinstein, we speculate about the implications when a potential power disturbance, like a capital market trend, is incorporated in our ultimatum bargaining setup. Since in our bargaining setup we treat the power to choose ct/pm separately from the power to define contracts' conditions, our four bargaining scenarios can be summarized in two "new" bargaining power scenarios given that the power disturbance assigns the bargaining power to determine contracts' conditions after ct/pm are fixed:

- In the "new" lender scenario we assume that the lender defines ct/pm. Given the lender's ct/pm-choice, the power disturbance assigns the power to define contracts' conditions either to the lender or to the borrower. Of course, the counterparty still has to accept the defined contracts' conditions, otherwise the game ends.
- In the "new" borrower scenario we assume that the borrower defines ct/pm. Given the borrower's ct/pm-choice, the power disturbance assigns the power to define contracts' conditions either to the lender or to the borrower. Of course, the counterparty still has to accept the defined contracts' conditions, otherwise the game ends.

Basically, the newly defined lender scenario is just a combination of our previous LS^a and LS^r scenarios, where once the lender (LS^a) and once the borrower (LS^r) had the right to set contracts' conditions given the lender's ct/pm-choice. Consequently, the lender's "new" preferred ct/pm-choice will be driven by the same underlying effects as in the "old" LS^a and LS^r scenarios. Depending on the lender's expectation about the probability distribution of the power disturbance (how likely is it that either the borrower receives the power to set contracts' conditions or that the lender keeps the power) either the effects previously observable for the "old" LS^a or for the "old" LS^r scenario will probably dominate the lender's "new" ct/pm-choice. The analog is true for the newly defined borrower scenarios, which results as a combination of the "old" BS^a and the "old" BS^r scenarios. Again, in the "new" borrower scenario the borrower's expectation about the probability distribution of the power disturbance is important to define which ct/pm-choice the borrower initially prefers.

Therefore, in principal all power effects observable for our "old" bargaining setup can probably be observable for the "new" bargaining setup as well depending on the lender's, respectively the borrower's, expectation about the probability distribution of the power disturbance. Given, for example, that the lender assumes that he definitively keeps the power to set the contracts' conditions in the "new" lender scenario and that the borrower believes that he also keeps the right to define contracts' conditions in the "new" borrower scenario, a power shift from the lender to the borrower ("new" lender scenario $\rightarrow$ "new" borrower scenario) will cause the same power effects as we observed for $LS^a \rightarrow BS^a$ given our "old" bargaining setup (see Proposition 1).

5.2.3 Debt Contracts in the Model Setup

To keep a clear distinction between both available contract types a *deterministic* verification mechanism à la Townsend (1979), Gale and Hellwig (1985), and Williamson (1987, 1986) is assumed for debt contracts, i. e. always verify the project return in the proclaimed default state $(\widehat{y} < h)$.

However, for debt contracts *stochastic* verification mechanisms are *Pareto-superior* since they reduce the associated costs and still prevent the borrower from cheating concerning y. These contracts exhibit substantially different and more complex features than contracts with deterministic verification mechanisms (cf. Border and Sobel (1987), Townsend (1988), Mookherjee and Png (1989), and Bernanke and Gertler (1989)). Stochastic verification mechanisms also cope with the issue of lacking subgame perfectness of the deterministic mechanisms, i. e. incurring verification costs vc even when the potential gains are lower (verifying y if $h - vc \leq \widehat{y} < h$).

But we apply the deterministic setting since it seems more realistic as already outlined in Section 3.1. Additionally, for example, Boyd and Smith (1994) point out that standard debt contracts with a deterministic verification scheme are the norm in debt financing and they are actually quite close to optimal when implementation costs are nontrivial.

Moreover, in our opinion the implications to consider debt contracts with a stochastic instead of a deterministic verification scheme will probably resemble the implications of a simple verification cost reduction in the developed principal-agent framework. Based on the results obtained in Chapters 3 and 4 and, in particular, in the robustness check in Section 4.5, we, therefore, expect the following implications due to the consideration of a stochastic instead of a deterministic verification schema:

- Public as well as private debt contracts become more attractive from the lender's as well as from the borrower's perspective for all four bargaining power scenarios.
- This effect is probably more clearly observable for the absolute power scenarios than for the restricted power scenarios since in the absolute power scenarios (only) agency cost considerations matter while in the restricted power scenarios additional strategic aspects have to be taken into account.
- Some bargaining power effects related to equity financing might no longer be observable since when debt financing becomes preferable to equity financing, the power effects related to equity financing vanish. Furthermore, "new" power effects related to debt financing can become observable when debt financing is now considered for project specifications where previously equity financing was preferable.
- Power effects related to the placement mode choice should be unaffected by the discussed variation in debt financing's verification mechanism.

6

Conclusion

The aim of this dissertation was to examine how the distribution of bargaining power between a lender and a borrower affects the contract type and the placement mode choice. Therefore, we developed a principal-agent framework where a lender (principal) and a borrower (agent) negotiate how to finance the borrower's risky project while informational asymmetries cause moral hazard and adverse selection problems. The borrower possesses an interim and ex-post informational advantage while the lender and the borrower are unaware whether their contractual partner has a positive outside option (A-type) or not (N-type) (ex-ante uncertainty). The contract negotiation game is analyzed for four alternative bargaining power scenarios where the power to determine contract type and placement mode as well as the right to set contracts' conditions is, each, either assigned to the lender or to the borrower. The lender and the borrower are risk neutral and they are maximizing their expected profits. Power effects are determined by a pair-wise (binary) comparison of the preferred contract type / placement mode (ct/pm)-choices in the alternative bargaining power scenarios. Section 6.1 summarizes the derived bargaining power effects, whereas Section 6.2 discusses their economic relevance.

6.1 Summary of Bargaining Power Effects

The bargaining power effects derived by the pair-wise (binary) comparison of the preferred ct/pm-choices in the alternative bargaining power scenarios (LS^a, LS^r, BS^r, BS^a), see Sections 3.5 to 4.4, are summarized below. The alternative bargaining power scenarios differ with respect to the distribution of bargaining power between the lender and the borrower since the lender and the borrower are jointly considered, i. e. the power obtained by one party is lost by the other and vice versa.

At first absolute power effects are considered since these effects only result from agency cost considerations. Additionally, strategic aspects have to be taken into account when the effects of the power to determine ct/pm or of the power to set contracts' conditions are examined. Strategic aspects affect the preferred ct/pm-choice in the restricted bargaining power scenarios (LS^r, BS^r) due to the separation of both bargaining power components. In the restricted power scenarios the ct/pm-optimizer tries to obtain a certain profit by his ct/pm-choice which the condition optimizer anticipates and tries to prevent.

The distribution of the *absolute bargaining power* to determine ct/pm and contracts' conditions affects the preferred ct/pm-choice $(LS^a \rightleftharpoons BS^a)$.

- The ct-choice is directly affected by the absolute power distribution since the negotiated contract conditions depend on the absolute power distribution and, therefore, the arising agency costs. For example, in the presented framework debt financing becomes more often preferable when the borrower obtains the absolute power and vice versa since when the borrower obtains (loses) the absolute power, debt financing's negotiated fixed repayment obligation (h) decreases (increases) and, consequently, the agency costs associated with debt financing decrease (increase). The agency costs of equity financing are constant, i. e. independent of the proportional return participation (q).
- The pm-choice is indirectly affected by the absolute power distribution for two reasons. Firstly, the borrower's informational rent in debt financing affects the cost differences of the alternative pms from the lender's perspective, whereas the borrower always has to bear the full costs of the pms since the lender never receives a rent due to interim and ex-post uncertainty. Hence, the preferred, i. e. cost optimal, pm-choice depends on the ct-choice and therefore on the power distribution. Secondly, the total surplus extractable from project financing ($\mu - AC - 1$) is probably lower in the lender than in the borrower scenario which provides the borrower with a stronger incentive to eliminate placement risk. Hence, the contract agreement probability can decrease or increase if the power is obtained by the borrower and vice versa.

See Propositions 1 and 2 for details (BM: p. 96, EM: pp. 220-221).

The distribution of the *power to determine ct/pm* also affects the preferred ct/pm-choice. The effects of the power to determine ct/pm depend on the distribution of the right to set contracts' conditions.

- In case the right to set contracts' conditions belongs to the lender, the distribution of the power to determine ct/pm affects both the ct and the pm decision directly $(LS^a \rightleftharpoons BS^r)$. The effects are driven by the borrower's informational rent seeking behavior in the BS^r scenario which the lender anticipates and tries to prevent. However, the only informational rent the lender can prevent by his strategy choice in public offerings is the borrower's rent due to his ex-ante advantage. For particular circumstances the lender is able to separate both borrower types, and, hence, their ex-ante informational advantage vanishes. The borrower's informational rent in debt financing due to his interim and ex-post advantage can not be prevented by the lender since the lender has no option to affect the borrower's potential rent in debt financing due to the latter's interim and ex-post informational advantage. Assuming, like we did (see Table 4.44), that the borrower's potential informational rent due to all three uncertainties is larger than his rent only due to his interim and ex-post advantage, and that the lender possesses the opportunity to separate both borrower types, a power shift from the lender to the N-type (A-type) borrower increases the favorability of public (private) debt and vice versa. However, if the lender does not possess this opportunity, the borrower can obtain an additional informational rent due to his ex-ante advantage. Hence, a power shift from the lender to the N-type (A-type) borrower causes all public (public and private) offerings with the same chance of contract agreement to become equally important since they all offer the same expected profit for the borrower. Assuming alternatively that the borrower's potential informational rent due to all three uncertainties is just as large as his rent due to his interim and ex-post advantage, a power shift from the lender to the borrower always favors public debt and vice versa.[1] See Propositions 3 and 4 for details (BM: pp. 98-98, EM: pp. 223-224).

- In case the right to set contracts' conditions belongs to the borrower, the distribution of the power to determine ct/pm again affects both the ct and the pm choice directly ($BS^a \rightleftharpoons LS^r$). Basically the effects are driven by the lender's objective to maximize the chance of contract agreement in the LS^r scenario since except in one situation the borrower can squeeze the lender's profit to the lender's profit alternative, i. e. the borrower can separate both lender types and therefore avoid an informational rent payment due to the lender's ex-ante advantage. Additionally, the lender never receives a rent from interim and ex-post uncertainty. In such circumstances a power shift from the borrower to the N-type (A-type) lender causes indifference between all public and private (only private) offerings with the same chance of contract agreement. The exception arises when only public offerings are feasible for the A-type lender since the A-type lender will choose a public placement and, therefore, confronts the borrower with a tradeoff in his strategy choice. The borrower has to choose among S1's disadvantage of paying an (ex-ante) informational rent to the N-type lender and S2's potential placement risk. If the expected placement risk, i. e. the resulting loss, exceeds the saved informational rent payment, the borrower prefers S1 to S2 which implies that the N-type lender obtains an (ex-ante)

---

[1] For example, this becomes obvious in the basic model where no ex-ante informational asymmetry and hence no potential rent due to this uncertainty exists. In such a situation the borrower prefers public debt whenever feasible to maximize his rent from his interim and ex-post advantage.

informational rent. Therefore, a power shift to the lender favors the respective public offering.

In total a power shift from the borrower to the lender increases the contract agreement probability and vice versa. See Propositions 5 and 6 for details (BM: pp. 99-99, EM: pp. 225-226).

Comparing the effects of the power to determine ct/pm given that either the lender or the borrower has the right to set contracts' conditions reveals that the strategic opportunities offered by this power to cope with informational asymmetries are more valuable for the borrower than for the lender. The strategic opportunities are more valuable for the borrower than for the lender since the borrower possesses an interim and ex-post informational advantage but not the lender. Due to these advantages the borrower obtains in debt financing an informational rent. This informational rent strengthens the borrower's position when bargaining with the lender about a potential (ex-ante) informational rent. Due to the rent from interim and ex-post uncertainty the borrower can (actively) oppose any attempt of the lender to separate both borrower types in the BS^r scenario, i. e. to avoid an ex-ante informational rent payment to the borrower. On the other side, the lender does not possess such an opportunity in the LS^r scenario since he has noting to "put at risk". The borrower can always squeeze the lender to his outside option when either the A-type lender does not participate at all or private financing is feasible for the latter. Therefore, the lender's and the borrower's valuations of the power to determine ct/pm differ since the power to choose ct/pm provides them with different strategic opportunities to cope with the informational asymmetries.

The distribution of the *power to set contracts' conditions* also affects the preferred ct/pm-choice. These power effects depend significantly on the distribution of the right to determine ct/pm. Since again strategic aspects outweigh agency cost considerations in the restricted power scenarios, the effects observable for $LS^a \rightleftharpoons LS^r$ resemble the effects for $BS^a \rightleftharpoons LS^r$; and the effects observable for $BS^a \rightleftharpoons BS^r$ resemble the effects for $LS^a \rightleftharpoons BS^r$. See Propositions 7 and 10 for details (BM: pp. 100-102, EM: pp. 228-231).

Finally, a *comparison of the power components* from the lender's and from the borrower's perspective reveals that the power to determine contracts' conditions is generally more important for profit maximization than the power to choose ct/pm. The latter is relevant for projects near their feasibility constraint where the potential informational rent can exceed the otherwise extractable surplus and contract agreement is not sure. See Results 2 and 3 for details (BM: pp. 103-104, EM: pp. 232-234).

6.2 Economic Relevance of Bargaining Power Effects

Clearly, the primary objective of this dissertation was to demonstrate that a firm's financing decision depends on the (ex-ante) distribution of bargaining power between the firm and potential fund suppliers. In particular, we were concerned about implications for the firm's ct/pm-choice. Therefore, we developed a model to examine how a firm's financing decision depends on the distribution of bargaining power.

Our model reveals several ways in which a firm's financing decision depends on the distribution of the bargaining power to choose ct/pm and on the distribution of the power to set contracts' conditions (see Propositions 1 to 10). The relevance of each power component depends on the particular project and uncertainty specification (see Results 2 and 3). In the absolute power scenarios only agency cost considerations affect the firm's ct/pm-choice, while in the restricted power scenarios strategic aspects gain importance. The firm's ct and pm choice are interrelated and have to be treated jointly.

Therefore, each firm with the intention to raise funds has to think about its own bargaining power (power distribution), its financing alternatives (own outside option), potential lenders' alternative investment opportunities (contractual partners' outside option), and the uncertainty in the market, e. g. how potential lenders evaluate the firm's financing alternatives, before entering contract negotiations. After realizing the status quo the firm can try to improve its bargaining position while potential lenders will optimize their position.

That firms care about bargaining power is empirically confirmed by Röell (1996) and Pagano et al. (1998). They find that firms going public, i. e. initially offering shares publicly, increase the competition among fund suppliers and experience a reduction in their costs of a bank loan even after controlling for firms' characteristics and the leverage reduction due to the equity issue. Assuming that going public increases a firm's power to determine ct/pm or generates a positive financing alternative (outside option) for bilateral bargaining situations, Röell's and Pagano's finding is consistent with our theoretical predictions, as a firm's, i. e. the borrower's, repayment obligation (h) reduces due to both variations.[2] Pagano et al. even observe that firms after going public borrow from a larger number of banks, and reduce the concentration of their borrowings. They also suggest that going public increases competition among fund suppliers in different capital market segments but also among

---

[2] Assuming the borrower obtains a positive outside option (A-type) while the lender still optimizes contracts' conditions (LS^a, BS^r) for private debt ($D^{\bullet}$) $h^*_{L,\text{N-type borrower}}$ is bounded above $h^*_{L,\text{A-type borrower}}$ since the lender can demand a higher repayment from a borrower without a financing alternative. Assuming alternatively that the power to choose ct/pm shifts to the borrower ($LS^a \rightarrow BS^r$ or $LS^r \rightarrow BS^a$), the borrower's profit increases for private debt financing ($D^{\bullet}$) which necessarily implies a reduction in h.

banks providing loans. The costs firms bear for an initial public offering can only realistically be explained taking the consequences of changes in the firm's bargaining position into account (cf. Pagano et al. (1998)).

Recently, Faulkender and Petersen (2003) even found that firms with a bond rating have significantly higher leverage ratios than firms without rating after controlling for firms' characteristics previously identified to explain firms' capital structure choice and for the possible endogenity of having a bond rating. Faulkender and Petersen observe that this discrepancy can partly be explained by credit rationing by firms' lenders à la Stiglitz and Weiss (cf. Stiglitz and Weiss (1981)) since the same type of market imperfection which affects a firm's capital structure choice affects potential lenders' willingness to supply funds. However, they can not explain the whole discrepancy. One possible explanation for the remaining discrepancy can be obtained from the developed bargaining power considerations. Obviously, firms with a bond rating are in a better bargaining position than firms without a rating since the former possess the opportunity to raise funds in different capital market segments.[3] Assuming therefore that a bond rating increases the borrower's bargaining power to choose ct/pm, Faulkender and Petersen (2003) actually observe that borrowers with more bargaining power favor debt financing. This finding is consistent with our theoretical predictions holding the power to set contracts' conditions constant, see Propositions 3 ($LS^a \to BS^r$) and 6 ($LS^r \to BS^a$).

Additionally, our analysis reveals that the capital structure predictions based on principal agent and asymmetric information considerations depend on the distribution of bargaining power between the lender and the borrower. This might be a potential explanation why empirical studies of capital structure, in particular, focusing on different countries obtain quite heterogeneous results (cf. Rajan and Zingales (1995), Booth et al. (2001), and Antoniou et al. (2002)). The capital market structure probably varies across the examined countries and hence the bargaining power of the firms. La Porta et al. (1997, 1998) suggest various legal environments in these countries as a potential source for the difference. They find that legal protection of investors matters for the availability of public debt and equity markets. However, Bancel and Mittoo (2003) find that the actual capital structure choice results from a complex interaction of many institutional features and business practices that is not fully captured in the different legal systems. Bancel and Mittoo even support the suggestion by Titman (2002, p. 114):

> "Corporate treasures do occasionally think about the kind of tradeoffs between tax savings and financial distress costs ... However, since these tradeoffs do not change much over time ... They spend much more thinking about changes in market conditions and the implications of these changes on how firms should be financed."

---

[3] Faulkender and Petersen (2003) interpret a bond rating as an indicator for firms' access to different debt market segments.

Therefore, we suggest to include bargaining power indicators, like the relation of fund supply and demand in empirical capital structure studies to overcome persistent gaps between theoretical predictions and empirical evidence (cf., e. g., Harris and Raviv (1991), Frank and Goyal (2003a) and Myers (2003)).

Such power indicators might also help to disentangle power from bankruptcy risk effects since a typical (inverse) proxy for bankruptcy risk is firm size (cf., e. g., Rajan and Zingales (1995)). Larger firms are supposed to be better diversified than smaller firms and hence larger firms are expected to bear less default risk. However, size is also a potential proxy for firms' bargaining power since e. g. larger firms often have access to capital market segments which smaller firms do not have. Larger firms can also offer potential lenders, like banks, further business opportunities, e. g. management of the firm's accounts, which smaller firms cannot do to this extend. The empirical observation that leverage is positively correlated with firm size (cf. Antoniou et al. (2002) and Frank and Goyal (2003a)) is consistent with both theoretical predictions.

Finally, the model's predictions can be useful for governments and regulators to assess their competition policy. Competition policy affects the distribution of bargaining power between firms and potential fund suppliers e. g. by influencing the competition among and within different capital market segments. For example, to abandon restrictions of institutional investors' business activities, such as the allowance of mergers between banks and insurance companies to form bankassurance groups, probably increases the competition among and within the related capital market segments and can shift the power to determine ct/pm from the lender to the borrower. Depending on the distribution of the power to set contracts' conditions this power redistribution has quite distinct effects as shown in Propositions 3 to 6. Assuming, for example, that the government's objective is to minimize agency costs, to abandon such restrictions seems to be reasonable if the power to set contracts' conditions belongs to the borrower (see Proposition 6 ($LS^r \to BS^a$)). However, if the government's objective is to maximize the chance of contract agreement, the power redistribution is counter productive. Therefore, the government's objective as well as the current bargaining power distribution are essential to evaluate regulatory changes.

A

Consistency Checks of Bargaining Power Effects (EM, PC1)

The consistency of the propositions about bargaining power effects in financial contracting, determined in Chapter 4, is examined in this appendix for an A-type lender / N-type borrower, N-type lender / A-type borrower and N-type lender / N-type borrower constellation, see Figures A.1 to A.6.

Fig. A.1: Power Effect's Consistency Check 1 for an A-Type Lender / N-Type Borrower (EM, PC1)

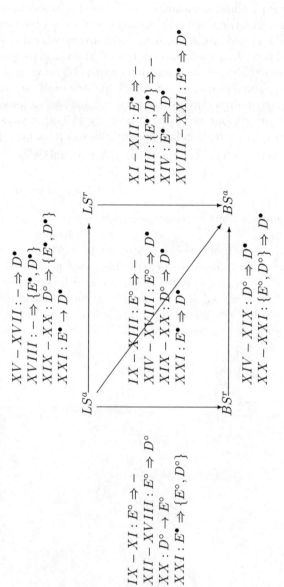

Note: This figure illustrates the consistency of the Propositions 1 ($LS^a \to BS^a$), 3 ($LS^a \to BS^r$), 6 ($LS^r \to BS^a$), 7 ($LS^a \to LS^r$) and 10 ($BS^r \to BS^a$) derived in Sections 4.4.1, 4.4.2 and 4.4.3 for the feasible contracting intervals. $\to$ refers to power effects which *can* occur given PC1, while $\Rightarrow$ indicates effects which *definitively* occur given PC1.

Fig. A.2: Power Effect's Consistency Check 2 for an A-Type Lender / N-Type Borrower (EM, PC1)

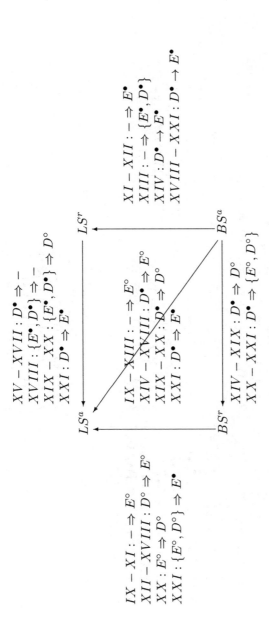

Note: This figure illustrates the consistency of the Propositions 2 ($BS^a \to LS^a$), 4 ($BS^r \to LS^a$), 5 ($BS^a \to LS^r$), 8 ($LS^r \to LS^a$) and 9 ($BS^a \to BS^r$) derived in Sections 4.4.1, 4.4.2 and 4.4.3 for the feasible contracting intervals. $\to$ refers to power effects which *can* occur given PC1, while $\Rightarrow$ indicates effects which *definitively* occur given PC1.

Fig. A.3: Power Effect's Consistency Check 1 for a N-Type Lender / A-Type Borrower (EM, PC1)

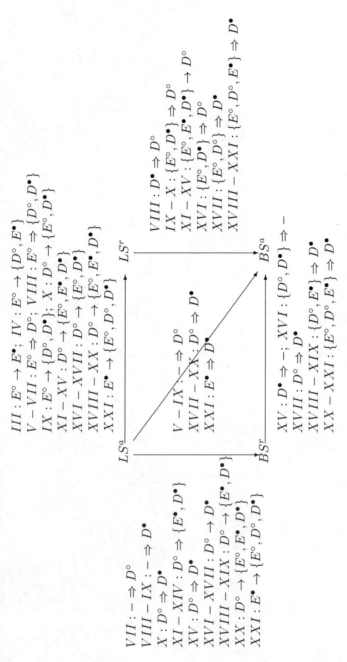

Note: This figure illustrates the consistency of the Propositions 1 ($LS^a \rightarrow BS^a$), 3 ($LS^a \rightarrow BS^r$), 6 ($LS^r \rightarrow BS^a$), 7 ($LS^r \rightarrow LS^r$) and 10 ($BS^r \rightarrow BS^a$) derived in Sections 4.4.1, 4.4.2 and 4.4.3 for the feasible contracting intervals. $\rightarrow$ refers to power effects which *can* occur given PC1, while $\Rightarrow$ indicates effects which *definitively* occur given PC1.

Fig. A.4: Power Effect's Consistency Check 2 for a N-Type Lender / A-Type Borrower (EM, PC1)

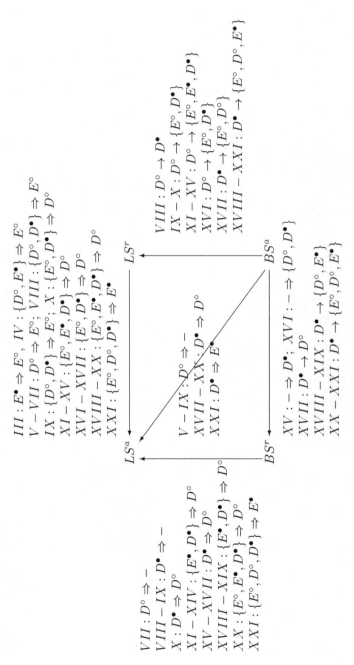

Note: This figure illustrates the consistency of the Propositions 2 ($BS^a \to LS^a$), 4 ($BS^r \to LS^a$), 5 ($LS^r \to LS^a$), 8 ($LS^r \to LS^a$) and 9 ($BS^a \to BS^r$) derived in Sections 4.4.1, 4.4.2 and 4.4.3 for the feasible contracting intervals. $\to$ refers to power effects which *can* occur given PC1, while $\Rightarrow$ indicates effects which *definitively* occur given PC1.

Fig. A.5: Power Effect's Consistency Check 1 for a N-Type Lender / N-Type Borrower (EM, PC1)

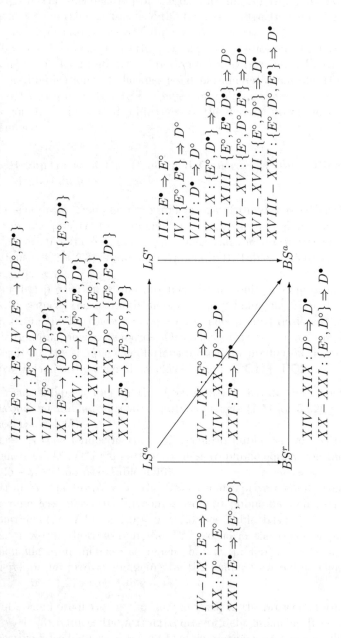

$III : E^\circ \to E^\bullet; IV : E^\circ \to \{D^\circ, E^\bullet\}$
$V - VII : E^\circ \Rightarrow D^\circ$
$VIII : E^\circ \Rightarrow \{D^\circ, D^\bullet\}$
$IX : E^\circ \to \{D^\circ, D^\bullet\}; X : D^\circ \to \{E^\circ, D^\bullet\}$
$XI - XV : D^\circ \to \{E^\bullet, E^\bullet, D^\bullet\}$
$XVI - XVII : D^\circ \to \{E^\circ, D^\bullet\}$
$XVIII - XX : D^\circ \to \{E^\circ, E^\bullet, D^\bullet\}$
$XXI : E^\bullet \to \{E^\circ, D^\circ, D^\bullet\}$

LS^r

$III : E^\bullet \Rightarrow E^\circ$
$IV : \{E^\circ, E^\bullet\} \Rightarrow D^\circ$
$VIII : D^\bullet \Rightarrow D^\circ$
$IX - X : \{E^\circ, D^\bullet\} \Rightarrow D^\circ$
$XI - XIII : \{E^\circ, E^\bullet, D^\bullet\} \Rightarrow D^\circ$
$XIV - XV : \{E^\circ, D^\circ, E^\bullet\} \Rightarrow D^\bullet$
$XVI - XVII : \{E^\circ, D^\circ\} \Rightarrow D^\bullet$
$XVIII - XXI : \{E^\circ, D^\circ, E^\bullet\} \Rightarrow D^\bullet$

BS^a

LS^a

$IV - IX : E^\circ \Rightarrow D^\circ$
$XIV - XX : D^\circ \Rightarrow D^\bullet$
$XXI : E^\bullet \Rightarrow D^\bullet$

BS^r

$XIV - XIX : D^\circ \Rightarrow D^\bullet$
$XX - XXI : \{E^\circ, D^\circ\} \Rightarrow D^\bullet$

$IV - IX : E^\circ \Rightarrow D^\circ$
$XX : D^\circ \to E^\circ$
$XXI : E^\bullet \Rightarrow \{E^\circ, D^\circ\}$

Note: This figure illustrates the consistency of the Propositions 1 ($LS^a \to BS^a$), 3 ($LS^a \to BS^r$), 6 ($LS^r \to BS^a$), 7 ($LS^r \to LS^a$) and 10 ($BS^r \to BS^a$) derived in Sections 4.4.1, 4.4.2 and 4.4.3 for the feasible contracting intervals. $\to$ refers to power effects which *can* occur given PC1, while $\Rightarrow$ indicates effects which *definitively* occur given PC1.

Fig. A.6: Power Effect's Consistency Check 2 for a N-Type Lender / N-Type Borrower (EM, PC1)

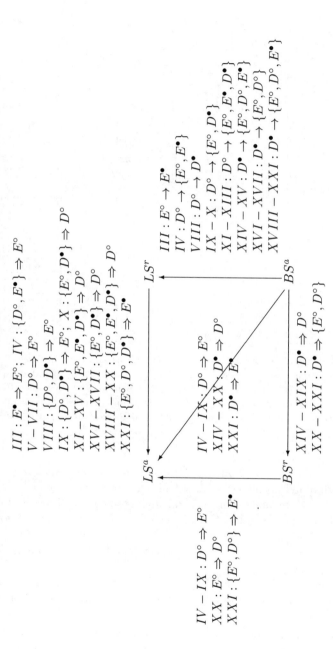

Note: This figure illustrates the consistency of the Propositions 2 ($BS^a \rightarrow LS^a$), 4 ($BS^r \rightarrow LS^a$), 5 ($BS^a \rightarrow LS^r$), 8 ($LS^r \rightarrow LS^a$) and 9 ($BS^a \rightarrow BS^r$) derived in Sections 4.4.1, 4.4.2 and 4.4.3 for the feasible contracting intervals. $\rightarrow$ refers to power effects which *can* occur given PC1, while $\Rightarrow$ indicates effects which *definitively* occur given PC1.

B

Derivation of Contract Type / Placement Mode-Choices for PC2 and PC3 (EM)

In this appendix the preferred ct/pm-choices for the four bargaining power scenarios are determined for the underlying parameter constellations PC2 and PC3 (see Table 4.44, p. 237) to examine the robustness of the propositions about bargaining power effects determined in Chapter 4 for PC1.

For the absolute bargaining power scenarios (LS^a, BS^a) the feasible contracting intervals given $\widehat{\lambda}_B = \lambda_B$, respectively $\widehat{\lambda}_L = \lambda_L$, are determined before a comparison of the expected profits reveals the lender's, respectively the borrower's, preferred ct/pm-choice.[1] For the scenarios with restricted bargaining power (LS^r, BS^r) given the lender's, respectively the borrower's, possible sets of consistent conditional expectations about the opponent's type, see Tables 4.31 (p. 187) and 4.37 (p. 202), firstly the consistency of the expectations is examined. Finally, based on the consistent conditional expectations the lender's, respectively the borrower's, preferred ct/pm-choice is predicted.

---

[1] See page 173 (209) for the derivation of $\widehat{\lambda}_B = \lambda_B$ $(\widehat{\lambda}_L = \lambda_L)$ in the LS^a (BS^a) scenario.

Table B.1: Expected Profits – Lender Optimization (EM, PC2, PC3)

ct/pm	$PL(\mu, \hat{\lambda}_B)$	$PB(\mu, \hat{\lambda}_B)$	BPCb/nb; CDr/nr; LCOC(AB)
		Parameter constellation 2 (for debt financing see parameter constellation 1, Table 4.25)	
$E^{\circ}_L\|S1$	$\mu - 1.6$	$\begin{cases} 0.3\|(1-\hat{\lambda}_L)0.3 \\ 0.3\|0.3 \end{cases}$	—
$E^{\circ}_L\|S2$	$(1-\hat{\lambda}_B)(\mu - 1.3) + \hat{\lambda}_B \overline{P}_L$	$0.3\|0$	—
$E^{\bullet(2)}_L$	$\mu - \hat{\lambda}_B 0.3 - 1.4$	$0.3\|0$	$\mu \geq 1.7 + \overline{P}_L$
$E^{\bullet(1)}_L$	$(1-\hat{\lambda}_B)(\mu - 1.4) + \hat{\lambda}_B \overline{P}_L$	$0.3\|0$	$\mu < 1.7 + \overline{P}_L$
		Parameter constellation 3 (for equity financing see parameter constellation 1, Table 4.25)	
$D^{\circ}_{L,Ia}$	$0.25\mu - 1$	$\dfrac{(1-\hat{\lambda}_L)0.75\mu + \hat{\lambda}_L 0.3\|(1-\hat{\lambda}_L)0.75\mu}{0.75\mu\|0.75\mu}$	$\mu \geq 0.4$
$D^{\circ}_{L,Ib}\|S1$	$\mu - 1.3$	$\dfrac{0.3\|(1-\hat{\lambda}_L)0.3}{0.3\|0.3}$	$\mu < 0.4$
$D^{\circ}_{L,Ib}\|S2$	$(1-\hat{\lambda}_B)(0.25\mu - 1) + \hat{\lambda}_B \overline{P}_L$	$\dfrac{0.3\|(1-\hat{\lambda}_L)0.75\mu}{0.3\|0.75\mu}$	$\mu < 0.4$
$D^{\circ}_{L,IIa}\dagger$	$\mu + \dfrac{0.05\overline{3}}{\mu} - 1.4$	$\dfrac{(1-\hat{\lambda}_L)\frac{0.05\overline{3}}{\mu}\big|\frac{0.05\overline{3}}{\mu} + \hat{\lambda}_L 0.3\|(1-\hat{\lambda}_L)\frac{0.05\overline{3}}{\mu}}{\frac{0.05\overline{3}}{\mu}\big|\frac{0.05\overline{3}}{\mu}}$	$0.2\overline{6} < \mu \leq 0.178$
$D^{\circ}_{L,IIb}\|S1$	$\mu + \dfrac{0.253}{\sqrt{\mu}} - 1.7$	$\dfrac{0.3\|(1-\hat{\lambda}_L)0.3}{0.3\|0.3}$	$\mu > 0.4$

table continued on next page

table continued from previous page

$D_{L,IIb}^{\circ}\|S2$	$(1-\hat{\lambda}_B)(\mu+\frac{0.05\overline{3}}{\mu}-1.4)+\hat{\lambda}_B\overline{P}_L$	$\left\{\begin{array}{l}0.3\|(1-\hat{\lambda}_L)\frac{0.05\overline{3}}{\mu}\\[2pt]0.3\|\frac{0.05\overline{3}}{\mu}\end{array}\right.$	$\mu>0.2\overline{6}$
$D_{L,Ia}^{\bullet(2)}$	$0.25\mu-1.05$	$\left\{\begin{array}{l}(1-\hat{\lambda}_L)(0.75\mu-0.05)+\hat{\lambda}_L0.3\|\\(1-\hat{\lambda}_L)(0.75\mu-0.05)\\0.75\mu-0.05\|0.75\mu-0.05\end{array}\right.$	$\mu\geq4.2+4\overline{P}_L$
$D_{L,Ib}^{\bullet(2)\,\dagger}$	$\hat{\lambda}_B(\mu-0.35)+(1-\hat{\lambda}_B)0.25\mu-1.05$	$\left\{\begin{array}{l}0.3\|(1-\hat{\lambda}_L)(0.75\mu-0.05)\\0.3\|0.75\mu-0.05\end{array}\right.$	$1.4+\overline{P}_L\leq\mu<0.4\overline{6}$
$D_{L,Ib}^{\bullet(1)}$	$(1-\hat{\lambda}_B)(0.25\mu-1.05)+\hat{\lambda}_B\overline{P}_L$	$\left\{\begin{array}{l}0.3\|(1-\hat{\lambda}_L)(0.75\mu-0.05)\\0.3\|0.75\mu-0.05\end{array}\right.$	$0.0\overline{6}\leq\mu<0.4\overline{6}$
$D_{L,Ic}^{\bullet(2)\,\dagger}$	$\mu-\hat{\lambda}_B0.3-1.1$	$0.3\|0$	$1.4+\overline{P}_L\leq\mu<0.4\overline{6}$
$D_{L,Ic}^{\bullet(1)\,\dagger}$	$(1-\hat{\lambda}_B)(\mu-1.1)+\hat{\lambda}_B\overline{P}_L$	$0.3\|0$	$1.1+\overline{P}_L\leq\mu<0.4\overline{6}$
$D_{L,IIa}^{\bullet(2)\,\dagger}$	$\hat{\lambda}_B0.25\mu+(1-\hat{\lambda}_B)(\mu+\frac{0.05\overline{3}}{\mu}-0.4)-1.05$	$\left\{\begin{array}{l}(1-\hat{\lambda}_L)(0.75\mu-0.05)+\hat{\lambda}_L0.3\|\\(1-\hat{\lambda}_L)(\frac{0.05\overline{3}}{\mu}-0.05)\\0.75\mu-0.05\|\frac{0.05\overline{3}}{\mu}-0.05\end{array}\right.$	$4.2+4\overline{P}_L\leq\mu\leq1.0\overline{6}$
$D_{L,IIa}^{\bullet(1)}$	$(1-\hat{\lambda}_B)(\mu+\frac{0.05\overline{3}}{\mu}-1.45)+\hat{\lambda}_B\overline{P}_L$	$\left\{\begin{array}{l}0.3\|(1-\hat{\lambda}_L)(\frac{0.05\overline{3}}{\mu}-0.05)\\0.3\|\frac{0.05\overline{3}}{\mu}-0.05\end{array}\right.$	$0.4\overline{6}\leq\mu\leq1.0\overline{6}$
$D_{L,IIb}^{\bullet(2)\,\dagger}$	$\mu+(1-\hat{\lambda}_B)\frac{0.05\overline{3}}{\mu}+\hat{\lambda}_B0.05-1.45$	$\left\{\begin{array}{l}0.3\|(1-\hat{\lambda}_L)(\frac{0.05\overline{3}}{\mu}-0.05)\\0.3\|\frac{0.05\overline{3}}{\mu}-0.05\end{array}\right.$	$1.4+\overline{P}_L\leq\mu<0.4\overline{6}$

table continued on next page

table continued from previous page

$D_{L,IIb}^{\bullet(1)}$	$(1-\widehat{\lambda}_B)(\mu + \frac{0.05\overline{3}}{\mu} - 1.45) + \widehat{\lambda}_B \overline{P}_L$	$\left\{\begin{array}{l} 0.3\|(1-\widehat{\lambda}_L)(\frac{0.05\overline{3}}{\mu} - 0.05) \\ 0.3\|\frac{0.05\overline{3}}{\mu} - 0.05 \end{array}\right.$	$0.2\overline{6} \le \mu < 0.4\overline{6}$
$D_{L,IIc}^{\bullet(2)}$	$\widehat{\lambda}_B 0.25\mu + (1-\widehat{\lambda}_B)(\mu + \frac{0.103}{\sqrt{\mu}} - 0.45) - 1.05$	$\left\{\begin{array}{l}(1-\widehat{\lambda}_L)(0.75\mu - 0.05) + \widehat{\lambda}_L 0.3\|0 \\ 0.75\mu - 0.05\|0 \end{array}\right.$	$\mu > 4.2 + 4\overline{P}_L$
$D_{L,IIc}^{\bullet(1)}$	$(1-\widehat{\lambda}_B)(\mu + \frac{0.103}{\sqrt{\mu}} - 1.5) + \widehat{\lambda}_B \overline{P}_L$	$0.3\|0$	$1.0\overline{6} < \mu < 4.2 + 4\overline{P}_L$
$D_{L,IId}^{\bullet(2)}$ †	$\mu + (1-\widehat{\lambda}_B)(\frac{0.103}{\sqrt{\mu}} - 0.4) - \widehat{\lambda}_B 0.3 - 1.1$	$0.3\|0$	$3.2 < \mu < 1.4$
$D_{L,IId}^{\bullet(1)}$ †	$(1-\widehat{\lambda}_B)(\mu + \frac{0.103}{\sqrt{\mu}} - 1.5) + \widehat{\lambda}_B \overline{P}_L$	$0.3\|0$	$3.2 < \mu < 1.4$
$D_{L,IIIa}^{\bullet(2)}$ †	$\mu + \frac{0.05\overline{3}}{\mu} - 1.45$	$\left\{\begin{array}{l}(1-\widehat{\lambda}_L)(\frac{0.05\overline{3}}{\mu} - 0.05) + \widehat{\lambda}_L 0.3\| \\ (1-\widehat{\lambda}_L)(\frac{\mu}{0.05\overline{3}} - 0.05) \\ \frac{0.05\overline{3}}{\mu} - 0.05\|\frac{0.05\overline{3}}{\mu} - 0.05 \end{array}\right.$	$0.2\overline{6} < \mu \le 0.152$
$D_{L,IIIb}^{\bullet(2)}$ †	$\mu + (1-\widehat{\lambda}_B)\frac{0.05\overline{3}}{\mu} + \widehat{\lambda}_B \frac{0.273}{\sqrt{\mu}} - \widehat{\lambda}_B 0.35 - 1.45$	$\left\{\begin{array}{l} 0.3\|(1-\widehat{\lambda}_L)(\frac{0.05\overline{3}}{\mu} - 0.05) \\ 0.3\|\frac{0.05\overline{3}}{\mu} - 0.05 \end{array}\right.$	$1.583 + 1.063\overline{P}_L <$ $\mu \le 1.0\overline{6}$
$D_{L,IIIb}^{\bullet(1)}$ †	$(1-\widehat{\lambda}_L)(\mu + \frac{0.05\overline{3}}{\mu} - 1.45) + \widehat{\lambda}_B \overline{P}_L$	$\left\{\begin{array}{l} 0.3\|(1-\widehat{\lambda}_L)(\frac{0.05\overline{3}}{\mu} - 0.05) \\ 0.3\|\frac{0.05\overline{3}}{\mu} - 0.05 \end{array}\right.$	$0.2\overline{6} < \mu \le 0.152$
$D_{L,IIIc}^{\bullet(2)}$	$\mu + \frac{(1-\widehat{\lambda}_B)0.103 + \widehat{\lambda}_B 0.273}{\sqrt{\mu}} - \widehat{\lambda}_B 0.3 - 1.5$	$0.3\|0$	$\mu \ge 1.583 + 1.063\overline{P}_L$
$D_{L,IIIc}^{\bullet(1)}$	$(1-\widehat{\lambda}_B)(\mu + \frac{0.103}{\sqrt{\mu}} - 1.5) + \widehat{\lambda}_B \overline{P}_L$	$0.3\|0$	$1.0\overline{6} < \mu < 1.583 + 1.063\overline{P}_L$

Note: The lender's and the borrower's expected profits are stated $(PL(\mu, \widehat{\lambda}_B), PB(\mu, \widehat{\lambda}_L))$ with the respective feasible contracting constraints. The power to determine contracts' conditions is given to the lender. The structure of the table is equivalent to Table 4.25.

Table B.2: Expected Profits – Borrower Optimization (EM, PC2, PC3)

ct/pm	$PB(\mu, \widehat{\lambda}_L)$	$PL(\mu, \widehat{\lambda}_L)$	NoN; BCOC(AL)	CDr/nr;
	Parameter constellation 2 (for debt financing see parameter constellation 1, Table 4.26)			
$E_B^{\circ}\vert S1$	$\mu - 1.6$	$\begin{cases} 0.3\vert(1-\widehat{\lambda}_B)0.3 \\ 0.3\vert0.3 \end{cases}$	—	
$E_B^{\circ}\vert S2$	$(1-\widehat{\lambda}_L)(\mu-1.3)+\widehat{\lambda}_L \overline{P}_B$	$0.3\vert0$		
$E_B^{\bullet(2)}$	$\mu - \widehat{\lambda}_L 0.3 - 1.4$	$0.3\vert0$	$\mu \geq 1.7 + \overline{P}_B$	
$E_B^{\bullet(1)}$	$(1-\widehat{\lambda}_L)(\mu-1.4)+\widehat{\lambda}_L \overline{P}_B$	$0.3\vert0$	$\mu < 1.7 + \overline{P}_B$	
	Parameter constellation 3 (for equity financing see parameter constellation 1, Table 4.26)			
$D_{B,I}^{\circ}\vert S1$	$\mu - 1.3$	$\begin{cases} 0.3\vert(1-\widehat{\lambda}_B)0.3 \\ 0.3\vert0.3 \end{cases}$	$\mu \geq 5.2$	
$D_{B,I}^{\circ}\vert S2$	$(1-\widehat{\lambda}_L)(\mu-1)+\widehat{\lambda}_L \overline{P}_B$	$0.3\vert0$	$\mu \geq 4$	
$D_{B,II}^{\circ}\vert S1$	$\mu + \frac{0.10\overline{6}}{\mu} + \frac{0.462}{\mu}\sqrt{(\mu-1.668)(\mu-0.032)} - 1.7$	$\begin{cases} 0.3\vert(1-\widehat{\lambda}_B)0.3 \\ 0.3\vert0.3 \end{cases}$	$1.668 \leq \mu < 5.2$	
$D_{B,II}^{\circ}\vert S2$	$(1-\widehat{\lambda}_L)\left[\mu + \frac{0.10\overline{6}}{\mu} + \frac{0.462}{\mu}\sqrt{(\mu-1.361)(\mu-0.039)} - 1.4\right] + \widehat{\lambda}_L \overline{P}_B$	$0.3\vert0$	$1.361 \leq \mu < 4$	
$D_{B,I}^{\bullet(2)}$	$\mu - \widehat{\lambda}_L 0.3 - 1.1$	$0.3\vert0$	$\mu \geq 5.4$	
$D_{B,I}^{\bullet(1)}$ †	$(1-\widehat{\lambda}_L)(\mu-1.1)+\widehat{\lambda}_L \overline{P}_B$	$0.3\vert0$	$5.4 \leq \mu < 1.4 + \overline{P}_B$	
$D_{B,II}^{\bullet(2)}$	$\mu + \widehat{\lambda}_L \frac{0.10\overline{6}}{\mu} - \widehat{\lambda}_L \frac{0.462}{\mu}\sqrt{(\mu-1.719)(\mu-0.031)} - \widehat{\lambda}_L 0.7 - 1.1$	$0.3\vert0$	$4.2 \leq \mu < 5.4$	

table continued on next page

table continued from previous page

$D^{\bullet(1)\dagger}_{B,II}$	$(1-\hat{\lambda}_L)(\mu-1.1)+\hat{\lambda}_L\overline{P}_B$	0.3|0	$4.2 \leq \mu < 1.721 + 0.602\overline{P}_B$
$D^{\bullet(2)}_{B,III}$	$\mu+\frac{0.10\overline{6}}{\mu}+\frac{0.462}{\mu}\cdot\left[\hat{\lambda}_L\sqrt{(\mu-1.719)(\mu-0.031)}+(1-\hat{\lambda}_L)\sqrt{(\mu-1.413)(\mu-0.038)}\right]-\hat{\lambda}_L 0.3 - 1.5$	0.3|0	$1.721 + 0.602\overline{P}_B \leq \mu < 4.2$
$D^{\bullet(1)}_{B,III}$	$(1-\hat{\lambda}_L)(\mu+\frac{0.10\overline{6}}{\mu}+\frac{0.462}{\mu}\sqrt{(\mu-1.413)(\mu-0.038)}-1.5)+\hat{\lambda}_L\overline{P}_B$	0.3|0	$1.413 \leq \mu < 1.721 + 0.602\overline{P}_B$

Note: The lender's and the borrower's expected profits ($PL(\mu,\hat{\lambda}_B)$, $PB(\mu,\hat{\lambda}_L)$) with the respective feasible contracting constraints are stated for the case that the borrower has the power to determine contracts' conditions. The structure of the table is equivalent to Table 4.26.

B.1 Lender Scenario with Absolute Power (LS^a)

B.1.1 Parameter Constellation 2

THE LENDER'S BEHAVIOR AS THE CONDITION OPTIMIZER ($\widehat{\lambda}_B = \lambda_B$)

Public equity financing

1. Feasible contracting intervals
   - A-type lender's perspective ($PL \geq P_L^+$)
     $E_L^\circ|S1 : [1.9, \infty); E_L^\circ|S2 : [1.6, \infty)$
   - N-type lender's perspective ($PL \geq 0$)
     $E_L^\circ|S1 : [1.6, \infty); E_L^\circ|S2 : [1.3, \infty)$
2. Profit comparison
   - A-type lender's perspective ($\overline{P}_L = P_L^+$)
     $E_L^\circ|S1$ is preferable to $E_L^\circ|S2$ for $\mu \geq 2.2$ and for $\mu < 2.2$ the other way round.
   - N-type lender's perspective ($\overline{P}_L = 0$)
     $E_L^\circ|S1$ is preferable to $E_L^\circ|S2$ for $\mu \geq 1.9$ and for $\mu < 1.9$ the other way round.
3. Preferred contract choice given public equity financing
   - A-type lender
     For $\mu < 1.6$: no choice; for $1.6 \leq \mu < 2.2 :$ $E_L^\circ|S2$ and for $\mu \geq 2.2 :$ $E_L^\circ|S1$.
   - N-type lender
     For $\mu < 1.3$: no choice; for $1.3 \leq \mu < 1.9 :$ $E_L^\circ|S2$ and for $\mu \geq 1.9 :$ $E_L^\circ|S1$.

Public debt financing (see derivation for PC1, Section 4.3)

3. Preferred contract choice given public debt financing
   - A-type lender
     For $\mu < 1.6$: no choice; for $1.6 \leq \mu < 1.726 :$ $D_{L,IIb}^\circ|S2$ and for $\mu \geq 1.726 :$ $D_{L,IIb}^\circ|S1$.
   - N-type lender
     For $\mu < 1.274$: no choice; for $1.274 \leq \mu < 1.368 :$ $D_{L,IIb}^\circ|S2$ and for $\mu \geq 1.368 :$ $D_{L,IIb}^\circ|S1$.

Private equity financing

1. Feasible contracting intervals
   - A-type lender's perspective ($PL \geq P_L^+$)
     $E_L^{\bullet(2)} : [2.0, \infty); E_L^{\bullet(1)} : [1.7, 2.0)$
   - N-type lender's perspective ($PL \geq 0$)
     $E_L^{\bullet(2)} : [1.7, \infty); E_L^{\bullet(1)} : [1.4, 1.7)$
2. No profit comparison required

3. Preferred contract choice given private equity financing
   - A-type lender
     For $\mu < 1.7$: no choice; for $1.7 \leq \mu < 2.0$: $E_L^{\bullet(1)}$ and for $\mu \geq 2.0$: $E_L^{\bullet(2)}$.
   - N-type lender
     For $\mu < 1.4$: no choice; for $1.4 \leq \mu < 1.7$: $E_L^{\bullet(1)}$ and for $\mu \geq 1.7$: $E_L^{\bullet(2)}$.

Private debt financing (see derivation for PC1, Section 4.3)

3. Preferred contract choice given private debt financing
   - A-type lender
     For $\mu < 1.653$: no choice; for $1.653 \leq \mu < 1.741$: $D_{L,IIa}^{\bullet(1)}$ or $D_{L,IIIb}^{\bullet(1)}$; for $1.741 \leq \mu \leq 3.2$: $D_{L,IIIb}^{\bullet(2)}$ and for $\mu > 3.2$: $D_{L,IIIc}^{\bullet(2)}$.
   - N-type lender
     For $\mu < 1.329$: no choice; for $1.329 \leq \mu < 1.4$: $D_{L,IIb}^{\bullet(1)}$; for $1.4 \leq \mu \leq 3.2$: $D_{L,IIIb}^{\bullet(2)}$ and for $\mu > 3.2$: $D_{L,IIIc}^{\bullet(2)}$.

THE LENDER'S BEHAVIOR AS THE ct/pm OPTIMIZER $(\widehat{\lambda}_B = \lambda_B)^2$

1. Feasible contracting intervals
   - A-type lender's perspective $(PL \geq P_L^+)$
     $E_L^\circ|S2 : [1.6, 2.2)$; $E_L^\circ|S1 : [2.2, \infty)$; $D_{L,IIb}^\circ|S2 : [1.6, 1.726)$; $D_{L,IIb}^\circ|S1 : [1.726, \infty)$; $E_L^{\bullet(1)} : [1.7, 2.0)$; $E_L^{\bullet(2)} : [2.0, \infty)$; $D_{L,IIa}^{\bullet(1)}$ and $D_{L,IIIb}^{\bullet(1)} : [1.653, 1.741)$; $D_{L,IIIb}^{\bullet(2)} : [1.741, 3.2]$; $D_{L,IIIc}^{\bullet(2)} : (3.2, \infty)$.
   - N-type lender's perspective $(PL \geq 0)$
     $E_L^\circ|S2 : [1.3, 1.9)$; $E_L^\circ|S1 : [1.9, \infty)$; $D_{L,IIb}^\circ|S2 : [1.274, 1.368)$; $D_{L,IIb}^\circ|S1 : [1.368, \infty)$; $E_L^{\bullet(1)} : [1.4, 1.7)$; $E_L^{\bullet(2)} : [1.7, \infty)$; $D_{L,IIIb}^{\bullet(2)} : [1.4, 3.2]$; $D_{L,IIIc}^{\bullet(2)} : (3.2, \infty)$.

2. Profit comparison
   - A-type lender's perspective $(\overline{P}_L = P_L^+)$
     $E_L^\circ|S2$ dominates: $E_L^{\bullet(1)}, D_{L,IIb}^\circ|S2, D_{L,IIa}^{\bullet(1)}$ and $D_{L,IIIb}^{\bullet(1)}$; $E_L^{\bullet(2)}$ dominates $E_L^\circ|S1$; $D_{L,IIb}^\circ|S1$ dominates $D_{L,IIIb}^{\bullet(2)}$; $D_{L,IIb}^\circ|S1$ is preferable to $E_L^\circ|S2$ for $\mu \geq 1.735$; $D_{L,IIb}^\circ|S1$ is preferable to $E_L^\circ|S1$ for $\mu < 19.2$; $D_{L,IIb}^\circ|S1$ is preferable to $E_L^{\bullet(2)}$ for $\mu < 8.533$; $D_{L,IIb}^\circ|S1$ is preferable to $d_{L,IIIc}^{\bullet(2)}$ for $\mu < 5.026$; $D_{L,IIIc}^{\bullet(2)}$ is preferable to $E_L^{\bullet(2)}$ for $\mu < 10.633$.
   - N-type lender's perspective $(\overline{P}_L = 0)$
     $D_{L,IIb}^\circ|S1$ dominates: $E_L^{\bullet(1)}, D_{L,IIIb}^{\bullet(2)}$; $E_L^{\bullet(2)}$ dominates $E_L^\circ|S1$; $E_L^\circ|S2$ is dominated by $D_{L,IIb}^\circ|S2$ and $D_{L,IIb}^\circ|S1$; $D_{L,IIb}^\circ|S2$ is preferable to

---

[2] We have to treat each ct/pm's contract choice interval-dependent since our profit comparison hinges on the assumption that the profit functions are continuously differentiable.

$E_L^\circ|S2$ for $\mu < 1.6$; $D_{L,IIb}^\circ|S1$ is preferable to $D_{L,IIb}^{\bullet(1)}$; $D_{L,IIb}^\circ|S1$ is preferable to $E_L^\circ|S1$ for $\mu < 19.2$; $D_{L,IIb}^\circ|S1$ is preferable to $E_L^{\bullet(2)}$ for $\mu < 8.533$; $D_{L,IIb}^\circ|S1$ is preferable to $D_{L,IIc}^{\bullet(2)}$ for $\mu < 5.026$; $D_{L,IIc}^{\bullet(2)}$ is preferable to $E_L^{\bullet(2)}$ for $\mu < 10.633$.

3. Preferred ct/pm-choices

Table B.3: Contract Type / Placement Mode-Choices and Agreement Probabilities in the Lender Scenario with Absolute Power (EM, PC2)

Interval	μ	Ct/pm-choice and agreement probabilities			
		A-type lender		N-type lender	
I	$[1; 1.274)$	$-$	(0)	$-$	(0)
$II - V$	$[1.274; 1.368)$	$-$	(0)	$D_{L,IIb}^\circ|S2$	$(\frac{1}{2})$
$VI - VII$	$[1.368; 1.6)$	$-$	(0)	$D_{L,IIb}^\circ|S1$	(1)
$VIII - XI$	$[1.6; 1.735)$	$E_L^\circ|S2$	$(\frac{1}{2})$	$D_{L,IIb}^\circ|S1$	(1)
$XII - XVII$	$[1.735; 5.026)$	$D_{L,IIb}^\circ|S1$	(1)	$D_{L,IIb}^\circ|S1$	(1)
XVIII	$[5.026; 10.633)$	$D_{L,IIIc}^{\bullet(2)}$	(1)	$D_{L,IIIc}^{\bullet(2)}$	(1)
XIX	$[10.633; \infty)$	$E_L^{\bullet(2)}$	(1)	$E_L^{\bullet(2)}$	(1)

Note: This table contains the lender's ct/pm-choice together with the related contract agreement probabilities stated in brackets.

B.1.2 Parameter Constellation 3

THE LENDER'S BEHAVIOR AS THE CONDITION OPTIMIZER $(\widehat{\lambda}_B = \lambda_B)$

Public equity financing (see derivation for PC1, Section 4.3)

3. Preferred contract choice given public equity financing
   - A-type lender
     For $\mu < 1.45$: no choice; for $1.45 \leq \mu < 2.05$: $E_L^\circ|S2$ and for $\mu \geq 2.05$: $E_L^\circ|S1$.
   - N-type lender
     For $\mu < 1.15$: no choice; for $1.15 \leq \mu < 1.75$: $E_L^\circ|S2$ and for $\mu \geq 1.75$: $E_L^\circ|S1$.

Public debt financing

1. Feasible contracting intervals
   - A-type lender's perspective $(PL \geq P_L^+)$
     $D_{L,Ia}^\circ : [5.2, \infty)$; $D_{L,IIb}^\circ|S1 : [1.812, \infty)$; $D_{L,IIb}^\circ|S2 : [1.668, \infty)$

- N-type lender's perspective $(PL \geq 0)$

  $D^{\circ}_{L,Ia} : [4, \infty); \ D^{\circ}_{L,IIb}|S1 : [1.493, \infty); \ D^{\circ}_{L,IIb}|S2 : [1.361, \infty)$

2. Profit comparison
   - A-type lender's perspective $(\overline{P}_L = P_L^+)$

     $D^{\circ}_{L,IIb}|S1$ dominates $D^{\circ}_{L,Ia}$; $D^{\circ}_{L,IIb}|S1$ is preferable to $D^{\circ}_{L,IIb}|S2$ for $\mu \geq 1.966$ and for $\mu < 1.966$ the other way round.
   - N-type lender's perspective $(\overline{P}_L = 0)$

     $D^{\circ}_{L,IIb}|S1$ dominates $D^{\circ}_{L,Ia}$; $D^{\circ}_{L,IIb}|S1$ is preferable to $D^{\circ}_{L,IIb}|S2$ for $\mu \geq 1.637$ and for $\mu < 1.637$ the other way round.

3. Preferred contract choice given public debt financing
   - A-type lender

     For $\mu < 1.668$: no choice; for $1.668 \leq \mu < 1.966 : \ D^{\circ}_{L,IIb}|S2$ and for $\mu \geq 1.966 : \ D^{\circ}_{L,IIb}|S1$.
   - N-type lender

     For $\mu < 1.361$: no choice; for $1.361 \leq \mu < 1.637 : \ D^{\circ}_{L,IIb}|S2$ and for $\mu \geq 1.637 : \ D^{\circ}_{L,IIb}|S1$.

Private equity financing (see derivation for PC1, Section 4.3)

3. Preferred contract choice given private equity financing
   - A-type lender

     For $\mu < 1.55$: no choice; for $1.55 \leq \mu < 1.85 : \ E^{\bullet(1)}_L$ and for $\mu \geq 1.85 : \ E^{\bullet(2)}_L$.
   - N-type lender

     For $\mu < 1.25$: no choice; for $1.25 \leq \mu < 1.55 : \ E^{\bullet(1)}_L$ and for $\mu \geq 1.55 : \ E^{\bullet(2)}_L$.

Private debt financing

1. Feasible contracting intervals
   - A-type lender's perspective $(PL \geq P_L^+)$

     $D^{\bullet(2)}_{L,Ia} : [5.4, \infty); \ D^{\bullet(2)}_{L,IIc} : [5.4, \infty); \ D^{\bullet(1)}_{L,IIc} : [1.721, 5.4); \ D^{\bullet(2)}_{L,IIIc} : [1.902, \infty); \ D^{\bullet(1)}_{L,IIIc} : [1.721, 1.902);$
   - N-type lender's perspective $(PL \geq 0)$

     $D^{\bullet(2)}_{L,Ia} : [4.2, \infty); \ D^{\bullet(2)}_{L,IIc} : [4.2, \infty); \ D^{\bullet(1)}_{L,IIc} : [1.413, 4.2); \ D^{\bullet(2)}_{L,IIIc} : [1.583, \infty); \ D^{\bullet(1)}_{L,IIIc} : [1.413, 1.583)$

2. Profit comparison[3]
   - A-type lender's perspective $(\overline{P}_L = P_L^+)$

     $D^{\bullet(2)}_{L,IIIc}$ dominates $D^{\bullet(2)}_{L,Ia}$ and $D^{\bullet(2)}_{L,IIc}$; $D^{\bullet(2)}_{L,IIIc}$ is preferable to $D^{\bullet(1)}_{L,IIc}$ for $\mu \geq 1.902$; $D^{\bullet(1)}_{L,IIc}$ and $D^{\bullet(1)}_{L,IIIc}$ generate the same expected profit.

---

[3] The results of this profit comparison coincide with the preferences stated for private debt financing on page 157.

- N-type lender's perspective $(\overline{P}_L = 0)$

  $D^{\bullet(2)}_{L,IIIc}$ dominates $D^{\bullet(2)}_{L,Ia}$ and $D^{\bullet(2)}_{L,IIc}$; $D^{\bullet(2)}_{L,IIIc}$ is preferable to $D^{\bullet(1)}_{L,IIc}$
  for $\mu \geq 1.583$; $D^{\bullet(1)}_{L,IIc}$ and $D^{\bullet(1)}_{L,IIIc}$ generate the same expected profit.

3. Preferred contract choice given private debt financing
   - A-type lender

     For $\mu < 1.721$: no choice; for $1.721 \leq \mu < 1.902$: $D^{\bullet(1)}_{L,IIc}$ or $D^{\bullet(1)}_{L,IIIc}$
     and for $\mu \geq 1.902$: $D^{\bullet(2)}_{L,IIIc}$.
   - N-type lender

     For $\mu < 1.413$: no choice; for $1.413 \leq \mu \leq 1.583$: $D^{\bullet(1)}_{L,IIc}$ or $D^{\bullet(1)}_{L,IIIc}$
     and for $\mu \geq 1.583$: $D^{\bullet(2)}_{L,IIIc}$.

THE LENDER'S BEHAVIOR AS THE ct/pm OPTIMIZER $\left(\widehat{\lambda}_B = \lambda_B\right)$[4]

1. Feasible contracting intervals
   - A-type lender's perspective $(PL \geq P_L^+)$

     $E^{\circ}_L|S2 : [1.45, 2.05)$; $E^{\circ}_L|S1 : [2.05, \infty)$; $D^{\circ}_{L,IIb}|S2 : [1.668, 1.966)$;
     $D^{\circ}_{L,IIb}|S1 : [1.966, \infty)$; $E^{\bullet(1)}_L : [1.55, 1.85)$; $E^{\bullet(2)}_L : [1.85, \infty)$; $D^{\bullet(1)}_{L,IIc}$
     and $D^{\bullet(1)}_{L,IIIc} : [1.722, 1.902)$; $D^{\bullet(2)}_{L,IIIc} : [1.902, \infty)$.
   - N-type lender's perspective $(PL \geq 0)$

     $E^{\circ}_L|S2 : [1.15, 1.75)$; $E^{\circ}_L|S1 : [1.75, \infty)$; $D^{\circ}_{L,IIb}|S2 : [1.361, 1.637)$;
     $D^{\circ}_{L,IIb}|S1 : [1.637, \infty)$; $E^{\bullet(1)}_L : [1.25, 1.55)$; $E^{\bullet(2)}_L : [1.55, \infty)$; $D^{\bullet(1)}_{L,IIc}$
     and $D^{\bullet(1)}_{L,IIIc} : [1.413, 1.583)$; $D^{\bullet(2)}_{L,IIIc} : [1.583, \infty)$.

2. Profit comparison
   - A-type lender's perspective $(\overline{P}_L = P_L^+)$

     $E^{\circ}_L|S2$ dominates: $E^{\bullet(1)}_L, D^{\bullet(1)}_{L,IIc}, D^{\bullet(1)}_{L,IIIc}, D^{\circ}_{L,IIb}|S2$;
     $E^{\bullet(2)}_L$ dominates: $E^{\circ}_L|S1, D^{\circ}_{L,IIb}|S1, D^{\bullet(2)}_{L,IIIc}$; $E^{\bullet(2)}_L$ is preferable to $E^{\circ}_L|S2$
     for $\mu \geq 1.95$; $E^{\circ}_L|S2$ is preferable to $D^{\bullet(2)}_{L,IIc}$ for $\mu < 2.196$.
   - N-type lender's perspective $(\overline{P}_L = 0)$

     $E^{\circ}_L|S2$ dominates: $E^{\bullet(1)}_L, D^{\circ}_{L,IIb}|S2, D^{\bullet(1)}_{L,IIc}, D^{\bullet(1)}_{L,IIIc}$;
     $E^{\bullet(2)}_L$ dominates: $E^{\circ}_L|S1, D^{\circ}_{L,IIIc}, D^{\circ}_{L,IIb}|S1$; $E^{\bullet(2)}_L$ is preferable to $E^{\circ}_L|S2$
     for $\mu \geq 1.65$; $E^{\circ}_L|S2$ is preferable to $D^{\bullet(2)}_{L,IIIc}$ for $\mu < 1.875$; $E^{\circ}_L|S2$ is
     preferable to $D^{\circ}_{L,IIb}|S1$ for $\mu < 1.881$.

---

[4] We have to treat each ct/pm's contract choice interval-dependent since our profit comparison hinges on the assumption that the profit functions are continuously differentiable.

3. Preferred ct/pm-choices

Table B.4: Contract Type / Placement Mode-Choices and Agreement Probabilities in the Lender Scenario with Absolute Power (EM, PC3)

Interval	μ	Ct/pm-choice and agreement probabilities			
		A-type lender		N-type lender	
I	$[1; 1.15)$	$-$	(0)	$-$	(0)
$II - V$	$[1.15; 1.45)$	$-$	(0)	$E_L^\circ \vert S2$	$(\frac{1}{2})$
$VI - XI$	$[145; 1.65)$	$E_L^\circ \vert S2$	$(\frac{1}{2})$	$E_L^\circ \vert S2$	$(\frac{1}{2})$
$XII - XIX$	$[1.65; 1.95)$	$E_L^\circ \vert S2$	$(\frac{1}{2})$	$E_L^{\bullet(2)}$	(1)
$XX - XXV$	$[1.95; \infty)$	$E_L^{\bullet(2)}$	(1)	$E_L^{\bullet(2)}$	(1)

Note: This table contains the lender's ct/pm-choice together with the related contract agreement probabilities stated in brackets.

B.2 Lender Scenario with Restricted Power (LS^r)

B.2.1 Parameter Constellation 2

EXPECTATION DERIVATION

Public equity financing ($\widehat{\lambda}_L \vert E^\circ \in \{0, \lambda_L\}$):

- for $\widehat{\lambda}_L \vert E^\circ = 0$
  the A-type borrower favors S2 to S1. Hence, he chooses $E_B^\circ \vert S2$ for $\mu \geq 1.6$.
  the N-type borrower favors S2 to S1. Hence, he chooses $E_B^\circ \vert S2$ for $\mu \geq 1.3$.
- for $\widehat{\lambda}_L \vert E^\circ = \lambda_L$
  the A-type borrower favors S1 to S2 for $\mu \geq 2.2$ and for $\mu < 2.2$ the other way round. Hence, he chooses $E_B^\circ \vert S2$ for $1.6 \leq \mu < 2.2$ and $E_B^\circ \vert S1$ for $\mu \geq 2.2$.
  the N-type borrower favors S1 to S2 for $\mu \geq 1.9$ and for $\mu < 1.9$ the other way round. Hence, he chooses $E_B^\circ \vert S2$ for $1.3 \leq \mu < 1.9$ and $E_B^\circ \vert S1$ for $\mu \geq 1.9$.

Public debt financing ($\widehat{\lambda}_L \vert D^\circ \in \{0, \lambda_L\}$) (see derivation for PC1, Section 4.3)

Private equity financing $(\widehat{\lambda}_L|E^\bullet \in \{\lambda_L, 1\}$ and $\widehat{\lambda}_L|E^\bullet = 0$ if the N-type borrower's BCOC(AL) is violated):

- for $\widehat{\lambda}_L|E^\bullet = 0^5$

  the N-type borrower chooses $E_B^{\bullet(1)}$ for $1.4 \le \mu < 1.7$.
- for $\widehat{\lambda}_L|E^\bullet = \lambda_L$

  the A-type borrower chooses $E_B^{\bullet(1)}$ for $1.7 \le \mu < 2.0$ and $E_B^{\bullet(2)}$ for $\mu \ge 2.0$.

  the N-type borrower chooses $E_B^{\bullet(1)}$ for $1.4 \le \mu < 1.7$ and $E_B^{\bullet(2)}$ for $\mu \ge 1.7$.
- for $\widehat{\lambda}_L|E^\bullet = 1^6$

  the A-type borrower only chooses $E_B^{\bullet(2)}$ for $\mu \ge 2.0$.

  the N-type borrower only chooses $E_B^{\bullet(2)}$ for $\mu \ge 1.7$.

Private debt financing $(\widehat{\lambda}_L|D^\bullet \in \{\lambda_L, 1\}$ and $\widehat{\lambda}_L|D^\bullet = 0$ if the N-type borrower's BCOC(AL) is not satisfied) (see derivation for PC1, Section 4.3)

---

[5] Since the consistency of $\widehat{\lambda}_L|E^\bullet = 0$ requires that the N-type borrower's BCOC(AL) is violated, i. e. the surplus from project financing $(\mu - 1)$ is not sufficient to cover the agency costs of private equity financing $(mc + 2nc)$ and the A-type lender's profit requirement (P_L^+), the A-type borrower never consistently expects $\widehat{\lambda}_L|E^\bullet = 0$. The A-type borrower's BPC can not be satisfied under these circumstances $(P_L^+ = P_B^+, \text{PC1})$.

[6] For $\widehat{\lambda}_L|E^\bullet = 1$ the borrower's alternative of offering $E^{\bullet(1)}$ is worthless since rejected by the (expected) A-type lender, hence only $E^{\bullet(2)}$ is feasible from the borrower's perspective.

Table B.5: Lender's Expected Profits and Contract Agreement Probabilities in the Lender Scenario with Restricted Power Depending on the Borrower's Expectation About the Lender's Type (EM, PC2)

Lender's ct/pm-choice	Borrower's expectation	Lender's expected profits and contract agreement probabilities			
		from	from	from	from
E°	$\hat{\lambda}_L|E^\circ = 0$	1.3 $0.3|0$ $(0|1-\lambda_B)$	1.6 $0.3|0$ $(0|1)$	1.9 $0.3|(1-\lambda_B)0.3$ $(1-\lambda_B|1)$	2.2 $0.3|0.3$ $(1|1)$
	$\hat{\lambda}_L|E^\circ = \lambda_L$	1.3 $0.3|0$ $(0|1-\lambda_B)$	1.6 $0.3|0$ $(0|1)$		
D°	$\hat{\lambda}_L|D^\circ = 0$	1.274 $0.3|0$ $(0|1-\lambda_B)$	1.318 $0.3|0$ $(0|1)$	1.680 $0.3|(1-\lambda_B)0.3$ $(1-\lambda_B|1)$	1.9 $0.3|0.3$ $(1|1)$
	$\hat{\lambda}_L|D^\circ = \lambda_L$	1.274 $0.3|0$ $(0|1-\lambda_B)$	1.318 $0.3|0$ $(0|1)$		
$E^\bullet$	$\hat{\lambda}_L|E^\bullet = 0$	1.4 $0.3|0$ $(0|1-\lambda_B)$	1.7 —		
	$\hat{\lambda}_L|E^\bullet = \lambda_L$	1.4 $0.3|0$ $(0|1-\lambda_B)$	1.7 $0.3|0$ $(1-\lambda_B|1)$	2.0 $0.3|0$ $(1|1)$	
	$\hat{\lambda}_L|E^\bullet = 1$	1.7 $0.3|0$ $(1-\lambda_B|1-\lambda_B)$	2.0 $0.3|0$ $(1|1)$		
$D^\bullet$	$\hat{\lambda}_L|D^\bullet = 0$	1.330 $0.3|0$ $(0|1-\lambda_B)$	1.4 $0.3|0$ $(0|1)$	1.653 —	
	$\hat{\lambda}_L|D^\bullet = \lambda_L$	1.330 $0.3|0$ $(0|1-\lambda_B)$	1.4 $0.3|0$ $(0|1)$	1.653 $0.3|0$ $(1-\lambda_B|1)$	1.741 $0.3|0$ $(1|1)$
	$\hat{\lambda}_L|D^\bullet = 1$	1.653 $0.3|0$ $(1-\lambda_B|1-\lambda_B)$	1.741 $0.3|0$ $(1|1)$		

Table B.6: Sets of Borrower's Consistent Conditional Expectations (SCCE) in the Lender Scenario with Restricted Power (EM, PC2)

μ	Sets of consistent conditional expectations
$[1.274, 1.3)$	$(-, 0, -, -)$
$[1.3, 1.318)$	$(0, 0, -, -)$
$[1.318, 1.4)$	$(-, 0, -, -)$
$[1.4, 1.6)$	$(-, 0, -, 0)$
$[1.6, 1.653)$	$(0, 0, -, 0)$
$[1.653, 1.7)$	$(0, 0, -, \lambda_L), (0, 0, -, 1), (-, \lambda_L, -, 1)$
$[1.7, 1.741)$	$(0, 0, \lambda_L, \lambda_L), (0, 0, 1, \lambda_L), (0, 0, \lambda_L, 1), (0, 0, 1, 1), (-, \lambda_L, 1, 1)$
$[1.741, 1.9)$	$(0, 0, -, \lambda_L), (-, \lambda_L, 1, 1)$
$[1.9, 2.0)$	$(0, 0, -, \lambda_L), (-, \lambda_L, -, 1)$
$[2.0, 2.2)$	$(0, 0, \lambda_L, \lambda_L), (-, \lambda_L, 1, 1)$
$[2.2, \infty)$	$(0, 0, \lambda_L, \lambda_L), (\lambda_L, \lambda_L, 1, 1)$

For $1.653 \leq \mu < 1.7$ $(0, 0, -, \lambda_L)$ is chosen by the borrower; for $1.7 \leq \mu < 1.741$: $(0, 0, \lambda_L, \lambda_L)$; for $1.741 \leq \mu < 2.0$: $(0, 0, -, \lambda_L)$ and for $\mu \geq 2.0$: $(0, 0, \lambda_L, \lambda_L)$.

CONTRACT TYPE / PLACEMENT MODE-CHOICE PREDICTIONS

Table B.7: Contract Type / Placement Mode-Choices and Agreement Probabilities in the Lender Scenario with Restricted Power (EM, PC2)

Interval	μ	Ct/pm-choice and agreement probabilities			
		A-type lender		N-type lender	
I	$[1, 1.274)$	$-$	(0)	$-$	(0)
II	$[1.274, 1.3)$	$-$	(0)	D°	$(\frac{1}{2})$
III	$[1.3, 1.318)$	$-$	(0)	E°, D°	$(\frac{1}{2})$
$IV - VI$	$[1.318, 1.4)$	$-$	(0)	D°	(1)
VII	$[1.4, 1.6)$	$-$	(0)	$D^\circ, D^\bullet$	(1)
VIII	$[1.6, 1.653)$	$-$	(0)	$E^\circ, D^\circ, D^\bullet$	(1)
$IX - X$	$[1.653, 1.7)$	$D^\bullet$	$(\frac{1}{2})$	$E^\circ, D^\circ, D^\bullet$	(1)
$XI - XII$	$[1.7, 1.741)$	$E^\bullet, D^\bullet$	$(\frac{1}{2})$	$E^\circ, D^\circ, E^\bullet, D^\bullet$	(1)
$XIII - XV$	$[1.741, 2.0)$	$D^\bullet$	(1)	$E^\circ, D^\circ, D^\bullet$	(1)
$XVI - XIX$	$[2.0, \infty)$	$E^\bullet, D^\bullet$	(1)	$E^\circ, D^\circ, E^\bullet, D^\bullet$	(1)

Note: This table contains the lender's ct/pm-choice together with the related contract agreement probabilities stated in brackets.

B.2.2 Parameter Constellation 3

EXPECTATION DERIVATION

Public equity financing ($\widehat{\lambda}_L | E^\circ \in \{0, \lambda_L\}$) (see derivation for PC1, Section 4.3)

Public debt financing ($\widehat{\lambda}_L | D^\circ \in \{0, \lambda_L\}$):

- for $\widehat{\lambda}_L | D^\circ = 0$
  the A-type borrower favors S2 to S1. Hence, he chooses $D^\circ_{B,II} | S2$ for $1.493 \leq \mu < 4$ and $D^\circ_{B,I} | S2$ for $\mu \geq 4$.
  the N-type borrower favors S2 to S1. Hence, he chooses $D^\circ_{B,II} | S2$ for $1.361 \leq \mu < 4$ and $D^\circ_{B,I} | S2$ for $\mu \geq 4$.

- for $\widehat{\lambda}_L | D^\circ = \lambda_L$
  the A-type borrower favors $D^\circ_{B,I} | S1$ to $D^\circ_{B,I} | S2$ for $\mu \geq 1.9$; $D^\circ_{B,II} | S1$ to $D^\circ_{B,I} | S2$ for $\mu \geq 2.163$; $D^\circ_{B,II} | S1$ to $D^\circ_{B,II} | S2$ for $\mu \geq 2.1$. Hence, he chooses $D^\circ_{B,II} | S2$ for $1.493 \leq \mu < 2.1$, $D^\circ_{B,II} | S1$ for $2.1 \leq \mu < 5.2$ and $D^\circ_{B,I} | S1$ for $\mu \geq 5.2$.
  the N-type borrower favors $D^\circ_{B,I} | S1$ to $D^\circ_{B,I} | S2$ for $\mu \geq 1.6$; $D^\circ_{B,II} | S1$ to $D^\circ_{B,I} | S2$ for $\mu \geq 1.945$; $D^\circ_{B,II} | S1$ to $D^\circ_{B,II} | S2$ for $\mu \geq 1.877$. Hence, he chooses $D^\circ_{B,II} | S2$ for $1.361 \leq \mu < 1.877$, $D^\circ_{B,II} | S1$ for $1.877 \leq \mu < 5.2$ and $D^\circ_{B,I} | S1$ for $\mu \geq 5.2$.

Private equity financing ($\widehat{\lambda}_L | E^\bullet \in \{\lambda_L, 1\}$ and $\widehat{\lambda}_L | E^\bullet = 0$ if the N-type borrower's BCOC(AL) is violated) (see derivation for PC1, Section 4.3)

Private debt financing ($\widehat{\lambda}_L | D^\bullet \in \{\lambda_L, 1\}$ and $\widehat{\lambda}_L | D^\bullet = 0$ if the N-type borrower's BCOC(AL) is not satisfied):

- for $\widehat{\lambda}_L | D^\bullet = 0$
  the A-type borrower chooses $D^{\bullet(1)}_{B,III}$ for $1.583 \leq \mu < 1.721$.[7]
  the N-type borrower chooses $D^{\bullet(1)}_{B,III}$ for $1.413 \leq \mu < 1.721$

- for $\widehat{\lambda}_L | D^\bullet = \lambda_L$
  the A-type borrower chooses $D^{\bullet(1)}_{B,III}$ for $1.583 \leq \mu < 1.902$, $D^{\bullet(2)}_{B,III}$ for $1.902 \leq \mu < 4.2$, $D^{\bullet(2)}_{B,II}$ for $4.2 \leq \mu < 5.4$ and $D^{\bullet(2)}_{B,I}$ for $\mu \geq 5.4$.
  the N-type borrower chooses $D^{\bullet(1)}_{B,III}$ for $1.413 \leq \mu < 1.721$, $D^{\bullet(2)}_{B,III}$ for $1.721 \leq \mu < 4.2$, $D^{\bullet(2)}_{B,II}$ for $4.2 \leq \mu < 5.4$ and $D^{\bullet(2)}_{B,I}$ for $\mu \geq 5.4$.

- for $\widehat{\lambda}_L | D^\bullet = 1$
  the A-type borrower chooses $D^{\bullet(2)}_{B,III}$ for $1.902 \leq \mu < 4.2$, $D^{\bullet(2)}_{B,II}$ for $4.2 \leq \mu < 5.4$ and $D^{\bullet(2)}_{B,I}$ for $\mu \geq 5.4$.
  the N-type borrower chooses $D^{\bullet(2)}_{B,III}$ for $1.721 \leq \mu < 4.2$, $D^{\bullet(2)}_{B,II}$ for $4.2 \leq \mu < 5.4$ and $D^{\bullet(2)}_{B,I}$ for $\mu \geq 5.4$.

---

[7] $D^{\bullet(1)}_{B,II}$ is restricted by the N-type borrower's BCOC(AL).

Table B.8: Lender's Expected Profits and Contract Agreement Probabilities in the Lender Scenario with Restricted Power Depending on the Borrower's Expectation About the Lender's Type (EM, PC3)

Lender's ct/pm-choice	Borrower's expectation	Lender's expected profits and contract agreement probabilities			
		from	from	from	from
E°	$\widehat{\lambda}_L \mid E^\circ = 0$	1.15 0.3|0 $(0\|1-\lambda_B)$	1.45 0.3|0 $(0\|1)$		
	$\widehat{\lambda}_L \mid E^\circ = \lambda_L$	1.15 0.3|0 $(0\|1-\lambda_B)$	1.45 0.3|0 $(0\|1)$	1.75 0.3|$(1-\lambda_B)$0.3 $(1-\lambda_B\|1)$	2.05 0.3|0.3 $(1\|1)$
D°	$\widehat{\lambda}_L \mid D^\circ = 0$	1.361 0.3|0 $(0\|1-\lambda_B)$	1.493 0.3|0 $(0\|1)$		
	$\widehat{\lambda}_L \mid D^\circ = \lambda_L$	1.361 0.3|0 $(0\|1-\lambda_B)$	1.493 0.3|0 $(0\|1)$	1.877 0.3|$(1-\lambda_B)$0.3 $(1-\lambda_B\|1)$	2.1 0.3|0.3 $(1\|1)$
$E^\bullet$	$\widehat{\lambda}_L \mid E^\bullet = 0$	1.25 0.3|0 $(0\|1-\lambda_B)$	1.55 —		
	$\widehat{\lambda}_L \mid E^\bullet = \lambda_L$	1.25 0.3|0 $(0\|1-\lambda_B)$	1.55 0.3|0 $(1-\lambda_B\|1)$	1.85 0.3|0 $(1\|1)$	
	$\widehat{\lambda}_L \mid E^\bullet = 1$	1.55 0.3|0 $(1-\lambda_B\|1-\lambda_B)$	1.85 0.3|0 $(1\|1)$		
$D^\bullet$	$\widehat{\lambda}_L \mid D^\bullet = 0$	1.413 0.3|0 $(0\|1-\lambda_B)$	1.583 0.3|0 $(0\|1)$	1.721 —	
	$\widehat{\lambda}_L \mid D^\bullet = \lambda_L$	1.413 0.3|0 $(0\|1-\lambda_B)$	1.583 0.3|0 $(0\|1)$	1.721 0.3|0 $(1-\lambda_B\|1)$	1.902 0.3|0 $(1\|1)$
	$\widehat{\lambda}_L \mid D^\bullet = 1$	1.721 0.3|0 $(1-\lambda_B\|1-\lambda_B)$	1.902 0.3|0 $(1\|1)$		

Table B.9: Sets of Borrower's Consistent Conditional Expectations (SCCE) in the Lender Scenario with Restricted Power (EM, PC3)

μ	Sets of consistent conditional expectations
$[1.15, 1.25)$	$(0, -, -, -)$
$[1.25, 1.361)$	$(0, -, 0, -)$
$[1.361, 1.413)$	$(0, 0, 0, -)$
$[1.413, 1.45)$	$(0, 0, 0, 0)$
$[1.45, 1.493)$	$(0, -, -, -)$
$[1.493, 1.55)$	$(0, 0, -, -)$
$[1.55, 1.574)$	$(0, 0, \lambda_L, -), (0, 0, 1, -)$
$[1.574, 1.721)$	$(0, 0, \lambda_L, 0), (0, 0, 1, 0)$
$[1.721, 1.75)$	$(0, 0, \lambda_L, \lambda_L), (0, 0, \lambda_L, 1), (0, 0, 1, 1)$
$[1.75, 1.85)$	$(0, 0, \lambda_L, \lambda_L), (\lambda_L, -, 1, 1)$
$[1.85, 1.902)$	$(0, 0, \lambda_L, -)$
$[1.902, 2.05)$	$(0, 0, \lambda_L, \lambda_L)$
$[2.05, 2.1)$	$(0, 0, \lambda_L, \lambda_L), (\lambda_L, -, 1, 1)$
$[2.1, \infty)$	$(0, 0, \lambda_L, \lambda_L), (\lambda_L, \lambda_L, 1, 1)$

In case of multiple sets of consistent conditional expectations the borrower chooses $(0, 0, \lambda_L, -)$ for $1.55 \leq \mu < 1.574$, $(0, 0, \lambda_L, 0)$ for $1.574 \leq \mu < 1.721$, $(0, 0, \lambda_L, \lambda_L)$ for $1.721 \leq \mu < 1.85$, $(0, 0, \lambda_L, -)$ for $1.85 \leq \mu < 1.902$ and $(0, 0, \lambda_L, \lambda_L)$ for $\mu \geq 1.902$.

CONTRACT TYPE / PLACEMENT MODE-CHOICE PREDICTIONS

Table B.10: Contract Type / Placement Mode-Choices and Agreement Probabilities in the Lender Scenario with Restricted Power (EM, PC3)

Interval	μ	Ct/pm-choice and agreement probabilities			
		A-type lender		N-type lender	
I	$[1,1.15)$	$-$	(0)	$-$	(0)
II	$[1.15,1.25)$	$-$	(0)	E°	$(\frac{1}{2})$
III	$[1.25,1.361)$	$-$	(0)	$E^\circ, E^\bullet$	$(\frac{1}{2})$
IV	$[1.361,1.413)$	$-$	(0)	$E^\circ, D^\circ, E^\bullet$	$(\frac{1}{2})$
V	$[1.413,1.45)$	$-$	(0)	$E^\circ, D^\circ, E^\bullet, D^\bullet$	$(\frac{1}{2})$
VI	$[1.45,1.493)$	$-$	(0)	E°	(1)
VII	$[1.493,1.55)$	$-$	(0)	E°, D°	(1)
VIII	$[1.55,1.574)$	$E^\bullet$	$(\frac{1}{2})$	$E^\circ, D^\circ, E^\bullet$	(1)
$IX - XIV$	$[1.574,1.721)$	$E^\bullet$	$(\frac{1}{2})$	$E^\circ, D^\circ, E^\bullet, D^\bullet$	(1)
$XV - XVI$	$[1.721,1.85)$	$E^\bullet, D^\bullet$	$(\frac{1}{2})$	$E^\circ, D^\circ, E^\bullet, D^\bullet$	(1)
XVII	$[1.85,1.902)$	$E^\bullet$	(1)	$E^\circ, D^\circ, E^\bullet$	(1)
$XVIII - XXV$	$[1.902,\infty)$	$E^\bullet, D^\bullet$	(1)	$E^\circ, D^\circ, E^\bullet, D^\bullet$	(1)

Note: This table contains the lender's ct/pm-choice together with the related contract agreement probabilities stated in brackets.

B.3 Borrower Scenario with Restricted Power (BS^r)

B.3.1 Parameter Constellation 2

EXPECTATION DERIVATION

Public equity financing $(\widehat{\lambda}_B|E^\circ = \lambda_B$ and $\widehat{\lambda}_B|E^\circ = 0$ if the N-type's LPC for D° with $\widehat{\lambda}_B|D^\circ \in \{0, \lambda_B\}$ is not satisfied):

- for $\widehat{\lambda}_B|E^\circ = 0$[8]
- for $\widehat{\lambda}_B|E^\circ = \lambda_B$
  the A-type lender favors S1 to S2 for $\mu \geq 2.2$ and for $\mu < 2.2$ the other

---

[8] The lender never consistently expects $\widehat{\lambda}_B|E^\circ = 0$ since once $E_B^\circ|S2$ becomes feasible ($\mu \geq 1.6$ from the A-type's and $\mu \geq 1.3$ from the N-type's perspective) the N-type's LPC for D° with $\widehat{\lambda}_B|D^\circ \in \{0, \lambda_B\}$ is satisfied. D° with $\widehat{\lambda}_B|D^\circ \in \{0, \lambda_B\}$ is feasible if $\mu \geq 1.274$.

way round. Hence, he chooses $E_L^\circ|S2$ for $1.6 \leq \mu < 2.2$ and $E_L^\circ|S1$ for $\mu \geq 2.2$.

the N-type lender favors S1 to S2 for $\mu \geq 1.9$ and for $\mu < 1.9$ the other way round. Hence, he chooses $E_L^\circ|S2$ for $1.3 \leq \mu < 1.9$ and $E_L^\circ|S1$ for $\mu \geq 1.9$.

Public debt financing $(\widehat{\lambda}_B|D^\circ \in \{0, \lambda_B\})$ (see derivation for PC1, Section 4.3)

Private equity financing $(\widehat{\lambda}_B|E^\bullet \in \{\lambda_B, 1\}$ and $\widehat{\lambda}_B|E^\bullet = 0$ if the N-type lender's participation constraint for D° and $\widehat{\lambda}_B|D^\circ \in \{0, \lambda_B\}$ is not satisfied):

- for $\widehat{\lambda}_B|E^\bullet = 0$
  not possible under the imposed constraint.
- for $\widehat{\lambda}_B|E^\bullet = \lambda_B$
  the A-type lender chooses $E_L^{\bullet(1)}$ for $1.7 \leq \mu < 2.0$ and $E_L^{\bullet(2)}$ for $\mu \geq 2.0$.
  the N-type lender chooses $E_L^{\bullet(1)}$ for $1.4 \leq \mu < 1.7$ and $E_L^{\bullet(2)}$ for $\mu \geq 1.7$.
- for $\widehat{\lambda}_B|E^\bullet = 1$
  the A-type lender only chooses $E_L^{\bullet(2)}$ for $\mu \geq 2.0$.
  the N-type lender only chooses $E_L^{\bullet(2)}$ for $\mu \geq 1.7$.

Private debt financing $(\widehat{\lambda}_B|D^\bullet \in \{\lambda_B, 1\}$ and $\widehat{\lambda}_B|D^\bullet = 0$ if the N-type lender's LCOC(AB) for $D^\bullet$ is violated) (see derivation for PC1, Section 4.3)

Table B.11: Borrower's Expected Profits and Contract Agreement Probabilities in the Borrower Scenario with Restricted Power Depending on the Lender's Expectation About the Borrower's Type (EM, PC2)

Borrower's ct/pm-choice	Lender's expectation	Borrower's expected profits and contract agreement probabilities		
		from	from	from
E°	$\widehat{\lambda}_B \mid E^\circ = \lambda_B$	1.3 $0.3\mid0$ $(0\mid1-\lambda_L)$ 2.2 $0.3\mid0.3$ $(1\mid1)$	1.6 $0.3\mid0$ $(0\mid1)$	1.9 $0.3\mid(1-\lambda_L)0.3$ $(1-\lambda_L\mid1)$
D°	$\widehat{\lambda}_B \mid D^\circ = 0$	1.274 $0.3\mid(1-\lambda_L)\frac{0.16}{\mu}$ $(0\mid1-\lambda_L)$	1.6 $0.3\mid\frac{0.16}{\mu}$ $(0\mid1)$	
	$\widehat{\lambda}_B \mid D^\circ = \lambda_B$	1.274 $0.3\mid(1-\lambda_L)\frac{0.16}{\mu}$ $(0\mid1-\lambda_L)$ 1.726 $0.3\mid0.3$ $(1\mid1)$	1.368 $0.3\mid(1-\lambda_L)0.3$ $(1-\lambda_L\mid1-\lambda_L)$	1.6 $0.3\mid(1-\lambda_L)0.3+\lambda_L\frac{0.16}{\mu}$ $(1-\lambda_L\mid1)$
$E^\bullet$	$\widehat{\lambda}_B \mid E^\bullet = \lambda_B$	1.4 $0.3\mid0$ $(0\mid1-\lambda_L)$	1.7 $0.3\mid0$ $(1-\lambda_L\mid1)$	2.0 $0.3\mid0$ $(1\mid1)$
	$\widehat{\lambda}_B \mid E^\bullet = 1$	1.7 $0.3\mid0$ $(1-\lambda_L\mid1-\lambda_L)$	2.0 $0.3\mid0$ $(1\mid1)$	
$D^\bullet$	$\widehat{\lambda}_B \mid D^\bullet = 0$	1.329 $0.3\mid(1-\lambda_L)\frac{0.16}{\mu}$ $(0\mid1-\lambda_L)$	1.4 —	1.653 $0.3\mid\frac{0.16}{\mu}-0.05$ $(1-\lambda_L\mid1)$
	$\widehat{\lambda}_B \mid D^\bullet = \lambda_B$	1.329 $0.3\mid(1-\lambda_L)\frac{0.16}{\mu}$ $(0\mid1-\lambda_L)$ 1.741 $0.3\mid\frac{0.16}{\mu}-0.05$ $(1\mid1)$	1.4 $0.3\mid(1-\lambda_L)\frac{0.16}{\mu}$ $(1-\lambda_L\mid1-\lambda_L)$ 3.2 $0.3\mid\frac{0.16}{\mu}-0.05$ $(1\mid1)$	
	$\widehat{\lambda}_B \mid D^\bullet = 1$	1.4 $0.3\mid(1-\lambda_L)\frac{0.16}{\mu}$ $(1-\lambda_L\mid1-\lambda_L)$	3.2 $0.3\mid\frac{0.16}{\mu}-0.05$ $(1\mid1)$	3.2 $0.3\mid0$ $(1\mid1)$

Table B.12: Sets of Lender's Consistent Conditional Expectations (SCCE) in the Borrower Scenario with Restricted Power (EM, PC2)

μ	Sets of consistent conditional expectations
$[1.274, 1.368)$	$(-, 0, -, -)$
$[1.368, 1.4)$	$(-, 0, -, -), (-, \lambda_B, -, -)$
$[1.4, 1.7)$	$(-, 0, -, 1), (-, \lambda_B, -, 1)$
$[1.7, 1.726)$	$(-, 0, 1, 1), (-, \lambda_B, 1, 1)$
$[1.726, 1.741)$	$(-, 0, 1, 1), (-, \lambda_B, -, -)$
$[1.741, 1.9)$	$(-, 0, -, 1), (-, \lambda_B, -, 1)$
$[1.9, 2.0)$	$(-, \lambda_B, -, 1)$
$[2.0, 2.2)$	$(-, \lambda_B, 1, 1)$
$[2.2, \infty)$	$(\lambda_B, \lambda_B, 1, 1)$

In case of multiple SCCEs the lender chooses $(-, \lambda_B, -, -)$ for $1.368 \leq \mu < 1.4$, $(-, 0, -, 1)$ for $1.4 \leq \mu < 1.7$, $(-, 0, 1, 1)$ for $1.7 \leq \mu < 1.741$ and $(-, 0, -, 1)$ for $1.741 \leq \mu < 1.9$.

CONTRACT TYPE / PLACEMENT MODE-CHOICE PREDICTIONS

Table B.13: Contract Type / Placement Mode-Choices and Agreement Probabilities in the Borrower Scenario with Restricted Power (EM, PC2)

Interval	μ	Ct/pm-choice and agreement probabilities			
		A-type borrower		N-type borrower	
I	$[1, 1.274)$	$-$	(0)	$-$	(0)
$II - V$	$[1.274, 1.368)$	$-$	(0)	D°	$(\frac{1}{2})$
VI	$[1.368, 1.4)$	D°	$(\frac{1}{2})$	D°	$(\frac{1}{2})$
VII	$[1.4, 1.6)$	$D^\bullet$	$(\frac{1}{2})$	D°	$(\frac{1}{2})$
$VIII - X$	$[1.6, 1.7)$	$D^\bullet$	$(\frac{1}{2})$	D°	(1)
$XI - XII$	$[1.7, 1.741)$	$E^\bullet, D^\bullet$	$(\frac{1}{2})$	D°	(1)
$XIII - XIV$	$[1.741, 1.9)$	$D^\bullet$	(1)	D°	(1)
XV	$[1.9, 2.0)$	$D^\circ, D^\bullet$	(1)	D°	(1)
XVI	$[2.0, 2.2)$	$D^\circ, E^\bullet, D^\bullet$	(1)	D°	(1)
$XVII - XIX$	$[2.2, \infty)$	$E^\circ, D^\circ, E^\bullet, D^\bullet$	(1)	E°, D°	(1)

Note: This table contains the borrower's ct/pm-choice together with the related contract agreement probabilities stated in brackets.

B.3.2 Parameter Constellation 3

EXPECTATION DERIVATION

Public equity financing ($\widehat{\lambda}_B|E^\circ = \lambda_B$ and $\widehat{\lambda}_B|E^\circ = 0$ if the N-type's LPC for D° with $\widehat{\lambda}_B|D^\circ \in \{0, \lambda_B\}$ is not satisfied) (see derivation for PC1, Section 4.3)[9]

Public debt financing ($\widehat{\lambda}_B|D^\circ \in \{0, \lambda_B\}$):

- for $\widehat{\lambda}_B|D^\circ = 0$
  the A-type lender favors S2 to S1 and $D^\circ_{L,IIb}|S1$ dominates $D^\circ_{L,Ia}$. Hence, he chooses $D^\circ_{L,IIb}|S2$ for $\mu \geq 1.667$.
  the N-type lender favors S2 to S1 and $D^\circ_{L,IIb}|S1$ dominates $D^\circ_{L,Ia}$. Hence, he chooses $D^\circ_{L,IIb}|S2$ for $\mu \geq 1.361$.

- for $\widehat{\lambda}_B|D^\circ = \lambda_B$
  the A-type lender favors $D^\circ_{L,IIb}|S1$ to $D^\circ_{L,IIb}|S2$ for $\mu \geq 1.966$ and $D^\circ_{L,IIb}|S1$ dominates $D^\circ_{L,Ia}$. Hence, he chooses $D^\circ_{L,IIb}|S2$ for $1.667 \leq \mu < 1.966$ and $D^\circ_{L,IIb}|S1$ for $\mu \geq 1.966$.
  the N-type lender favors $D^\circ_{L,IIb}|S1$ to $D^\circ_{L,IIb}|S2$ for $\mu \geq 1.637$ and $D^\circ_{L,IIb}|S1$ dominates $D^\circ_{L,Ia}$. Hence, he chooses $D^\circ_{L,IIb}|S2$ for $1.361 \leq \mu < 1.637$ and $D^\circ_{L,IIb}|S1$ for $\mu \geq 1.637$.

Private equity financing ($\widehat{\lambda}_B|E^\bullet \in \{\lambda_B, 1\}$ and $\widehat{\lambda}_B|E^\bullet = 0$ if the N-type lender's participation constraint for D° and $\widehat{\lambda}_B|D^\circ \in \{0, \lambda_B\}$ is not satisfied) (see derivation for PC1, Section 4.3)

Private debt financing ($\widehat{\lambda}_B|D^\bullet \in \{\lambda_B, 1\}$ and $\widehat{\lambda}_B|D^\bullet = 0$ if the N-type lender's LCOC(AB) for $D^\bullet$ is violated):

- for $\widehat{\lambda}_B|D^\bullet = 0$
  the A-type lender never consistently expects $\widehat{\lambda}_B|D^\bullet = 0$; the N-type lender chooses $D^{\bullet(1)}_{L,IIIc}$ for $1.427 \leq \mu < 1.583$.

- for $\widehat{\lambda}_L|D^\bullet = \lambda_L$.
  the lender can either offer both borrower types a contract, only the N-type borrower, or no contract at all. Due to the lender's preference analysis for private debt financing (see page 157) we realize that the A-type lender chooses $D^{\bullet(1)}_{L,IIc}$ or $D^{\bullet(1)}_{L,IIIc}$ for $1.741 \leq \mu < 1.902$, and $D^{\bullet(2)}_{L,IIIc}$ for $\mu \geq 1.902$.
  the N-type lender chooses $D^{\bullet(1)}_{L,IIc}$ or $D^{\bullet(1)}_{L,IIIc}$ for $1.42 \leq \mu < 1.583$ and $D^{\bullet(2)}_{L,IIIc}$ for $\mu \geq 1.583$.

---

[9] In comparison to PC1, the restriction of $\widehat{\lambda}_L|E^\circ = 0$ to be consistent lifts from $\mu < 1.274$ to $\mu < 1.361$.

- for $\widehat{\lambda}_B | D^\bullet = 1$

  the A-type lender either finances the expected A-type borrower with risk-less or risky debt, $D^{\bullet(2)}_{L,Ia}, D^{\bullet(2)}_{L,IIa}, D^{\bullet(2)}_{L,IIc}$ dominated by $D^{\bullet(2)}_{L,IIIb}, D^{\bullet(2)}_{L,IIIc}$. Hence, he chooses $D^{\bullet(2)}_{L,IIIc}$ for $\mu \geq 1.902$.

  the N-type lender chooses $D^{\bullet(2)}_{L,IIIc}$ for $\mu \geq 1.902$.

Table B.14: Borrower's Expected Profits and Contract Agreement Probabilities in the Borrower Scenario with Restricted Power Depending on the Lender's Expectation About the Borrower's Type (EM, PC3)

Borrower's ct/pm-choice	Lender's expectation	Borrower's expected profits and contract agreement probabilities			
		from	from	from	from
E°	$\widehat{\lambda}_B\mid E^{\circ} = 0$	1.15 $\frac{0.3\mid 0}{(0\mid 1-\lambda_L)}$	1.361 —		
	$\widehat{\lambda}_B\mid E^{\circ} = \lambda_B$	1.15 $\frac{0.3\mid 0}{(0\mid 1-\lambda_L)}$	1.45 $\frac{0.3\mid 0}{(0\mid 1)}$	1.75 $\frac{0.3\mid(1-\lambda_L)0.3}{(1-\lambda_L\mid 1)}$	2.05 $\frac{0.3\mid 0.3}{(1\mid 1)}$
D°	$\widehat{\lambda}_B\mid D^{\circ} = 0$	1.361 $\frac{0.3\mid(1-\lambda_L)\frac{0.16}{\mu}}{(0\mid 1-\lambda_L)}$	1.667 $\frac{0.3\mid\frac{0.16}{\mu}}{(0\mid 1)}$		
	$\widehat{\lambda}_B\mid D^{\circ} = \lambda_B$	1.361 $\frac{0.3\mid(1-\lambda_L)\frac{0.16}{\mu}}{(0\mid 1-\lambda_L)}$	1.637 $\frac{0.3\mid(1-\lambda_L)0.3}{(1-\lambda_L\mid 1-\lambda_L)}$	1.667 $\frac{0.3\mid(1-\lambda_L)0.3 + \lambda_L\frac{0.16}{\mu}}{(1-\lambda_L\mid 1)}$	1.966 $\frac{0.3\mid 0.3}{(1\mid 1)}$
$E^{\bullet}$	$\widehat{\lambda}_B\mid E^{\bullet} = \lambda_B$	1.25 $\frac{0.3\mid 0}{(0\mid 1-\lambda_L)}$	1.55 $\frac{0.3\mid 0}{(1-\lambda_L\mid 1)}$	1.85 $\frac{0.3\mid 0}{(1\mid 1)}$	
	$\widehat{\lambda}_B\mid E^{\bullet} = 1$	1.55 $\frac{0.3\mid 0}{(1-\lambda_L\mid 1-\lambda_L)}$	1.85 $\frac{0.3\mid 0}{(1\mid 1)}$		
$D^{\bullet}$	$\widehat{\lambda}_B\mid D^{\bullet} = 0$	1.427 $\frac{0.3\mid 0}{(0\mid 1-\lambda_L)}$	1.583 —		
	$\widehat{\lambda}_B\mid D^{\bullet} = \lambda_B$	1.42 $\frac{0.3\mid 0}{(0\mid 1-\lambda_L)}$	1.583 $\frac{0.3\mid 0}{(1-\lambda_L\mid 1-\lambda_L)}$	1.741 $\frac{0.3\mid 0}{(1-\lambda_L\mid 1)}$	1.902 $\frac{0.3\mid 0}{(1\mid 1)}$
	$\widehat{\lambda}_B\mid D^{\bullet} = 1$	1.902 $\frac{0.3\mid 0}{(1\mid 1)}$			

Table B.15: Sets of Lender's Consistent Conditional Expectations (SCCE) in the Borrower Scenario with Restricted Power (EM, PC3)

μ	Sets of consistent conditional expectations
$[1.15, 1.361)$	$(0, -, -, -)$
$[1.361, 1.55)$	$(-, 0, -, -)$
$[1.55, 1.637)$	$(-, 0, 1, -)$
$[1.637, 1.75)$	$(-, 0, 1, -), (-, \lambda_B, 1, -)$
$[1.75, 1.85)$	$(\lambda_B, -, 1, -)$
$[1.85, 1.966)$	$(\lambda_B, \lambda_B, 1, 1)$
$[1.966, 2.05)$	$(-, \lambda_B, 1, 1)$
$[2.05, \infty)$	$(\lambda_B, \lambda_B, 1, 1)$

For $1.637 \leq \mu < 1.75$ the lender chooses $(-, 0, 1, -)$.

CONTRACT TYPE / PLACEMENT MODE-CHOICE PREDICTIONS

Table B.16: Contract Type / Placement Mode-Choices and Agreement Probabilities in the Borrower Scenario with Restricted Power (EM, PC3)

Interval	μ	Ct/pm-choice and agreement probabilities			
		A-type borrower		N-type borrower	
I	$[1,1.15)$	$-$	(0)	$-$	(0)
$II - III$	$[1.15,1.361)$	$-$	(0)	E°	$(\frac{1}{2})$
$IV - VII$	$[1.361,1.55)$	$-$	(0)	D°	$(\frac{1}{2})$
$VIII - XIII$	$[1.55,1.667)$	$E^\bullet$	$(\frac{1}{2})$	D°	$(\frac{1}{2})$
$XIV - XV$	$[1.667,1.75)$	$E^\bullet$	$(\frac{1}{2})$	D°	(1)
XVI	$[1.75,1.85)$	$E^\circ, E^\bullet$	$(\frac{1}{2})$	E°	(1)
$XVII - XX$	$[1.85,1.966)$	$E^\circ, D^\circ, E^\bullet, D^\bullet$	(1)	E°, D°	(1)
$XXI - XXII$	$[1.966,2.05)$	$D^\circ, E^\bullet, D^\bullet$	(1)	D°	(1)
$XXIII - XXV$	$[2.05,\infty)$	$E^\circ, D^\circ, E^\bullet, D^\bullet$	(1)	E°, D°	(1)

Note: This table contains the borrower's ct/pm-choice together with the related contract agreement probabilities stated in brackets.

B.4 Borrower Scenario with Absolute Power (BS^a)

B.4.1 Parameter Constellation 2

THE BORROWER'S BEHAVIOR AS THE CONDITION OPTIMIZER ($\widehat{\lambda}_L = \lambda_L$)

Public equity financing

1. Feasible contracting intervals
    - A-type borrower's perspective ($PB \geq P_B^+$)
      $E_B^\circ|S1 : [1.9, \infty)$; $E_B^\circ|S2 : [1.6, \infty)$
    - N-type borrower's perspective ($PB \geq 0$)
      $E_B^\circ|S1 : [1.6, \infty)$; $E_B^\circ|S2 : [1.3, \infty)$
2. Profit comparison
    - A-type borrower's perspective ($\overline{P}_B = P_B^+$)
      $E_B^\circ|S1$ is preferable to $E_B^\circ|S2$ for $\mu \geq 2.2$ and for $\mu < 2.2$ the other way round.
    - N-type borrower's perspective ($\overline{P}_B = 0$)
      $E_B^\circ|S1$ is preferable to $E_B^\circ|S2$ for $\mu \geq 1.9$ and for $\mu < 1.9$ the other way round.
3. Preferred contract choice given public equity financing
    - A-type borrower
      For $\mu < 1.6$: no choice; for $1.6 \leq \mu < 2.2$: $E_B^\circ|S2$ and for $\mu \geq 2.2$: $E_B^\circ|S1$.
    - N-type borrower
      For $\mu < 1.3$: no choice; for $1.3 \leq \mu < 1.9$: $E_B^\circ|S2$ and for $\mu \geq 1.9$: $E_B^\circ|S1$.

Public debt financing (see derivation for PC1, Section 4.3)

3. Preferred contract choice given public debt financing
    - A-type borrower
      For $\mu < 1.318$: no choice; for $1.318 \leq \mu < 1.\overline{3}$: $D_{B,II}^\circ|S2$; for $1.\overline{3} \leq \mu < 1.9$: $D_{B,I}^\circ|S2$ and for $\mu \geq 1.9$: $D_{B,I}^\circ|S1$.
    - N-type borrower
      For $\mu < 1.274$: no choice; for $1.274 \leq \mu < 1.\overline{3}$: $D_{B,II}^\circ|S2$ for $1.\overline{3} \leq \mu < 1.68$: $D_{B,I}^\circ|S2$; for $1.68 \leq \mu < 1.7\overline{3}$: $D_{B,II}^\circ|S1$ and for $\mu \geq 1.7\overline{3}$: $D_{B,I}^\circ|S1$.

Private equity financing

1. Feasible contracting intervals
    - A-type borrower's perspective ($PB \geq P_B^+$)
      $E_B^{\bullet(2)} : [2.0, \infty)$; $E_B^{\bullet(1)} : [1.7, 2.0)$
    - N-type borrower's perspective ($PB \geq 0$)
      $E_B^{\bullet(2)} : [1.7, \infty)$; $E_B^{\bullet(1)} : [1.4, 1.7)$
2. No profit comparison required

3. Preferred contract choice given private equity financing
   - A-type borrower
     For $\mu < 1.7$: no choice; for $1.7 \leq \mu < 2.0$: $E_B^{\bullet(1)}$ and for $\mu \geq 2.0$: $E_B^{\bullet(2)}$.
   - N-type borrower
     For $\mu < 1.4$: no choice; for $1.4 \leq \mu < 1.7$: $E_B^{\bullet(1)}$ and for $\mu \geq 1.7$: $E_B^{\bullet(2)}$.

Private debt financing (see derivation for PC1, Section 4.3)

3. Preferred choice given private debt financing
   - A-type borrower
     For $\mu < 1.4$: no choice; for $1.4 \leq \mu < 1.741$: $D_{B,II}^{\bullet(1)}$; for $1.741 \leq \mu < 1.8$: $D_{B,II}^{\bullet(2)}$ and for $\mu \geq 1.8$: $D_{B,I}^{\bullet(2)}$.
   - N-type borrower
     For $\mu < 1.330$: no choice; for $1.330 \leq \mu < 1.4$: $D_{B,III}^{\bullet(1)}$; for $1.4 \leq \mu < 1.653$: $D_{B,II}^{\bullet(1)}$; for $1.653 \leq \mu < 1.8$: $D_{B,II}^{\bullet(2)}$ an for $\mu \geq 1.8$: $D_{B,I}^{\bullet(2)}$.

THE BORROWER'S BEHAVIOR AS THE ct/pm OPTIMIZER $(\widehat{\lambda}_L = \lambda_L)$[10]

1. Feasible contracting intervals
   - A-type borrower's perspective $(PB \geq P_B^+)$
     $E_B^\circ|S2 : [1.6, 2.2)$; $E_B^\circ|S1 : [2.2, \infty)$; $D_{B,II}^\circ|S2 : [1.318, 1.\bar{3})$; $D_{B,I}^\circ|S2 : [1.\bar{3}, 1.9)$; $D_{B,I}^\circ|S1 : [1.9, \infty)$; $E_B^{\bullet(1)} : [1.7, 2.0)$; $E_B^{\bullet(2)} : [2.0, \infty)$; $D_{B,II}^{\bullet(1)} : [1.4, 1.741)$; $D_{B,II}^{\bullet(2)} : [1.741, 1.8)$; $D_{B,I}^{\bullet(2)} : [1.8, \infty)$
   - N-type borrower's perspective $(PB \geq 0)$
     $E_B^\circ|S2 : [1.3, 1.9)$; $E_B^\circ|S1 : [1.9, \infty)$; $D_{B,II}^\circ|S2 : [1.274, 1.\bar{3})$; $D_{B,I}^\circ|S2 : [1.\bar{3}, 1.68)$; $D_{B,II}^\circ|S1 : [1.68, 1.7\bar{3})$; $D_{B,I}^\circ|S1 : [1.7\bar{3}, \infty)$; $E_B^{\bullet(1)} : [1.4, 1.7)$; $E_B^{\bullet(2)} : [1.7, \infty)$; $D_{B,III}^{\bullet(1)} : [1.330, 1.4)$; $D_{B,II}^{\bullet(1)} : [1.4, 1.653)$; $D_{B,II}^{\bullet(2)} : [1.653, 1.8)$; $D_{B,I}^{\bullet(2)} : [1.8, \infty)$
2. Profit comparison
   - A-type borrower's perspective $(\overline{P}_B = P_B^+)$
     $D_{B,I}^\circ|S2$ dominates: $D_{B,II}^{\bullet(1)}, D_{B,II}^{\bullet(2)}$; $D_{B,I}^{\bullet(2)}$ dominates: $E_B^\circ|S1, D_{B,I}^\circ|S1$, $E_B^{\bullet(2)}$; $D_{B,I}^\circ|S2$ is preferable to $E_B^\circ|S2, E_B^{\bullet(1)}$; $D_{B,I}^{\bullet(2)}$ is preferable to $D_{B,I}^\circ|S2$ for $\mu \geq 1.8$; $D_{B,I}^{\bullet(2)}$ is preferable to $E_B^\circ|S2$.
   - N-type borrower's perspective $(\overline{P}_B = 0)$
     $D_{B,I}^\circ|S2$ dominates $D_{B,I}^{\bullet(1)}$; $D_{B,II}^{\bullet(2)}$ dominates $D_{B,II}^\circ|S1$; $D_{B,I}^{\bullet(2)}$ dominates $E_B^\circ|S1$; $D_{B,II}^\circ|S2$ is preferable to $E_B^\circ|S2, D_{B,III}^{\bullet(1)}$; $D_{B,I}^\circ|S2$ is preferable

---

[10] We have to treat each ct/pm's contract choice interval-dependent since our profit comparison hinges on the assumption that the profit functions are continuously differentiable.

to $E^\circ_B|S2, D^{\bullet(1)}_{B,III}, E^{\bullet(1)}_B$; $D^{\bullet(2)}_{B,II}$ is preferable to $D^\circ_{B,I}|S2$ for $\mu \geq 1.659$; $D^{\bullet(2)}_{B,II}$ is preferable to $E^\circ_B|S2, E^{\bullet(2)}_B, D^\circ_{B,I}|S1$; $D^{\bullet(2)}_{B,I}$ is preferable to $D^\circ_{B,I}|S1, E^{\bullet(2)}_B$.

3. Preferred ct/pm-choices

Table B.17: Contract Type / Placement Mode-Choices and Agreement Probabilities in the Borrower Scenario with Absolute Power (EM, PC2)

Interval	μ	Ct/pm-choices and agreement probabilities			
		A-type borrower		N-type borrower	
I	$[1; 1.274)$	$-$	(0)	$-$	(0)
$II - III$	$[1.274; 1.318)$	$-$	(0)	$D^\circ_{B,II}|S2$	$(\frac{1}{2})$
IV	$[1.318; 1.\bar{3})$	$D^\circ_{B,II}|S2$	$(\frac{1}{2})$	$D^\circ_{B,II}|S2$	$(\frac{1}{2})$
$V - IX$	$[1.\bar{3}; 1.659)$	$D^\circ_{B,I}|S2$	$(\frac{1}{2})$	$D^\circ_{B,I}|S2$	$(\frac{1}{2})$
$X - XIII$	$[1.659; 1.8)$	$D^\circ_{B,I}|S2$	$(\frac{1}{2})$	$D^{\bullet(2)}_{B,II}$	(1)
$XIV - XIX$	$[1.8; \infty)$	$D^{\bullet(2)}_{B,I}$	(1)	$D^{\bullet(2)}_{B,I}$	(1)

Note: This table contains the borrower's ct/pm-choice together with the related contract agreement probabilities stated in brackets.

B.4.2 Parameter Constellation 3

THE BORROWER'S BEHAVIOR AS THE CONDITION OPTIMIZER $(\widehat{\lambda}_L = \lambda_L)$

Public equity financing (see derivation for PC1, Section 4.3)

3. Preferred contract choice given public equity financing
   - A-type borrower
     For $\mu < 1.45$: no choice; for $1.45 \leq \mu < 2.05$: $E^\circ_B|S2$ and for $\mu \geq 2.05$: $E^\circ_B|S1$.
   - N-type borrower
     For $\mu < 1.15$: no choice; for $1.15 \leq \mu < 1.75$: $E^\circ_B|S2$ and for $\mu \geq 1.75$: $E^\circ_B|S1$.

Public debt financing

1. Feasible contracting intervals
   - A-type borrower's perspective $(PB \geq P^+_B)$
     $D^\circ_{B,I}|S1 : [5.2, \infty)$; $D^\circ_{B,I}|S2 : [4, \infty)$; $D^\circ_{B,II}|S1 : [1.812, 5.2)$; $D^\circ_{B,II}|S2 : [1.493, 4)$

- N-type borrower's perspective ($PB \geq 0$)
  $D^{\circ}_{B,I}|S1 : [5.2, \infty); D^{\circ}_{B,I}|S2 : [4, \infty); D^{\circ}_{B,II}|S1 : [1.668, 5.2); D^{\circ}_{B,II}|S2 : [1.361, 4)$

2. Profit comparison
   - A-type borrower's perspective ($\overline{P}_B = P^+_B$)
     $D^{\circ}_{B,II}|S2$ is dominated by $D^{\circ}_{B,I}|S1$ and $D^{\circ}_{B,II}|S1$; $D^{\circ}_{B,II}|S1$ is preferable to $D^{\circ}_{B,II}|S2$ for $\mu \geq 2.104$.
   - N-type borrower's perspective ($\overline{P}_B = 0$)
     $D^{\circ}_{B,II}|S2$ is dominated by $D^{\circ}_{B,I}|S1$ and $D^{\circ}_{B,II}|S1$; $D^{\circ}_{B,II}|S1$ is preferable to $D^{\circ}_{B,II}|S2$ for $\mu \geq 1.877$.

3. Preferred contract choice given public debt financing
   - A-type borrower
     For $\mu < 1.493$: no choice; for $1.493 \leq \mu < 2.104 :$ $D^{\circ}_{B,II}|S2$; for $2.104 \leq \mu < 5.2 :$ $D^{\circ}_{B,II}|S1$ and for $\mu \geq 5.2 :$ $D^{\circ}_{B,I}|S1$.
   - N-type borrower
     For $\mu < 1.361$: no choice; for $1.361 \leq \mu < 1.877 :$ $D^{\circ}_{B,II}|S2$; for $1.877 \leq \mu < 5.2 :$ $D^{\circ}_{B,II}|S1$ and for $\mu \geq 5.2 :$ $D^{\circ}_{B,I}|S1$.

Private equity financing (see derivation for PC1, Section 4.3)

3. Preferred contract choice given private equity financing
   - A-type borrower
     For $\mu < 1.55$: no choice; for $1.55 \leq \mu < 1.85 :$ $E^{\bullet(1)}_B$ and for $\mu \geq 1.85 :$ $E^{\bullet(2)}_B$.
   - N-type borrower
     For $\mu < 1.25$: no choice; for $1.25 \leq \mu < 1.55 :$ $E^{\bullet(1)}_B$ and for $\mu \geq 1.55 :$ $E^{\bullet(2)}_B$.

Private debt financing

1. Feasible contracting intervals
   - A-type borrower's perspective ($PB \geq P^+_B$)
     $D^{\bullet(1)}_{B,III} : [1.583, 1.902); D^{\bullet(2)}_{B,III} : [1.902, 4.2); D^{\bullet(2)}_{B,II} : [4.2, 5.4); D^{\bullet(2)}_{B,I} : [5.4, \infty)$.
   - N-type borrower's perspective ($PB \geq 0$)
     $D^{\bullet(1)}_{B,III} : [1.413, 1.721); D^{\bullet(2)}_{B,III} : [1.721, 4.2); D^{\bullet(2)}_{B,II} : [4.2, 5.4); D^{\bullet(2)}_{B,I} : [5.4, \infty)$.

2. No profit comparison required
3. Preferred choice given private debt financing
   - A-type borrower
     For $\mu < 1.583$: no choice; for $1.583 \leq \mu < 1.902 :$ $D^{\bullet(1)}_{B,III}$; for $1.902 \leq \mu < 4.2 :$ $D^{\bullet(2)}_{B,III}$; for $4.2 \leq \mu < 5.4 :$ $D^{\bullet(2)}_{B,II}$ and for $\mu \geq 5.4 :$ $D^{\bullet(2)}_{B,I}$.
   - N-type borrower
     For $\mu < 1.413$: no choice; for $1.413 \leq \mu < 1.721 :$ $D^{\bullet(1)}_{B,III}$; for $1.721 \leq \mu < 4.2 :$ $D^{\bullet(2)}_{B,III}$; for $4.2 \leq \mu < 5.4 :$ $D^{\bullet(2)}_{B,II}$ and for $\mu \geq 5.4 :$ $D^{\bullet(2)}_{B,I}$.

THE BORROWER'S BEHAVIOR AS THE ct/pm OPTIMIZER $\left(\widehat{\lambda}_L = \lambda_L\right)^{11}$

1. Feasible contracting intervals
   - A-type borrower's perspective $(PB \geq P_B^+)$
     $E_B^\circ|S2 : [1.45, 2.05)$; $E_B^\circ|S1 : [2.05, \infty)$; $D_{B,II}^\circ|S2 : [1.493, 2.104)$;
     $D_{B,II}^\circ|S1 : [2.104, 5.2)$; $D_{B,I}^\circ|S1 : [5.2, \infty)$; $E_B^{\bullet(1)} : [1.55, 1.85)$; $E_B^{\bullet(2)} :$
     $[1.85, \infty)$; $D_{B,III}^{\bullet(1)} : [1.583, 1.902)$; $D_{B,III}^{\bullet(2)} : [1.902, 4.2)$; $D_{B,II}^{\bullet(2)} : [4.2, 5.4)$;
     $D_{B,I}^{\bullet(2)} : [5.4, \infty)$.
   - N-type borrower's perspective $(PB \geq 0)$
     $E_B^\circ|S2 : [1.15, 1.75)$; $E_B^\circ|S1 : [1.75, \infty)$; $D_{B,II}^\circ|S2 : [1.361, 1.877)$;
     $D_{B,II}^\circ|S1 : [1.877, 5.2)$; $D_{B,I}^\circ|S1 : [5.2, \infty)$; $E_B^{\bullet(1)} : [1.25, 1.55)$; $E_B^{\bullet(2)} :$
     $[1.55, \infty)$; $D_{B,III}^{\bullet(1)} : [1.413, 1.721)$; $D_{B,III}^{\bullet(2)} : [1.721, 4.2)$; $D_{B,II}^{\bullet(2)} : [4.2, 5.4)$;
     $D_{B,I}^{\bullet(2)} : [5.4, \infty)$.
2. Profit comparison
   - A-type borrower's perspective $(\overline{P}_B = P_B^+)$
     $E_B^\circ|S2$ dominates $E_B^{\bullet(1)}, D_{B,III}^{\bullet(1)}$; $D_{B,II}^\circ|S2$ is preferable to $E_B^\circ|S2$ for
     $\mu \geq 1.627$; $D_{B,II}^\circ|S2$ is preferable to $E_B^\circ|S1$; $E_B^{\bullet(2)}$ is preferable to
     $D_{B,II}^\circ|S2$ for $\mu \geq 2.013$; $D_{B,III}^{\bullet(2)}$ is preferable to $D_{B,II}^\circ|S2$ for $\mu \geq 1.988$;
     $D_{B,III}^{\bullet(2)}$ is preferable to $E_B^{\bullet(2)}$ for $\mu \geq 1.933$; $D_{B,III}^{\bullet(2)}$ is preferable
     to $D_{B,II}^\circ|S1$; $D_{B,I}^{\bullet(2)}$ and $D_{B,II}^{\bullet(2)}$ are preferable to $E_B^\circ|S1$, $D_{B,II}^\circ|S1$,
     $D_{B,I}^\circ|S1$, $E_B^{\bullet(2)}$.
   - N-type borrower's perspective $(\overline{P}_B = 0)$
     $E_B^\circ|S2$ dominates $E_B^{\bullet(1)}, D_{B,III}^{\bullet(1)}$; $E_B^{\bullet(2)}$ dominates $E_B^\circ|S1$; $D_{B,II}^\circ|S2$ is
     preferable to $E_B^\circ|S2$ for $\mu \geq 1.627$; $E_B^{\bullet(2)}$ is preferable to $D_{B,II}^\circ|S2$
     for $\mu \geq 1.659$; $D_{B,II}^\circ|S1$ is preferable to $E_B^{\bullet(2)}$ for $\mu \geq 2.422$; $D_{B,II}^{\bullet(2)}$
     is preferable to $E_B^{\bullet(2)}$ for $\mu \geq 1.933$; $D_{B,I}^{\bullet(2)}$, $D_{B,II}^{\bullet(2)}$ and $D_{B,III}^{\bullet(2)}$ are
     preferable to $D_{B,II}^\circ|S1$, $D_{B,I}^\circ|S1$, $E_B^{\bullet(2)}$.

---

[11] We have to treat each ct/pm's contract choice interval-dependent since our profit comparison hinges on the assumption that the profit functions are continuously differentiable.

3. Preferred ct/pm-choices

Table B.18: Contract Type / Placement Mode-Choices and Agreement Probabilities in the Borrower Scenario with Absolute Power (EM, PC3)

Interval	μ	Ct/pm-choices and agreement probabilities			
		A-type borrower		N-type borrower	
I	$[1; 1.15)$	$-$	(0)	$-$	(0)
$II - V$	$[1.15; 1.45)$	$-$	(0)	$E_B^\circ \lvert S2$	$(\frac{1}{2})$
$VI - IX$	$[1.45; 1.627)$	$E_B^\circ \lvert S2$	$(\frac{1}{2})$	$E_B^\circ \lvert S2$	(1)
$X - XII$	$[1.627; 1.659)$	$D_{B,II}^\circ \lvert S2$	$(\frac{1}{2})$	$D_{B,II}^\circ \lvert S2$	$(\frac{1}{2})$
$XIII - XVIII$	$[1.659; 1.933)$	$D_{B,II}^\circ \lvert S2$	$(\frac{1}{2})$	$E_B^{\bullet(2)}$	(1)
$XIX - XXI$	$[1.933; 1.988)$	$D_{B,II}^\circ \lvert S2$	$(\frac{1}{2})$	$D_{B,III}^{\bullet(2)}$	(1)
$XXII - XXIII$	$[1.988; 4.2)$	$D_{B,III}^{\bullet(2)}$	(1)	$D_{B,III}^{\bullet(2)}$	(1)
XXIV	$[4.2; 5.4)$	$D_{B,II}^{\bullet(2)}$	(1)	$D_{B,II}^{\bullet(2)}$	(1)
XXV	$[5.4; \infty)$	$D_{B,I}^{\bullet(2)}$	(1)	$D_{B,I}^{\bullet(2)}$	(1)

Note: This table contains the borrower's ct/pm-choice together with the related contract agreement probabilities stated in brackets.

C

Robustness Checks of Bargaining Power Effects
Derived for PC1 – Propositions 2 to 10 (EM)

Table C.1: Robustness of the Power Effect Predictions of Proposition 2
$(BS^a \rightarrow LS^a$, EM)

Effect	PC1	PC2	PC3
• A-type borrower → A-type lender			
$D^\bullet \Rightarrow -$	$XVII - XVIII$	$-$	$-$
$D^\bullet \Rightarrow D^\circ$	$XIX - XX$	$XIV - XVII$	$-$
$D^\bullet \Rightarrow E^\bullet$	XXI	XIX	$XXII - XXV$
$- \Rightarrow D^\circ$	$-$	$XII - XIII$	$-$
$- \Rightarrow E^\bullet$	$-$	$-$	$XX - XXI$
• A-type borrower → N-type lender			
$D^\circ \Rightarrow -$	$V - IX$	$IV - V$	$X - XII$
$D^\bullet \Rightarrow D^\circ$	$XVII - XX$	$XIV - XVII$	$-$
$D^\bullet \Rightarrow E^\bullet$	XXI	XIX	$XIX - XXV$
$E^\circ \Rightarrow -$	$-$	$-$	$VI - IX$
• N-type borrower → A-type lender			
$- \Rightarrow E^\circ$	$IX - XIII$	$VIII - IX$	$VI - XII$
$D^\bullet \Rightarrow E^\circ$	$XIV - XVIII$	$X - XI$	XIX
$D^\bullet \Rightarrow D^\circ$	$XIX - XX$	$XII - XVII$	$-$
$D^\bullet \Rightarrow E^\bullet$	XXI	XIX	$XX - XXV$
$E^\bullet \Rightarrow E^\circ$	$-$	$-$	$XIII - XVIII$
• N-type borrower → N-type lender			
$D^\circ \Rightarrow E^\circ$	$IV - IX$	$-$	$X - XII$
$D^\bullet \Rightarrow D^\circ$	$XIV - XX$	$IX - XVII$	$-$
$D^\bullet \Rightarrow E^\bullet$	XXI	XIX	$XIX - XXV$

Note: This table contains all bargaining power effects which are observable for an absolute power shift from the borrower to the lender $(BS^a \rightarrow LS^a)$ given PC1, PC2 or PC3.

Table C.2: Robustness of the Power Effect Predictions of Proposition 3
$(LS^a \to BS^r, \text{EM})$

Effect	PC1	PC2	PC3
• A-type lender $\to$ A-type borrower			
$- \Rightarrow D^\bullet$	XV	$-$	$-$
$- \Rightarrow \{D^\circ, D^\bullet\}$	$XVI - XVII$	$-$	$-$
$- \Rightarrow \{D^\circ, E^\bullet, D^\bullet\}$	XVIII	$-$	$-$
$D^\circ \to \{E^\bullet, D^\bullet\}$	XIX	XVI	$-$
$D^\circ \to \{E^\circ, E^\bullet, D^\bullet\}$	XX	XVII	$-$
$E^\bullet \to \{E^\circ, D^\circ, D^\bullet\}$	XXI	XIX	$XX,$ $XXIII - XXV$
$D^\circ \Rightarrow -$	$-$	XII	$-$
$D^\circ \Rightarrow D^\bullet$	$-$	XIII	$-$
$D^\circ \to D^\bullet$	$-$	XV	$-$
$D^\bullet \to \{E^\circ, D^\circ, E^\bullet\}$	$-$	XVIII	$-$
$- \Rightarrow \{E^\circ, D^\circ, E^\bullet, D^\bullet\}$	$-$	$-$	$XVII - XIX$
$E^\bullet \to \{D^\circ, D^\bullet\}$	$-$	$-$	$XXI - XXII$
• A-type lender $\to$ N-type borrower			
$E^\circ \Rightarrow -$	$IX - XI$	$-$	$VI - XIII$
$E^\circ \Rightarrow D^\circ$	$XII - XVIII$	$VIII - XI$	$XIV - XV$
$D^\circ \to E^\circ$	XX	XVII	$-$
$E^\bullet \Rightarrow \{E^\circ, D^\circ\}$	XXI	XIX	$XX,$ $XXIII - XXV$
$D^\bullet \Rightarrow \{E^\circ, D^\circ\}$	$-$	XVIII	$-$
$E^\circ \to D^\circ$	$-$	$-$	$XVII - XIX$
$E^\bullet \Rightarrow D^\circ$	$-$	$-$	$XXI - XXII$

table continued on next page

table continued from previous page			
• N-type lender $\to$ A-type borrower			
$- \Rightarrow D^\circ$	VII	$-$	$-$
$- \Rightarrow D^\bullet$	$VIII - IX$	$-$	$-$
$D^\circ \Rightarrow D^\bullet$	X, XV	$VII - X,$ $XIII - XIV$	$-$
$D^\circ \Rightarrow \{E^\bullet, D^\bullet\}$	$XI - XIV$	$XI - XII$	$-$
$D^\circ \to D^\bullet$	$XVI - XVII$	XV	$-$
$D^\circ \to \{E^\bullet, D^\bullet\}$	$XVIII - XIX$	XVI	$-$
$D^\circ \to \{E^\circ, E^\bullet, D^\bullet\}$	XX	XVII	$-$
$E^\bullet \to \{E^\circ, D^\circ, D^\bullet\}$	XXI	XIX	$XVII - XX,$ $XXIII - XXV$
$D^\bullet \to \{E^\circ, D^\circ, E^\bullet\}$	$-$	XVIII	$-$
$- \Rightarrow E^\bullet$	$-$	$-$	$VIII - XI$
$E^\bullet \to E^\circ$	$-$	$-$	XVI
$E^\bullet \to \{D^\circ, D^\bullet\}$	$-$	$-$	$XXI - XXII$
• N-type lender $\to$ N-type borrower			
$E^\circ \Rightarrow D^\circ$	$IV - IX$	$-$	$IV - XI$
$D^\circ \to E^\circ$	XX	XVII	$-$
$E^\bullet \Rightarrow \{E^\circ, D^\circ\}$	XXI	XIX	$XVII - XX,$ $XXIII - XXV$
$D^\bullet \Rightarrow \{E^\circ, D^\circ\}$	$-$	XVIII	$-$
$E^\bullet \Rightarrow D^\circ$	$-$	$-$	$XIII - XV,$ $XXI - XXII$
$E^\bullet \Rightarrow E^\circ$	$-$	$-$	XVI

Note: This table contains all bargaining power effects which are observable for a redistribution of the bargaining power to determine ct/pm from the lender to the borrower while the former keeps the power to set the contracts' conditions ($LS^a \to BS^r$) given PC1, PC2 or PC3.

Table C.3: Robustness of the Power Effect Predictions of Proposition 4
($BS^r \rightarrow LS^a$, EM)

Effect	PC1	PC2	PC3
• A-type borrower $\rightarrow$ A-type lender			
$D^{\bullet} \Rightarrow -$	XV	$-$	$-$
$\{D^{\circ}, D^{\bullet}\} \Rightarrow -$	$XVI - XVII$	$-$	$-$
$\{D^{\circ}, E^{\bullet}, D^{\bullet}\} \Rightarrow -$	XVIII	$-$	$-$
$\{E^{\bullet}, D^{\bullet}\} \Rightarrow D^{\circ}$	XIX	XVI	$-$
$\{E^{\circ}, E^{\bullet}, D^{\bullet}\} \Rightarrow D^{\circ}$	XX	XVII	$-$
$\{E^{\circ}, D^{\circ}, D^{\bullet}\} \Rightarrow E^{\bullet}$	XXI	XIX	$XX,$ $XXIII - XXV$
$- \Rightarrow D^{\circ}$	$-$	XII	$-$
$D^{\bullet} \Rightarrow D^{\circ}$	$-$	$XIII - XV$	$-$
$\{E^{\circ}, D^{\circ}, E^{\bullet}\} \Rightarrow D^{\bullet}$	$-$	XVIII	$-$
$\{E^{\circ}, D^{\circ}, E^{\bullet}, D^{\bullet}\} \Rightarrow -$	$-$	$-$	$XVI - XIX$
$\{D^{\circ}, D^{\bullet}\} \Rightarrow E^{\bullet}$	$-$	$-$	$XXI - XXII$
• A-type borrower $\rightarrow$ N-type lender			
$D^{\circ} \Rightarrow -$	VII	$-$	$-$
$D^{\bullet} \Rightarrow -$	$VIII - IX$	$-$	$-$
$D^{\bullet} \Rightarrow D^{\circ}$	$X, XV - XVII$	$VII - X,$ $XIII - XV$	$-$
$\{E^{\bullet}, D^{\bullet}\} \Rightarrow D^{\circ}$	$XI - XIV,$ $XVIII - XIX$	$XI - XII,$ XVI	$-$
$\{E^{\circ}, D^{\circ}, D^{\bullet}\} \Rightarrow E^{\bullet}$	XXI	XIX	$XVII - XX,$ $XXIII - XXV$
$\{E^{\circ}, D^{\bullet}, E^{\bullet}\} \Rightarrow D^{\circ}$	XX	XVII	$-$
$\{E^{\circ}, D^{\circ}, E^{\bullet}\} \Rightarrow D^{\bullet}$	$-$	XVIII	$-$
$E^{\bullet} \Rightarrow -$	$-$	$-$	$VIII - XI$
$E^{\circ} \Rightarrow E^{\bullet}$	$-$	$-$	XVI
$\{D^{\circ}, D^{\bullet}\} \Rightarrow E^{\bullet}$	$-$	$-$	$XXI - XXII$

table continued on next page

table continued from previous page			
• N-type borrower → A-type lender			
$- \Rightarrow E^\circ$	$IX - XI$	$-$	$VI - XIII$
$D^\circ \Rightarrow E^\circ$	$XII - XVIII$	$VIII - XI$	$XIV - XV,$ $XVII - XIX$
$E^\circ \to D^\circ$	XX	XVII	$-$
$\{E^\circ, D^\circ\} \Rightarrow E^\bullet$	XXI	XIX	$XX, XXIII - XXV$
$\{E^\circ, D^\circ\} \Rightarrow D^\bullet$	$-$	XVIII	$-$
$D^\circ \Rightarrow E^\bullet$	$-$	$-$	$XXI - XXII$
• N-type borrower → N-type lender			
$D^\circ \Rightarrow E^\circ$	$IV - IX$	$-$	$IV - XI$
$E^\circ \Rightarrow D^\circ$	XX	XVII	$-$
$\{E^\circ, D^\circ\} \Rightarrow E^\bullet$	XXI	XIX	$XVII - XX,$ $XXIII - XXV$
$\{E^\circ, D^\circ\} \Rightarrow D^\bullet$	$-$	XVIII	$-$
$D^\circ \Rightarrow E^\bullet$	$-$	$-$	$XII - XV,$ $XXI - XXII$
$E^\circ \Rightarrow E^\bullet$	$-$	$-$	XVI

Note: This table contains all bargaining power effects which are observable for a redistribution of the bargaining power to determine ct/pm from the borrower to the lender while the latter keeps the power to set the contracts' conditions ($BS^r \to LS^a$) given PC1, PC2 or PC3.

Table C.4: Robustness of the Power Effect Predictions of Proposition 5
$(BS^a \to LS^r$, EM)

Effect	PC1	PC2	PC3
• A-type borrower $\to$ A-type lender			
$- \Rightarrow D^\bullet$	$XV - XVI$	XIII	$-$
$D^\bullet \to E^\bullet$	$XVIII - XXI$	$XVI - XIX$	$XXII - XXV$
$- \Rightarrow E^\bullet$	$-$	$-$	XVII
$- \Rightarrow \{E^\bullet, D^\bullet\}$	$-$	$-$	$XVIII - XXI$
• A-type borrower $\to$ N-type lender			
$D^\circ \to D^\bullet$	VIII	VII	$-$
$D^\circ \to \{E^\circ, D^\bullet\}$	$IX - X, XVI$	XIII	$-$
$D^\circ \to \{E^\circ, E^\bullet, D^\bullet\}$	$XI - XV$	$XI - XII$	$X - XVI,$ $XVIII - XXI$
$D^\bullet \to \{E^\circ, D^\circ\}$	XVII	$XIV - XV$	$-$
$D^\bullet \to \{E^\circ, D^\circ, E^\bullet\}$	$XVIII - XXI$	$XVI - XIX$	$XXII - XXV$
$D^\circ \to E^\circ$	$-$	$VIII - X$	$-$
$E^\circ \to D^\circ$	$-$	$-$	VII
$E^\circ \to \{D^\circ, E^\bullet\}$	$-$	$-$	VIII
$E^\circ \to \{D^\circ, E^\bullet, D^\bullet\}$	$-$	$-$	IX
$D^\circ \to \{E^\circ, E^\bullet\}$	$-$	$-$	XVII
• N-type borrower $\to$ A-type lender			
$- \Rightarrow E^\bullet$	$XI - XII$	$-$	$VIII - XIII$
$- \Rightarrow \{E^\bullet, D^\bullet\}$	XIII	$-$	$-$
$D^\bullet \to E^\bullet$	$XIV,$ $XVIII - XXI$	$XI - XII,$ $XVI - XIX$	$XIX - XXV$
$- \Rightarrow D^\bullet$	$-$	IX	$-$
$E^\bullet \to D^\bullet$	$-$	$-$	$XV - XVI,$ XVIII
			table continued on next page

table continued from previous page

- N-type borrower → N-type lender

$E^\circ \to E^\bullet$	III	–	III
$D^\circ \to \{E^\circ, E^\bullet\}$	IV	–	–
$D^\circ \to D^\bullet$	VIII	VII	–
$D^\circ \to \{E^\circ, D^\bullet\}$	$IX - X$	–	–
$D^\circ \to \{E^\circ, E^\bullet, D^\bullet\}$	$XI - XIII$	–	$X - XII$
$D^\bullet \to \{E^\circ, D^\circ, E^\bullet\}$	$XIV - XV,$ $XVIII - XXI$	$XI - XII,$ $XVI - XIX$	$XIX - XXV$
$D^\bullet \to \{E^\circ, D^\circ\}$	$XVI - XVII$	$XIII - XV$	–
$D^\circ \to E^\circ$	–	$III, VIII - IX$	–
$D^\bullet \Rightarrow \{E^\circ, D^\circ\}$	–	X	–
$E^\circ \to D^\circ$	–	–	VII
$E^\circ \to \{D^\circ, E^\bullet\}$	–	–	$IV, VIII$
$E^\circ \to \{D^\circ, E^\bullet, D^\bullet\}$	–	–	V, IX
$E^\bullet \to \{E^\circ, D^\circ, D^\bullet\}$	–	–	$XIII - XVI,$ XVIII
$E^\bullet \to \{E^\circ, D^\circ\}$	–	–	XVII

Note: This table contains all bargaining power effects which are observable for a redistribution of the bargaining power to determine ct/pm from the borrower to the lender while the former keeps the power to set contracts' conditions ($BS^a \to LS^r$) given PC1, PC2 or PC3.

Table C.5: Robustness of the Power Effect Predictions of Proposition 6
$(LS^r \rightarrow BS^a$, EM)

Effect	PC1	PC2	PC3
• A-type lender → A-type borrower			
$D^\bullet \Rightarrow -$	$XV - XVI$	XIII	$-$
$E^\bullet \Rightarrow D^\bullet$	$XVIII - XXI$	$XVI - XIX$	$XXII - XXV$
$E^\bullet \Rightarrow -$	$-$	$-$	XVII
$\{E^\bullet, D^\bullet\} \Rightarrow -$	$-$	$-$	$XVIII - XXI$
• A-type lender → N-type borrower			
$E^\bullet \Rightarrow -$	$XI - XII$	$-$	$VIII - XII$
$\{E^\bullet, D^\bullet\} \Rightarrow -$	XIII	$-$	$-$
$E^\bullet \Rightarrow D^\bullet$	$XIV,$	$XI - XII,$	$XIX - XXV$
	$XVIII - XXI$	$XVI - XIX$	
$D^\bullet \Rightarrow -$	$-$	IX	$-$
$D^\bullet \Rightarrow E^\bullet$	$-$	$-$	$XV - XVI,$
			XVIII
• N-type lender → A-type borrower			
$D^\bullet \Rightarrow D^\circ$	VIII	VII	$-$
$\{E^\circ, D^\bullet\} \Rightarrow D^\circ$	$IX - X, XVI$	XIII	$-$
$\{E^\circ, E^\bullet, D^\bullet\} \Rightarrow D^\circ$	$XI - XV$	$XI - XII$	$X - XVI,$
			$XVIII - XXI$
$\{E^\circ, D^\circ\} \Rightarrow D^\bullet$	XVII	$XIV - XV$	$-$
$\{E^\circ, D^\circ, E^\bullet\} \Rightarrow D^\bullet$	$XVIII - XXI$	$XVI - XIX$	$XXII - XXV$
$E^\circ \Rightarrow D^\circ$	$-$	$VIII - X$	$-$
$D^\circ \Rightarrow E^\circ$	$-$	$-$	VII
$\{D^\circ, E^\bullet\} \Rightarrow E^\circ$	$-$	$-$	VIII
$\{D^\circ, E^\bullet, D^\bullet\} \Rightarrow E^\circ$	$-$	$-$	IX
$\{E^\circ, E^\bullet\} \Rightarrow D^\circ$	$-$	$-$	XVII

table continued on next page

table continued from previous page			
• N-type lender → N-type borrower			
$E^{\bullet} \Rightarrow E^{\circ}$	III	$-$	III
$\{E^{\circ}, E^{\bullet}\} \Rightarrow D^{\circ}$	IV	$-$	$-$
$D^{\bullet} \Rightarrow D^{\circ}$	VIII	VII	$-$
$\{E^{\circ}, D^{\bullet}\} \Rightarrow D^{\circ}$	$IX - X$	$-$	$-$
$\{E^{\circ}, E^{\bullet}, D^{\bullet}\} \Rightarrow D^{\circ}$	$XI - XIII$	$-$	$X - XII$
$\{E^{\circ}, D^{\circ}, E^{\bullet}\} \Rightarrow D^{\bullet}$	$XIV - XV,$ $XVIII - XXI$	$XI - XII,$ $XVI - XIX$	$XIX - XXV$
$\{E^{\circ}, D^{\circ}\} \Rightarrow D^{\bullet}$	$XVI - XVII$	$X, XIII - XV$	$-$
$E^{\circ} \Rightarrow D^{\circ}$	$-$	$III, VIII - IX$	$-$
$\{D^{\circ}, E^{\bullet}\} \Rightarrow E^{\circ}$	$-$	$-$	$IV, VIII$
$\{D^{\circ}, E^{\bullet}, D^{\bullet}\} \Rightarrow E^{\circ}$	$-$	$-$	V, IX
$D^{\circ} \Rightarrow E^{\circ}$	$-$	$-$	VII
$\{E^{\circ}, D^{\circ}, D^{\bullet}\} \Rightarrow E^{\bullet}$	$-$	$-$	$XIII - XVI,$ XVIII
$\{E^{\circ}, D^{\circ}\} \Rightarrow E^{\bullet}$	$-$	$-$	XVII

Note: This table contains all bargaining power effects which are observable for a redistribution of the bargaining power to determine ct/pm from the lender to the borrower, while the latter keeps the power to set contracts' conditions ($LS^r \to BS^a$) given PC1, PC2 or PC3.

Table C.6: Robustness of the Power Effect Predictions of Proposition 7
($LS^a \rightarrow LS^r$, EM)

Effect	PC1	PC2	PC3
• A-type lender			
$- \Rightarrow D^\bullet$	$XV - XVII$	$-$	$-$
$- \Rightarrow \{E^\bullet, D^\bullet\}$	XVIII	$-$	$XVIII - XIX$
$D^\circ \Rightarrow \{E^\bullet, D^\bullet\}$	$XIX - XX$	$XVI - XVII$	$-$
$E^\bullet \rightarrow D^\bullet$	XXI	XIX	$XX - XXV$
$D^\circ \Rightarrow -$	$-$	XII	$-$
$D^\circ \Rightarrow D^\bullet$	$-$	$XIII - XV$	$-$
$D^\bullet \rightarrow E^\bullet$	$-$	XVIII	$-$
$- \Rightarrow E^\bullet$	$-$	$-$	XVII
• N-type lender			
$E^\circ \rightarrow E^\bullet$	III	$-$	III
$E^\circ \rightarrow \{D^\circ, E^\bullet\}$	IV	$-$	$IV, VIII$
$E^\circ \Rightarrow D^\circ$	$V - VII$	$-$	$-$
$E^\circ \Rightarrow \{D^\circ, D^\bullet\}$	VIII	$-$	$-$
$E^\circ \rightarrow \{D^\circ, D^\bullet\}$	IX	$-$	$-$
$D^\circ \rightarrow \{E^\bullet, D^\bullet\}$	$X, XVI - XVII$	$XIII - XV$	$-$
$D^\circ \rightarrow \{E^\circ, E^\bullet, D^\bullet\}$	$XI - XV,$ $XVIII - XX$	$X - XII,$ $XVI - XVII$	$-$ $-$
$E^\circ \rightarrow \{E^\circ, D^\circ, D^\bullet\}$	XXI	$-$	$V, IX - XII$
$D^\circ \rightarrow E^\circ$	$-$	$III, VIII - X$	$-$
$D^\circ \rightarrow D^\bullet$	$-$	VII	$-$
$D^\bullet \rightarrow \{E^\circ, D^\circ, E^\bullet\}$	$-$	XVIII	$-$
$E^\bullet \rightarrow \{E^\circ, D^\circ, D^\bullet\}$	$-$	XIX	$XIII - XVI,$ $XVIII - XXV$
$E^\circ \rightarrow D^\circ$	$-$	$-$	VII
$E^\bullet \rightarrow \{E^\circ, D^\circ\}$	$-$	$-$	XVII

Note: This table contains all bargaining power effects which are observable for a redistribution of the bargaining power to set contracts' conditions from the lender to the borrower while the former keeps the power to determine ct/pm ($LS^a \rightarrow LS^r$) given PC1, PC2 or PC3.

Table C.7: Robustness of the Power Effect Predictions of Proposition 8
$(LS^r \to LS^a$, EM)

Effect	PC1	PC2	PC3
• A-type lender			
$D^\bullet \Rightarrow -$	$XV - XVII$	$-$	$-$
$\{E^\bullet, D^\bullet\} \Rightarrow -$	XVIII	$-$	$XVIII - XIX$
$\{E^\bullet, D^\bullet\} \Rightarrow D^\circ$	$XIX - XX$	$XVI - XVII$	$-$
$D^\bullet \to E^\bullet$	XXI	XIX	$XX - XXV$
$- \Rightarrow D^\circ$	$-$	XII	$-$
$D^\bullet \Rightarrow D^\circ$	$-$	$XIII - XV$	$-$
$E^\bullet \to D^\bullet$	$-$	XVIII	$-$
$E^\bullet \Rightarrow -$	$-$	$-$	XVII
• N-type lender			
$E^\bullet \Rightarrow E^\circ$	III	$-$	III
$\{D^\circ, E^\bullet\} \Rightarrow E^\circ$	IV	$-$	$IV, VIII$
$D^\circ \Rightarrow E^\circ$	$V - VII$	$-$	VII
$\{D^\circ, D^\bullet\} \Rightarrow E^\circ$	$VIII, IX$	$-$	$-$
$\{E^\circ, D^\bullet\} \Rightarrow D^\circ$	$X, XVI - XVII$	$XIII - XV$	$-$
$\{E^\circ, E^\bullet, D^\bullet\} \Rightarrow D^\circ$	$XI - XV,$ $XVIII - XX$	$X - XII,$ $XVI - XVII$	$-$ $-$
$\{E^\circ, D^\circ, D^\bullet\} \Rightarrow E^\bullet$	XXI	XIX	$XII - XVI,$ $XVIII - XXV$
$E^\circ \Rightarrow D^\circ$	$-$	$III, VIII - X$	$-$
$D^\bullet \Rightarrow D^\circ$	$-$	VII	$-$
$\{E^\circ, D^\circ, E^\bullet\} \Rightarrow D^\bullet$	$-$	XVIII	$-$
$\{D^\circ, E^\bullet, D^\bullet\} \Rightarrow E^\circ$	$-$	$-$	$V, IX - XI$
$\{E^\circ, D^\circ\} \Rightarrow E^\bullet$	$-$	$-$	XVII

Note: This table contains all bargaining power effects which are observable for a redistribution of the bargaining power to set contracts' conditions from the borrower to the lender while the latter keeps the power to determine ct/pm $(LS^r \to LS^a)$ given PC1, PC2 or PC3.

Table C.8: Robustness of the Power Effect Predictions of Proposition 9
    ($BS^a \rightarrow BS^r$, EM)

Effect	PC1	PC2	PC3
• A-type borrower			
$- \Rightarrow D^\bullet$	XV	XIII	–
$- \Rightarrow \{D^\circ, D^\bullet\}$	XVI	–	–
$D^\bullet \rightarrow D^\circ$	XVII	XV	–
$D^\bullet \rightarrow \{D^\circ, E^\bullet\}$	$XVIII - XIX$	XVI	XXII
$D^\bullet \rightarrow \{E^\circ, D^\circ, E^\bullet\}$	$XX - XXI$	$XVII - XIX$	$XXIII - XXV$
$- \Rightarrow \{E^\circ, D^\circ, E^\bullet, D^\bullet\}$	–	–	$XVII - XX$
$- \Rightarrow \{D^\circ, E^\bullet, D^\bullet\}$	–	–	XXI
• N-type borrower			
$D^\bullet \Rightarrow D^\circ$	$XIV - XIX$	$X - XVI$	$XXI - XXII$
$D^\bullet \Rightarrow \{E^\circ, D^\circ\}$	$XX - XXI$	$XVII - XIX$	$XIX - XX,$
			$XXIII - XXV$
$E^\circ \Rightarrow D^\circ$	–	–	$IV - IX$
$E^\bullet \Rightarrow D^\circ$	–	–	$XIII - XV$
$E^\bullet \Rightarrow E^\circ$	–	–	XVI
$E^\bullet \Rightarrow \{E^\circ, D^\circ\}$	–	–	$XVII - XVIII$

Note: This table contains all bargaining power effects which are observable for a redistribution of the bargaining power to set contracts' conditions from the borrower to the lender while the former keeps the power to determine ct/pm ($BS^a \rightarrow BS^r$) given PC1, PC2 or PC3.

Table C.9: Robustness of the Power Effect Predictions of Proposition 10 $(BS^r \to BS^a$, EM)

Effect	PC1	PC2	PC3
• A-type borrower			
$D^\bullet \Rightarrow -$	XV	XIII	$-$
$\{D^\circ, D^\bullet\} \Rightarrow -$	XVI	$-$	$-$
$D^\circ \Rightarrow D^\bullet$	XVII	XV	$-$
$\{D^\circ, E^\bullet\} \Rightarrow D^\bullet$	$XVIII - XIX$	XVI	XXII
$\{E^\circ, D^\circ, E^\bullet\} \Rightarrow D^\bullet$	$XX - XXI$	$XVII - XIX$	$XXIII - XXV$
$\{E^\circ, D^\circ, E^\bullet, D^\bullet\} \Rightarrow -$	$-$	$-$	$XVII - XX$
$\{D^\circ, E^\bullet, D^\bullet\} \Rightarrow -$	$-$	$-$	XXI
• N-type borrower			
$D^\circ \Rightarrow D^\bullet$	$XIV - XIX$	$X - XVI$	$XXI - XXII$
$\{E^\circ, D^\circ\} \Rightarrow D^\bullet$	$XX - XXI$	$XVII - XIX$	$XIX - XX,$ $XXIII - XXV$
$D^\circ \Rightarrow E^\circ$	$-$	$-$	$IV - IX$
$D^\circ \Rightarrow E^\bullet$	$-$	$-$	$XIII - XV$
$E^\circ \Rightarrow E^\bullet$	$-$	$-$	XVI
$\{E^\circ, D^\circ\} \Rightarrow E^\bullet$	$-$	$-$	$XVII - XVIII$

Note: This table contains all bargaining power effects which are observable for a redistribution of the bargaining power to set contracts' conditions from the lender to the borrower while the latter keeps the power to determine ct/pm $(BS^r \to BS^a)$ given PC1, PC2 or PC3.

References

Admati, A., Pfleiderer, P., 2000. Forcing firms to talk: Financial disclosure regulation and externalities. Review of Financial Studies 13, 479–519.

Aghion, P., Tirole, J., 1994. The management of innovation. Quarterly Journal of Economics 109, 1185–1209.

Akerlof, G., 1970. Market for lemons: Quality uncertainty and the market mechanism. Quarterly Journal of Economics 84, 488–500.

Allen, L., Jagtiani, J., 1996. Risk and market segmentation in financial intermediaries' returns. Tech. Rep. 36, Wharton School, University of Pennsylvania, Philadelphia.

Amihud, Y., Lev, B., Travlos, N., 1990. Corporate control and the choice of investment financing: The case of corporate acquisitions. Journal of Finance 45, 603–616.

Antoniou, A., Guney, Y., Paudyal, K., 2002. Determinants of corporate capital structure: Evidence from european countries. Tech. rep., Centre for Empirical Research in Finance, University of Durham, Durham.

Bancel, F., Mittoo, U., 2003. The determinants of capital structure choice: A survey of european firms. Tech. rep., Asper School of Business, University of Manitoba, Manitoba.

Barry, C., Mascarella, C., Vertsuypens, M., 1991. Underwriter warrants, underwriter compensation, and the costs of going public. Journal of Financial Economics 29, 113–135.

Bebchuk, L., 1999. A rent-protection theory of corporate ownership and control. Tech. Rep. 7203, National Bureau of Economic Research, Cambridge, MA.

Bebchuk, L., Fried, J., Walker, D., 2002. Managerial power and rent extraction in the design of executive compensation. Tech. Rep. 9068, National Bureau of Economic Research, Cambridge, MA.

Becht, M., Bolton, P., Röell, A., 2003. Corporate governance and control. In: Constantinides, G., Harris, M., Stulz, R. (Eds.), Handbook of the Economics of Finance. North-Holland, Amsterdam, 1–110.

Beck, T., Demirgüc-Kunt, A., Maksimovic, V., 2004. Bank competition and access to finance: International evidence. Journal of Money, Credit, and Banking 36, 627–648.

Berglöf, E., von Thadden, E.-L., 1994. Short-term versus long-term interests: Capital structure with multiple investors. Quarterly Journal of Economics 109, 1055–1084.

Bergman, Y., Callen, J., 1991. Opportunistic underinvestment in debt renegotiations and capital structure. Journal of Financial Economics 29, 137–171.

Berkovitch, E., Israel, R., Spiegel, Y., 2000. Managerial compensation and capital structure. Journal of Economics and Management Strategy 9, 549–584.

Bernanke, B., Gertler, M., 1989. Agency costs, net worth, and business fluctuations. American Economic Review 79, 14–31.

Besanko, D., Thakor, A., 1987. Collateral and rationing: Sorting equilibria in monopolistic and competitive credit markets. International Economic Review 28, 671–689.

Best, R., Zhang, H., 1993. Alternative information sources and the information content of bank loans. Journal of Finance 48, 1507–1522.

Blackwell, D., Kidwell, D., 1988. An investigation of cost differences between public sales and private placements of debt. Journal of Financial Economics 22, 253–278.

Bolton, P., Freixas, X., 2000. Equity, bonds, and bank debt: Capital structure and financial market equilibrium under asymmetric information. Journal of Political Economy 108, 324–351.

Bolton, P., Jeanne, O., 2005. Structuring and restructuring sovereign debt: The role of seniority. Tech. Rep. 11071, National Bureau of Economic Research, Cambridge, MA.

Bolton, P., Scharfstein, D., 1996. Optimal debt structure and the number of creditors. Journal of Political Economy 104, 1–25.

Bonaccorsi di Patti, E., Dell'Ariccia, G., 2004. Bank competition and firm creation. Journal of Money, Credit, and Banking 36, 225–252.

Boot, A., 2000. Relationship banking: What do we know? Journal of Financial Intermediation 9, 7–25.

Booth, L., Aivazian, V., Demirguc-Kunt, A., Maksimovic, V., 2001. Capital structure in developing countries. Journal of Finance 56, 87–130.

Border, K., Sobel, J., 1987. Samurai accountant: A theory of auditing and plauder. Review of Economic Studies 54, 525–540.

Boyd, J., Smith, B., 1994. How good are standard debt contracts? Stochastic versus nonstochastic monitoring in a costly state verification environment. Journal of Business 67, 539–561.

Brander, J., Lewis, T., 1986. Oligopoly and financial structure: The limited liability effect. American Economic Review 76, 956–970.

Bronars, S., Deere, D., 1991. The threat of unionization, the use of debt, and the preservation of shareholder wealth. Quarterly Journal of Economics 106, 231–254.

Campbell, T., Chan, Y.-S., 1992. Optimal financial contracting with ex-post and ex-ante observability problems. Quarterly Journal of Economics 107, 785–795.

Canning, D., 1989. Bargaining theory. In: Hahn, F. (Ed.), The Economics of Missing Markets, Information, and Games. Clarendon Press, Oxford, 163–187.

Carey, M., Rosen, R., 2001. Public debt as a punching bag: An agency model of the mix of public and private debt. Tech. rep., Federal Reserve Board, Washington.

Chaplinsky, S., Niehaus, G., 1993. Do inside ownership and leverage share common determinants? Quarterly Journal of Business and Economics 32, 51–65.

Chatterjee, K., 1985. Disagreement in bargaining: Models with incomplete information. In: Roth, A. (Ed.), Game-Theoretic Models of Bargaining. Cambridge University Press, Cambridge, UK, 9–26.

Chemmanur, T., Fulhieri, P., 1994. Reputation, renegotiation, and the choice between bank loans and publicly traded debt. Review of Financial Studies 7, 475–506.

Choe, C., 1998. A mechanism design approach to an optimal contract under ex-ante and ex-post private information. Review of Economic Design 3, 237–255.

Coase, R., 1937. The nature of the firm. Economica 4, 386–405.

Cramton, P., 1991. Dynamic bargaining with transaction costs. Management Science 37, 1221–1233.

Crawford, V. P., Sobel, J., 1982. Strategic information transmission. Econometrica 50, 1431–1451.

Cronqvist, H., Nilsson, M., 2003. The choice between rights offerings and private equity placements. Tech. rep., University of Chicago, Chicago, IL.

Dasgupta, S., Nanda, V., 1993. Bargaining and brinkmanship: Capital structure choice by regulated firms. International Journal of Industrial Organization 11, 475–497.

Dasgupta, S., Sengupta, K., 1993. Sunk investment, bargaining and choice of capital structure. International Economic Review 34, 203–220.

Denis, D., Mihov, V., 2003. The choice among bank debt, non-bank private debt and public debt: Evidence from new corporate borrowings. Journal of Financial Economics 70, 3–28.

Dhaliwal, D., Khurana, I., Pereira, R., 2003. Costly public disclosure and the choice between private and public debt. Tech. rep., University of Missouri-Columbia, Columbia, MO.

Diamond, D., 1984. Financial intermediation and delegated monitoring. Review of Economic Studies 51, 393–414.

Diamond, D., 1991. Monitoring and reputation: The choice between bank loans and directly placed debt. Journal of Political Economy 99, 689–721.

Dowd, K., 1992. Optimal financial contracts. Oxford Economic Papers 44, 672–693.

Eckbo, E., Masulis, R., 1992. Adverse selection and the rights offer paradox. Journal of Financial Economics 32, 293–332.

Eckbo, E., Norli, O., 2005. The choice of seasoned-equity selling mechanisms: Theory and evidence. Tech. Rep. DP4833, Centre for Economic Policy Research, London.

Edgeworth, F., 1881. Mathematical Psychics. Kegan Paul & Publishers, London.

Ellingsen, T., Rydqvist, K., 1997. The stock market as a screening device and the decision to go public. Tech. Rep. 174, Stockholm School of Economics, Stockholm.

Elsas, R., Krahnen, J., 2000. Collateral, default risk and relationship lending: An empirical study on financial contracting. Tech. rep., Center for Financial Studies, Frankfurt am Main.

Fairchild, R., 2004. Financial contracting between managers and venture capitalists: The role of value-added services, reputation seeking and bargaining power. Journal of Financial Research 27, 481–495.

Fama, E., Jensen, M., 1983. Separation of ownership and control. Journal of Law and Economics 26, 301–325.

Faulkender, M., Petersen, M., 2003. Does the source of capital affect capital structure? Tech. Rep. 9930, National Bureau of Economic Research, Cambridge, MA.

Fernández, R., Özler, S., 1999. Debt concentration and bargaining power: Large banks, small banks, and secondary market prices. International Economic Review 40, 333–355.

Frank, M., Goyal, V., 2003a. Capital structure decisions. Tech. rep., University of British Columbia, Vancouver.

Frank, M., Goyal, V., 2003b. Testing the pecking order theory of capital structure. Journal of Financial Economics 67, 217–248.

Frank, M., Goyal, V., 2005. Trade off and pecking order theories of debt. Tech. rep., Center for Corporate Governance, Tuck School of Business at Dartmouth, Hanover, NH.

Freixas, X., Rochet, J.-C., 1997. Microeconomics of Banking. MIT Press, Cambridge, MA.

Fudenberg, D., Levine, D., Tirole, J., 1985. Infinite-horizon models of bargaining with one-sided incomplete information. In: Roth, A. (Ed.), Game-Theoretic Models of Bargaining. Cambridge University Press, Cambridge, UK, 73–98.

Fulghieri, P., Lukin, D., 2001. Information production, dilution costs, and optimal security design. Journal of Financial Economics 61, 3–42.

Galai, D., Masulis, R., 1976. The option pricing model and the risk factor of stock. Journal of Financial Economics 3, 53–81.

Gale, D., Hellwig, M., 1985. Incentive-compatible debt contracts: The one-period problem. Review of Economic Studies 52, 647–663.

Gertner, R., Scharfstein, D., 1991. A theory of workouts and the effects of reorganization law. Journal of Finance 46, 1189–1222.

Goldberg, L., Rai, A., 1996. The structure-performance relationship for European banking. Journal of Banking and Finance 20, 745–771.

Gomes, A., Phillips, G., 2005. Why do public firms issue private and public equity, convertibles and debt? Tech. Rep. 11294, National Bureau of Economic Research, Cambridge, MA.

Gompers, P., Lerner, J., 1996. The use of covenants: An empirical analysis of venture partnership agreements. Journal of Law and Economics 39, 463–498.

Gompers, P., Lerner, J., 1999. An analysis of compensation in the U.S. venture capital partnership. Journal of Financial Economics 51, 3–44.

Gompers, P., Lerner, J., 2000. Money chasing deals? The impact of fund inflows on private equity valuations. Journal of Financial Economics 55, 281–325.

Gorton, G., Kahn, J., 2000. The design of bank loan contracts. Review of Financial Studies 13, 331–364.

Graham, J., Harvey, C., 2001. The theory and practice of corporate finance: Evidence from the field. Journal of Financial Economics 60, 187–243.

Grossman, S., Hart, O., 1988. One share-one vote and the market for corporate control. Journal of Financial Economics 20, 175–202.

Guzman, M., 2000. Bank structure, capital accumulation and growth: A simple macroeconomic model. Economic Theory 16, 421–455.

Habib, M. A., Johnsen, D. B., 2000. The private placement of debt and outside equity as an information revelation mechanism. Review of Financial Studies 13, 1017–1055.

Hanka, G., 1998. Debt and the terms of employment. Journal of Financial Economics 48, 245–282.

Harris, M., Raviv, A., 1988a. Corporate control contests and capital structure. Journal of Financial Economics 20, 55–86.

Harris, M., Raviv, A., 1988b. Corporate governance: Voting rights and majority rule. Journal of Financial Economics 20, 203–235.

Harris, M., Raviv, A., 1989. The design of securities. Journal of Financial Economics 24, 255–287.

Harris, M., Raviv, A., 1991. The theory of capital structure choice. Journal of Finance 46, 297–355.

Harris, M., Raviv, A., 1992. Financial contracting theory. In: Laffont, J. (Ed.), Advances in Economic Theory. 6th World Congress of the Econometric Society. Cambridge University Press, Cambridge, UK, 64–150.

Hart, O., 2001. Financial contracting. Journal of Economic Literature 39, 1079–1100.

Hart, O., Holmström, B., 1987. Theory of contracts. In: Bewley, T. (Ed.), Advances in Economic Theory. 5th World Congress of the Econometric Society. Cambridge University Press, Cambridge, UK, 71–155.

Hart, O., Moore, J., 1995. Debt and seniority: An analysis of the role of hard claims in constraining management. American Economic Review 85, 567–585.

Hege, U., Mella-Barral, P., 2000. Bargaining power and optimal leverage. Tech. rep., Centre for Economic Policy and Research, London.

Heinkel, R., Schwartz, E., 1986. Rights versus underwritten offerings: An asymmetric information approach. Journal of Finance 41, 1–18.

Hellwig, M., 2000. Costly state verification: The choice between ex-ante and ex-post verification mechanisms. Tech. Rep. 6, University of Mannheim, Mannheim.

Heron, R., Lie, E., 2004. A comparison of the motivations for and the information content of different types of equity offerings. Journal of Business 77, 605–632.

Hertzel, M., Smith, R., 1993. Market discounts and shareholder gains for placing equity privately. Journal of Finance 48, 459–485.

Hovakimian, A., Opler, T., Titman, S., 2001. The debt-equity choice. Journal of Financial and Quantitative Analysis 36, 1–24.

Hvide, H., Leite, T., 2002. Debt, outside equity, and capital structure under costly state verification. Tech. rep., Norwegian School of Economics and Business, Bergen.

Inderst, R., 2002. Contract design and bargaining power. Economics Letters 74, 171–176.

Inderst, R., Müller, H., 2004. The effect of capital market characteristics on the value of start-up firms. Journal of Financial Economics 72, 319–356.

Israel, R., Ofer, A., Siegel, D., 1989. The information content of equity for debt swaps: An investigation of analysts' forecasts of firm cash flows. Journal of Financial Economics 25, 349–370.

Jensen, M., 1986. Agency costs of free cash flow, corporate finance and takeovers. American Economic Review 76, 323–339.

Jensen, M., Meckling, W., 1976. Theory of the firm: Managerial behavior, agency costs and ownership structure. Journal of Financial Economics 3, 305–360.

Kennan, J., Wilson, R., 1993. Bargaining with private information. Journal of Economic Literature 61, 45–104.

Klimenko, M., 2002. Trade interdependence, the international financial institutions, and the recent evolution of sovereign-debt renegotiations. Journal of International Economics 58, 177–209.

Kose, J., Dilip, M., Badih, S., 1993. Debt as an engine of creative innovation. Tech. Rep. S-93-37, New York University, New York.

Krishnaswami, S., Spindt, P., Subramanian, V., 1999. Information asymmetry, monitoring, and the placement structure of corporate debt. Journal of Financial Economics 54, 407–434.

Kwan, S., Carleton, W., 2004. Financial contracting and the choice between private placement and publicly offered bonds. Tech. Rep. 20, Federal Reserve Bank of San Francisco, San Francisco.

La Porta, R., de Silanes, F. L., Shleifer, A., Vishny, R., 1997. Legal determinants of external finance. Journal of Finance 52, 1131–1152.

La Porta, R., de Silanes, F. L., Shleifer, A., Vishny, R., 1998. Law and finance. Journal of Political Economy 106, 1113–1155.

Leland, H., Pyle, D., 1977. Informational asymmetries, financial structure and financial intermediation. Journal of Finance 32, 371–387.

Mackie-Mason, J., 1990. Do firms care who provides their financing? In: Hubbard, G. (Ed.), Asymmetric Information, Corporate Finance and Investment. Chicago University Press, Chicago, IL, 63–104.

Mas-Colell, A., Whinston, M., Green, J., 1995. Microeconomic Theory. Oxford University Press, New York, Oxford.

Maskin, E., Tirole, J., 1990. The principal-agent relationship with an informed principal, I: The case of private values. Econometrica 58, 379–409.

Maskin, E., Tirole, J., 1992. The principal-agent relationship with an informed principal, II: Common values. Econometrica 60, 1–42.

Mason, C., Gottesman, A., Prevost, A., 2003. Shareholder intervention, managerial resistance, and corporate control: A Nash equilibrium approach. Quarterly Review of Economics and Finance 43, 466–482.

Masulis, R., 1983. The impact of capital structure change on firm value: Some estimates. Journal of Finance 38, 107–126.

Masulis, R., Korwar, A., 1986. Seasoned equity offerings: An empirical investigation. Journal of Financial Economics 15, 91–118.

Maug, E., 1998. Large shareholders as monitors: Is there a trade-off between liquidity and control? Journal of Finance 53, 65–98.

Modigliani, F., Miller, M., 1958. The cost of capital, corporate finance and the theory of investment. American Economic Review 48, 261–297.

Mookherjee, D., Png, I., 1989. Optimal auditing, insurance, and redistribution. Quarterly Journal of Economics 104, 399–415.

Mukhopadhyay, B., 2002. Costly state verification and optimal investment. Journal of Economics and Finance 26, 233–248.

Myers, S., 1977. Determinants of corporate borrowing. Journal of Financial Economics 5, 145–175.

Myers, S., 1984. The capital structure puzzle. Journal of Finance 39, 575–592.

Myers, S., 2000. Outside equity. Journal of Finance 45, 1005–1037.

Myers, S., 2003. Financing of corporations. In: Constantinides, G., Harris, M., Stulz, R. (Eds.), Handbook of the Economics of Finance. North-Holland, Amsterdam, 215–254.

Myers, S., Majluf, N., 1984. Corporate financing and investment decisions when firms have information that investors do not have. Journal of Financial Economics 13, 187–221.

Nakamura, L., 1993. Recent research in commercial banking, information and lending. Financial Markets, Institutions, and Instruments 2, 73–88.

Nash, J., 1950. Equilibrium points in n-person games. Proceedings of the National Academy of Sciences U.S.A. 36, 48–49.

Nash, J., 1951. Non-cooperative games. Annals of Mathematics 54, 286–295.

Nash, J., 1953. Two-person cooperative games. Econometrica 21, 128–140.

Nenova, T., 2003. The value of corporate voting rights and control: A cross-country analysis. Journal of Financial Economics 68, 325–351.

Osano, H., 2003. Managerial bargaining power and board independence. Tech. Rep. 559, Kyoto Institut of Economic Research, Kyoto.

Pagano, M., Panetta, F., Zingales, L., 1998. Why do companies go public? An empirical analysis. Journal of Finance 53, 27–64.

Petersen, M., Rajan, R., 1995. The effect of credit market competition on lending relationships. Quarterly Journal of Economics 110, 407–443.

Rajan, R., 1992. Insiders and outsiders: The choice between informed and arm's-length debt. Journal of Finance 47, 1367–1400.

Rajan, R., Zingales, L., 1995. What do we know about capital structure? Some evidence from international data. Journal of Finance 50, 1421–1460.

Ramakrishnan, R., Thakor, A., 1984. Information reliability and a theory of financial intermediation. Review of Economic Studies 51, 415–432.

Ritter, J., 1987. The costs of going public. Journal of Financial Economics 19, 269–281.

Röell, A., 1996. The decision to go public: An overview. European Economic Review 40, 1071–1081.

Romer, D., 2000. Advanced Macroeconomics. McGraw-Hill, Irwin.

Ross, S., 1977. The determination of financial structure: The incentive-signaling approach. Bell Journal of Economics 8, 23–40.

Roth, A., 1985a. Bargaining with Inomplete Information. Cambridge University Press, Cambridge, UK.

Roth, A., 1985b. Game-Theoretic Models of Bargaining. Cambridge University Press, Cambridge, UK.

Rothschild, M., Stiglitz, J., 1976. Equilibrium in competitive insurance markets: An essay in the economics of imperfect information. Quarterly Journal of Economics 80, 629–649.

Rubinstein, A., 1982. Perfect equilibrium in a bargaining model. Econometrica 50, 97–109.

Sarig, O., 1998. The effect of leverage on bargaining with a corporation. Financial Review 33, 1–16.

Schäfer, D., 2002. Restructuring know-how and collateral. Kredit und Kapital 35, 572–597.

Schmid-Klein, L., O'Brien, T., Peters, S., 2002. Debt vs. equity and asymmetric information: A review. Financial Review 37, 317–349.

Schmidt, D., Wahrenburg, M., 2003. Contractual relations between European VC-Funds and investors: The impact of reputation and bargaining power on contractual design. Tech. Rep. 2003/15, Center for Financial Studies, Frankfurt am Main.

Schmitz, P., 2001. The hold-up problem and incomplete contracts: A survey of recent topics in contract theory. Bulletin of Economic Research 53, 1–17.

Serrano, R., 2004. Fifty years of the Nash Program, 1953 - 2003. Tech. Rep. 21, Department of Economics, Brown University, Providence, RI.

Shleifer, A., Vishny, R., 1986. Larger shareholders and corporate control. Journal of Political Economy 94, 461–488.

Shleifer, A., Vishny, R., 1997. A survey of corporate governance. Journal of Finance 52, 737–783.

Shleifer, A., Wolfenson, D., 2002. Investor protection and equity markets. Journal of Financial Economics 66, 3–28.

Shyam-Sunder, L., Myers, S., 1999. Testing static tradeoff against pecking order models of capital structure. Journal of Financial Economics 51, 219–244.

Smith, C., Warner, J., 1979. On financial contracting: An analysis of bond covenants. Journal of Financial Economics 7, 117–161.

Spence, M., 1973. Job market signaling. Quarterly Journal of Economics 87, 355–374.

Spier, K., Sykes, A., 1998. Capital structure, priority rules, and the settlement of civil claims. International Review of Law and Economics 18, 187–200.

Stiglitz, J., Weiss, A., 1981. Credit rationing in markets with imperfect information. American Economic Review 71, 393–410.

Stoughton, N., Talmor, E., 1999. Managerial bargaining power in the determination of compensation contracts and corporate investment. International Economic Review 40, 69–93.

Subramaniam, V., 1996. Underinvestment, debt financing, and long-term supplier relations. Journal of Law, Economics and Organization 12, 461–479.

Titman, S., 2002. The Modigliani and Miller theorem and market efficiency. Financial Management 31, 101–115.

Titman, S., Wessels, R., 1988. The determinants of capital structure choice. Journal of Finance 43, 1–18.

Townsend, R., 1979. Optimal contracts and competitive markets with costly state verification. Journal of Economic Theory 21, 265–293.

Townsend, R., 1988. Information constrained insurance: The relation principle extended. Journal of Monetary Economics 21, 411–450.

Varian, H., 1992. Microeconomic Analysis. Norton & Company, New York, London.

Verrecchia, R., 1983. Discretionary disclosure. Journal of Accounting and Economics 5, 179–194.

von Neumann, J., Morgenstern, O., 1944. Theory of Games and Economic Behavior. Princeton University Press, Princeton, NJ.

Wang, C., Williamson, S. D., 1993. Adverse selection in credit markets with costly screening. Tech. Rep. 9310001, Economics Working Paper Archive at WUSTL, Series Finance, Washington University, St. Louis, MO.

Williamson, S., 1986. Costly monitoring, financial intermediation, and equilibrium credit rationing. Journal of Monetary Economics 18, 159–179.

Williamson, S., 1987. Costly monitoring, loan contracts, and equilibrium credit rationing. Quarterly Journal of Economics 102, 135–145.

Wilson, C., 1977. A model of insurance markets with incomplete information. Journal of Economic Theory 16, 167–207.

Wu, Y.-L., 2004. The choice of equity selling mechanisms. Journal of Financial Economics 74, 93–119.

Yosha, O., 1995. Information disclosure costs, and the choice of financing source. Journal of Financial Intermediation 4, 3–20.

Zarutskie, R., 2005. Evidence on the effects of bank competition on firm borrowing and investment. Journal of Financial Economics , forthcoming.

List of Figures

List of Tables

Lecture Notes in Economics and Mathematical Systems

For information about Vols. 1–483
please contact your bookseller or Springer-Verlag

Printing: Krips bv, Meppel
Binding: Stürtz, Würzburg